# Quantitative Methods in Biological and Medical Sciences

H.O. Lancaster

# Quantitative Methods in Biological and Medical Sciences

A Historical Essay

Springer-Verlag

New York Berlin Heidelberg London Paris
Tokyo Hong Kong Barcelona Budapest

H.O. Lancaster
Emeritus Professor of Mathematical Statistics
University of Sydney
Sydney, NSW 2006, Australia

With one figure.

Library of Congress Cataloging-in-Publication Data
Lancaster, H.O. (Henry Oliver), 1913–
    Quantitative methods in biological and medical sciences: a
historical essay / H.O. Lancaster.
        p.    cm.
    Includes bibliographical references and index.
    ISBN 0-387-94279-3.—ISBN 3-540-94279-3
    1. Medicine—Mathematics—History.    2. Biomathematics—History.
I. Title.
    [DNLM: 1. Meta-Analysis.    2. Biometry—methods.    WA 950 1994]
R853.M3L36    1994
574′.0151—dc20
DNLM/DLC
for Library of Congress                                          94-8063

Printed on acid-free paper.

Production managed by Hal Henglein, manufacturing supervised by Jacqui Ashri.
Typeset by Asco Trade Typesetting Ltd., Hong Kong.
Printed and bound by Braun-Brumfield, Ann Arbor, MI.
Printed in the United States of America.

9 8 7 6 5 4 3 2 1

ISBN 0-387-94279-3 Springer-Verlag New York Berlin Heidelberg
ISBN 3-540-94279-3 Springer-Verlag Berlin Heidelberg New York

# Dedication

I dedicate this book to Ernst Mayr (b. 1904), who has contributed greatly to the practice and history of biology, considered as science rather than as technology, and has done much toward the unification of the history and philosophy of the natural sciences.

# Contents

# Acknowledgments

I thank Professor Ernst Mayr for allowing me to dedicate this book to him and for permission to make a number of quotations from his authoritative book *The Growth of Biological Thought*.

I thank the Australian Research Grants Scheme for its financial support over four years in the preparation of this book.

I also thank the University of Sydney for accommodation and other assistance.

Many members of the University of Sydney have provided help and advice, among whom should be especially mentioned the University Librarian, Dr. Neil Radford, and his staff. I thank particularly Dr. Neville Weber of the School of Mathematics and Statistics for reading through the text and for clarifying some details of the mathematics, and Professor E. Seneta and Associate Professor E.D. Fackerell for various discussions in mathematical statistics and applied mathematics, respectively. From other schools, I thank also Emeritus Professor J.L. Still for some discussions on biochemistry, Professor M.R. Bennett for generalities on mathematics in biology, Dr. Julius Rocca for reading of the first two chapters on Greek science, Dr. Ursula Bygott for her constant encouragement, and Mrs. Elsie Adler for help with the manuscript.

Among individuals outside the University of Sydney, I thank Miss Brenda Heagney, Librarian to the Royal Australasian College of Physicians for searches of the literature and other research. I thank also the Mosman (Municipal) Library, Professor C.C. Heyde of the Australian National University for various assistance, and Dr. David Christie and Professor Annette Dobson for reading some chapters.

Among overseas correspondents, I thank Sir Michael Atiyah of Trinity College, Cambridge, and Dr. H. Davies of University College, London, for information on H.T.J. Norton; Freie Universität Berlin for data on F.H. Mueller; Professor Michael Healey of the London School of Hygiene for data on W.J. Martin; Professor R.L. Plackett for data on Bryan Robinson; Professor Dr. K. Holubar for reading sections on sepsis; Ms. Ingrid Radkey, University of California Biology Library, for material on Wilhelm Weinberg;

Dr. R.J. Wolfe, Curator of Rare Books and Manuscripts, Boston Medical Library, for help on the identity of O.W. Holmes's "competent person"; Dr. J.M. Tanner for suggestions on anthropometry; Professor Geoffrey Dean for advice on the porphyrias; and Professor Peter Armitage for his help and encouragement over the years of writing.

I also thank Mrs. Philippa Holy, my secretary and research assistant, for her steady and accurate work and for many kindnesses and help.

# Introduction

My original intention was to write a history of medical statistics, used in its prewar sense, expanding the writings on the subject by Major Greenwood, from which I formed many of my ideas in the early days immediately after the Second World War. In later years, I decided that the scope of his works was narrower than what I think is appropriate now, for he was writing in an era before the acceptance and use of the Fisherian methods and he was probably not aware of the mathematization of many parts of biological theory.

Further, the boundary between the medical and biological sciences has largely disappeared. Many texts have now been written on branches of the theory and practice inspired by R.A. Fisher (see §4.13).

I discuss the history of the use of quantitative methods in the biological sciences, defined after the style of Peller (1967) as that branch of science that uses a quantitative approach to, or quantitative logical reasoning on, any issue having to do with medicine or biology. The mathematical techniques are various and not classified here. Within the book I use "biological sciences" to include medicine but use the longer phrase in its title to avoid misunderstandings as to content. Moreover, most of the experimental work carried out in medical research laboratories is performed on animals other than man. I call the book a historical essay, as it would be difficult to give a proper account of all the history of mathematics applicable to biology.

In general, I do not develop mathematical proofs except to give a relatively complete description of the history of the normal distributions. A knowledge of matrices is common to all users of electronic machines, for example in applied biometry. No detailed knowledge of biological anatomy or physiology or disease is required. For diseases, the reader might use my text, *Expectations of Life* (Springer-Verlag, New York, 1990).

I have reduced the biographical writing to a minimum because most readers will have access to the *Dictionary of Scientific Biography*, and so I have appended an asterisk in the text and in the person index next to the name of those scientists who have a full biography in the Dictionary.

This book discusses applications of mathematics, in particular counting and measurement, to biological problems that have brought about changes in the way scientists describe the world. Those applications that do not affect the world picture and so may be regarded as technological, for example, agricultural crop yields, are not discussed.

Many branches of mathematics applied to the technology of biology have often a large, informative literature that the reader could be referred to, for example, Latin (or Eulerian) squares in biometry.

I have no illusions about references to mathematics as "queen of the sciences" because only the theory of numbers refers directly to the natural world. Indeed, it is better to think of mathematics as a language applicable to many branches of science. A great physicist has said that science consists of physics and stamp collecting; but biology is no longer a mere extension of taxonomy and biologists are attacking difficult and deep problems. Mayr (1982) has written:

Where the study of man is concerned, it is biology that provides methodology and conceptualization. The President of France has recently formulated this conviction in these words: "There is no doubt that mathematics, physics, and other sciences rather ill-advisedly referred to as 'exact' ... will continue to afford surprising discoveries— yet, I cannot help feeling that the real scientific revolution of the future must come from biology." (p. 42)

I hope that this book will help to fill gaps between the traditional history and philosophy of science and the new biology.

Let us consider the needs for mathematics in the modern world. The mathematics varies from the most elementary up to the most difficult and modern. Further, the proportion of stochastic content varies. Some problems have little stochastic content, for example, similitude; the flight of birds; the hydrodynamics of the circulatory system; and the physics of vision and of hearing; this latter requires applications of Fourier analysis.

With the development of biological research toward the molecular level, there are many applications of sciences that have a heavy mathematical background. Thus difficult mathematical problems continue to arise in genetics; for example, we find articles studying the movement of the chromosomes at meiosis or mitosis using advanced topology. The reducing process, which yields one-half the chromosomes for the next generation, is analogous to the tossing of a coin, and some philosophers believe that it may be the only place in nature where a probabilistic ratio exists. The theory of the gene positions on the chromosomes is again mathematical and continues to play a leading role in the current work on the human genome. Mathematics again crops up in the problems of ascertainment.

In population genetics, there are fields of knowledge where experiments must be supplemented by mathematical methods, for the time required to obtain answers by experiment is too long. Nevertheless, there are observations that help to verify the notions of evolution and natural selection, examples of which are given in Chapters 4 through 7. In this general field, there are good relations between mathematician and geneticist. Sometimes they are the same person, as was the case with R.A. Fisher.

To whom is the book directed? First, statisticians of any field. Second, applied mathematicians. Third, medical or biological historians and teachers of the subjects mentioned. Fourth, the educated layman with an interest in evolution. Fifth, students of the subjects mentioned. Sixth, teachers at the prematriculation level who often advise their students to drop mathematics, incorrectly so if later the student is to undertake a biological career. The mathematics is in general not developed beyond quite elementary levels, with the exceptions of the treatment of the multivariate normal distribution, which brings in Arthur Cayley's derivation, and the matrix theory of population growth.

Examples of the theory must be given; of course, some of the peas and *Drosophila melanogaster* will come into the picture but many readers will welcome the choice of man, which means we can avoid the difficulties of introducing many species with unfamiliar Latin names and structure and yet give examples of a number of quite general genetic phenomena.

# 1
# Greek Science

## 1.1 Atomism

The first two chapters consider several themes dating back to Greek times that have relevance to modern science. The opinions of biologists and philosophers of biology, for example, D'Arcy Wentworth Thompson, William Bateson, and Ernst Mayr, are often based on their readings or traditions of Greek history. Further, Greek science was more advanced than the science of the authors writing, say, in the early 1500s. Some themes considered by the Greeks have been fully clarified only in the twentieth century. A good example of this is *atomism*, with which we begin.

Leucippus* (floruit† Greece, fifth century B.C.), probably the founder of the School of Abdera, may be mentioned with Democritus* (floruit late fifth century B.C.) as early Greek atomists, although authors suggest other workers in the second millennium B.C. According to the atomists, the universe is explained by the existence of solid corporeal elements, infinite in number, shape, and size but otherwise lacking in sensible qualities, which were originally scattered through the infinite; their size can vary from the invisible up to the size of the cosmos. They are physically indivisible, and so are referred to as *atomos*. Although indivisible, these atoms are certainly extended and indestructible. They are homogeneous in substance and contain no void and no interstices and are in perpetual movement. These atoms exist in a void. They combine to form the elements of the cosmos and the properties of their combinations depend on the geometric relations between them. The modern general theory asserts that matter is not infinitely divisible (see §2.1).

Incidentally, Mayr (1982) writes of Democritus that

he was the first to have posed a problem that has split philosophers ever since: Does organization of phenomena, particularly the world of life, result purely from chance

---

*The asterisk next to a name means that there is a full biographical entry in the *Dictionary of Scientific Biography*, Charles Scribner's Sons, New York (1970–1990), which is abbreviated throughout this text as *DSB*.
† flourished.

or is it necessity, owing to the structure of the elementary components, the atoms? Chance or necessity has ever since been the theme of controversies among philosophers.... It was Darwin, more than 2,200 years later, who showed that chance and necessity are not the only two options, and that the two-step process of natural selection [namely "genetic recombination and reproductive success," Mayr (1982, p. 519)] avoids Democritus's dilemma. (p. 86)

## 1.2 Forms and Ideas

Pythagoras* (*ca* 560–*ca* 480 B.C.) founded a school in Croton, in southern Italy. Members of the school were the first to develop a branch of the science of mathematics, the theory of numbers as it is now called. Under their influence, the notion arose that number was essential to the study of the world, whether physical or biological. As a by-product of their thought, number mysticism became popular. One of their great achievements was the discovery of incommensurability, that not all numbers could be expressed as a ratio of integers, in particular $\sqrt{2}$.

Remarkably, according to Heath (1921), Eudoxus of Cnidus* (*ca* 400–*ca* 347 B.C.) had already derived a solution to this problem of definition of a real number corresponding to a given equation. Indeed, Heath (pp. 326–327) states: "The greatness of the new theory itself needs no further argument when it is remembered that the definition of equal ratios in Eucl. V, Def. 5 corresponds exactly to the modern theory of irrationals due to Dedekind, and that it is word for word the same as Weierstrass's definition of equal numbers."[1]

In other words Eudoxus had given a discussion of the "Dedekind cut" of Julius Wilhelm Richard Dedekind* (1831–1916). The set of rational numbers can be split into two subsets, namely those that are less than or equal to a given rational number and those that are greater than it. Now $\sqrt{2}$ is not rational but it can be determined by squaring any rational number, $A$ say, whether $A \leqslant \sqrt{2}$ or $A > \sqrt{2}$ and so a "cut" is achieved and the point is said to be at the Dedekind cut.

Plato* (427–348/347 B.C.) may be taken as representing Greek thought on the doctrines of Forms. Others before him had held that there was no stable substance in the sense world and that the sense world was in a state of flux. Plato was anxious to continue the Socratic search for universals. It was Plato who, wishing to separate universals from sensible particulars, called them Forms and Ideas. The Greeks believed that causation in nature was regulated by laws that can be stated in mathematical terms. They were successful in solving problems in physics and astronomy but were rather less successful in biology. They thought that the variable world of the phenomena was but a

---

[1] From Heath, T.L. *A History of Greek Mathematics.* Copyright © The Clarendon Press, Oxford, 1921. Reprinted by permission of Oxford University Press.

reflection of a limited number of fixed forms, *eide*, as Plato called them (see Mayr, 1982, p. 38). The effects of this line of thought are still with us, as we show in §2.3. Mayr (1982) believes that other theses of Plato were damaging to scientific biology:

The second was the concept of an animate cosmos, a living, harmonious whole (Hall, 1969, p. 93), which made it so difficult in later periods to explain how evolution could have taken place, because any change would disturb the harmony. Third, he replaced spontaneous generation by a creative power, a demiurge. Since Plato was a polytheist and pagan, his demiurge was something of a less concrete person than the creator-god of the great monotheistic religions. Yet, it was in terms of monotheism that the demiurge, the craftsman who made the world, was later interpreted. And it was this interpretation which led to the later Christian tradition that "it is the task of the philosopher to reveal the blueprint of the creator," a tradition still powerful up to the middle of the nineteenth century (natural theology, Louis Agassiz). The fourth of Plato's influential dogmas was his great stress on "soul." References to noncorporeal principles can be found also in pre-Socratic philosophers but nowhere as specific, detailed, and all-pervasive as in Plato. (p. 305)

Plato is now regarded as the antihero of science, especially of biology, from later Roman times up to the present.

## 1.3 Aristotle*

Mayr (1982) writes:

No one prior to Darwin has made a greater contribution to our understanding of the living world than Aristotle (384–322 B.C.). His knowledge of biological matters was vast and had diverse sources. In his youth he was educated by Asclepiadic physicians; later he spent three years of his life on the island of Lesbos, where he evidently devoted much time to the study of marine organisms. In almost any portion of the history of biology one has to start with Aristotle. He was the first to distinguish various of the disciplines of biology and to devote to them monographic treatments (*De partibus animalium, De generatione animalium*, and so forth). He was the first to discover the great heuristic value of comparison and is rightly celebrated as the founder of the comparative method. He was the first to give detailed life histories of a large number of species of animals. He devoted an entire book to reproductive biology and life histories.... He was intensely interested in the phenomenon of organic diversity, as well as in the meaning of the differences among animals and plants. Although he did not propose a formal classification, he classified animals according to certain criteria and his arrangement of the invertebrates was superior to that of Linnaeus two thousand years later. (pp. 87–88)

Clearly, Aristotle was a great and versatile scientist.

## 1.4 Canons of Experimentation

A noteworthy feature of the work of Strato of Lampsacus* (d. 271/268 B.C.) was his extensive use of experiment. Good popularizing accounts of his work and attitudes are given by Magner (1979) and by H.B. Gottschalk (*DSB*,

XIII, pp. 91–95). Strato can be seen as working in the traditions of the school of Aristotle and of Theophrastus. Let us consider him as the prototype of the "good scientist" of modern times. Observations are made and then some hypothesis constructed. Tests are made on the hypothesis and perhaps new observations are suggested. Strato's experiments often form a progressive series; he would look for the existence of similar phenomena occurring under natural conditions; here he was trying to avoid the charge that his experiments distorted nature. (See §7.5 for examples from population genetics.)

From ancient Greek days, there are many examples of workers testing their hypotheses in the field; notably in astronomy, the stars and planets continue in their courses and yield tests of hypotheses. The same is not true in biology, for special investigations will usually have to be undertaken. There is a temptation to be *Platonic* and stay with the written word. We might well call the opposite behavior *Stratonic*.

There seems to be a revival of interest in Greek philosophy and science, the title of his chair at Cambridge University, according to G.E.R. Lloyd (1991), who believes these subjects have been greatly neglected. Some modern scientists deserve the title *Stratonic*, for example the list of geneticists responsible for the *evolutionary synthesis* (see §7.7).

# 2
# Later Influences of the Greek Authors

## 2.1 Atomism and Modern Biology

John Dalton* (1766–1844), in the early years of the nineteenth century [see his biography by Thackray (*DSB*, III, pp. 537–547) and Thackray (1970)], observed that in a chemical reaction the chemical combining ratios of weights between two different elements appeared to be simple ratios of integers. From this observation, he deduced the existence of atoms of weights, all multiples of some unknown weight, peculiar to the atoms of the particular element. However, the physicists did not regard Dalton's work as a complete proof of the existence of atoms (see Maiocchi, 1990).

Robert Brown* (1773–1858) had observed the movements of microscopic particles in the fluid parts of preparations of plants; he saw that they persisted in these preparations even after the death of the plants. Further, the phenomenon appeared in any material that could be ground down to form particles of microscopic size and so be suspended in fluid. Marian Smoluchowski* (1872–1917) in 1906 and Albert Einstein* (1879–1955) in 1905 and 1922 were able to explain the motion as the result of constant batterings of molecules on the particles in the fluid; these were stochastic events and sometimes there would result sufficient impacts in a short time to cause a motion visible microscopically. Einstein possibly had little knowledge of Brown's work but was working on the existence of molecules, Dalton's atoms, and giving an idea of size. The method has been used to determine the Avogadro number; it also gave firm evidence of the existence of molecules.

[For a discussion of the difficulties of interpreting Brownian movement, see S.G. Brush's account of Ludwig Boltzmann* (1844–1906) (*DSB*, II, pp. 260–268).]

## 2.2 Solids and Liquids in Living Beings

We may apply commonsense ideas on atomism to the study of the necessity of structure to allow physiological functions to occur. An example can be given. Samuel Thomas Soemmerring* (1755–1830), the famous German anatomist, claimed in 1799 that the intraventricular cerebrospinal fluid was the seat of the analysis of observations of the body or "organ of intelligence." Kruta, in his biography of Karl Asmund Rudolphi* (1771–1832) (*DSB*, XI, pp. 592–593), notes that Rudolphi considered that such processes as thought and reasoning could not arise from such a structureless organ as the cerebrospinal fluid. Rudolphi examined the development of many plants and concluded that each plant in its early stages consists wholly of cellular tissue. Fluids can only retain memories in a very limited sense, nor can they originate information; in any case (nonfluid) structures would be needed to measure the content of the fluid. The same argument applies to the humoralist theory in §3.3. [See Karl Ernst von Baer* (1792–1876), who stated that every animal that springs from a sexual union is developed from an ovum.]

A similar argument on the necessity for discrete organs or organelles applies to signalling within a living body. Chemical signalling is possible within a small living body, say, an ameba. Some signalling is used in the larger bodies, for example, the hormones, where the message is simple and general, as in the adrenal carrying the warning "be prepared," but there are obvious difficulties in signalling more precise statements if the message is to be received only by certain cells. The larger forms have therefore developed a nerve system in which electrical impulses are carried from site to site (that is, organ to organ). Here discrete channels are being used and we can say that solid or particulate structures are present, namely neurons and fibres. As Kandel, Schwartz, and Jessell (1991, p. 28) point out, the brain uses stereotyped electrical signals to process all the information it receives and analyzes. The signals are virtually identical in all nerve cells. There are excellent diagrams and explanatory matter in the book. The physiology involved is a superb example of reductionism, that is, fixing attention on a single organ rather that the whole body.

## 2.3  Platonic Science in Modern Biology

We may ask what was the cause of failure of Platonic traditions when mathematics was applied to biology in modern times, whereas they were so successful in astronomy. Perhaps the diversity of life is such that its features cannot be imagined but must be observed; some scientists maintain that there are more parameters to be reckoned with in an ameba than in the sun. It would be difficult to work from "a limited number of fixed forms" as in §1.2 to the whole of life, without the necessary observations as a guide. Plato's influence also encouraged a belief in spontaneous generation and vitalism.

William Bateson* (1861–1926) describes in detail the differences between the essentialists (or theoretical biologists) and the experimental or observational biologists in a memorable passage (Bateson, 1894, pp. 574–575). In this passage, Bateson is comparing the fates (or indeed failures) of Platonists on the one hand and the narrow-minded naturalists on the other. He concludes that it is by the study of variation alone that the problem of the modes of evolution can be attacked.

In §4.4 we give an example of a Platonic type of solution by Galton to the problem of ancestral heredity. Indeed, we can see that this solution led to a loss of reputation by both Galton and his biographer, Karl Pearson.

In taxonomy, the word *type* had an established meaning and so Quetelet (1845) had no necessity to refer to Plato; for it is easy to identify the type, Quetelet's "average man," with the Platonic *eidos*. The essentialist notion of type was to persist. Thus Pearson (1892, p. 484 of the 1900 edition) writes that "Nature aims at a type."

For an extreme example of Platonic thought, see Remmert (1991):

Leibniz believed that a continuity principle underlay all the laws of nature. The law of continuity "*Natura non facit saltus*" runs like a red thread through all his work in philosophy, physics and mathematics. In the *Initia rerum Mathematicarum metaphysica* (*Math. Schriften* VII, pp. 17–29) it says: "... Continuity however is attributable to time as much as to spatial extension, to qualities just as to motion, actually to every transition in nature, since these never proceed by leaps." Leibniz applied his continuity principle also, for example, to *biology* and in this seems to have anticipated Darwin somewhat; in a letter to Varignon he writes: "The continuity principle carries such conviction for me that I would not in the least be astonished at the discovery of intermediate life-forms many of whose characteristics, like their methods of feeding and reproduction, would give them equal claim to being plants or animals...." The continuity postulate later became known as Leibniz' dogma. (p. 38)[1]

## 2.4 Classification and Comparative Method

From the point of view of this book, Aristotle should be remembered for classification and the comparative method. (See Chapter 16 for classification and many examples throughout the book for comparative methods.)

Comparative methods seem to lead inevitably to quantitative methods [see Harvey and Pagel (1991), reviewed in *Science* 254:134–136, and writing on the method in evolutionary genetics]. W. Farr used the comparative method in the nineteenth century with skill to show that the death rates in the English counties varied greatly, from which he concluded that the lifestyle varied between counties and that remedial measures could be applied with profit.

The comparative method has been useful in suggesting hints on the causes of some noninfective diseases. (See §13.1 for the researches on the etiology of pellagra; §13.3 for the etiology of cancers such as scrotal cancer; §13.4 for the etiology of melanoma; §13.5 for disease caused by ionizing radiation; and §13.7 for the association of tobacco with lung cancer.)

---

[1] From Remmert, R. *Theory of Complex Functions.* Copyright © Springer-Verlag, New York, 1991. Reprinted with permission.

## 2.5 Canonical or Stratonic Science

A general discussion of this problem has been given by Mayr (1982) in his epilogue, "Toward a Science of Science." We confine our attention to some special points on the applications of mathematics to biology.

For a mathematical model to be appropriate, we have to ensure that there exists a living agent or agents having, at least approximately, the required properties. In mathematics, when a new field is axiomatized, it is required to prove that there exists a set of "elements" such that they obey the proposed set of axioms. In §12.8, we find that there is no disease that has been found to have the required properties for some of the differential equation models that are widely used; in particular, measles does not have the required property.

An interesting case is given by the following: Hermon Carey Bumpus* (1862–1943), in Bumpus (1899) cited by C.J. Bajema (*DSB*, XVII, p. 131*b*), observed on 1 February 1898 a severe winter storm with snow, sleet, and rain in Providence, Rhode Island, causing many sparrows to be blown down. He believed that this could be taken as a test of natural selection and compared various parameters of those 74 birds that survived with those of 64 that died. He concluded that "selective elimination is most severe with extremely variable individuals, no matter in what direction the variations may occur."

(See §7.5 for a more important example of work, especially that of Četverikov, verifying in nature theory that had been developed as mathematics or laboratory science.)

## 2.6 Mathematization of Other Sciences

In some branches of science other than biology, the need to give an overall picture of the use of mathematics has been realized in Vol. 52 (1961) of *Isis*, pp. 133–352, on the progress of quantification in the sciences generally. Woolf (1961a) gives an introduction on p. 133 and is followed by S.S. Wilks, science, p. 135; A.C. Crombie, medieval physics, p. 143; T.S. Kuhn, modern physical science, p. 161; H. Guerlac, chemistry, p. 194; R.H. Shryock, medieval science, p. 215; E.G. Boring, psychology, p. 238; J.J. Spengler, economics, p. 258; P.F. Lazarsfeld, sociology, p. 277; G.W. Gerard, biology, p. 334. [See also Woolf (1961b). For psychology see also Brown and Smith (1991).]

It does not seem proper to give a long discussion on scientific theory in this book, although we have already made a note on the harm that Platonic theory had done to scientific thought whereby the scientist hoped to arrive at the truth by thought processes rather than by observation of the outside world. Viner (1952) notes that:

Theory is always simpler than reality. Even when it seems terribly complex, it is still "simpliste," as compared to the range of factors operating as conditions, as means, or as ends, in any actual concrete situation.... It is a great temptation for the theorist to work from a few premises. It simplifies his task of analysis. It makes it easier for him to reach definite and precise answers, and to reach them by rigorous and technically elegant procedures. He has, moreover, to support him in this practice the doctrine of ancient origin, but nevertheless of highly questionable validity for the social sciences, that the progress of scientific analysis is marked by the substitution of simple for complex solutions to problems and of precise and definite for qualified and contingent answers to questions. (p. 12)[2]

Of course, there are some relatively simple problems in biology, such as the growth of populations as treated below in our Chapters 6 and 7. Satisfactory conclusions can be obtained that accord with observations from human populations and that can be generalized to other species.

---

[2] Viner, Jacob. *International Trade and Economic Development*. Copyright ©1952 by The Free Press, a Division of Macmillan, Inc.; copyright renewed 1980 by Princeton Bank & Trust Company. Reprinted with the permission of the publisher.

# 3
# Microscopic World and the Structure of Living Organisms

## 3.1 Biodiversity

Biodiversity makes the modelling of the life sciences and the application of mathematics to them very difficult. The reader is referred to the commentary of Mayr (1982, pp. 99–103 and 864).

It can be noted that the Platonists had little interest in diversity (see §1.2). For them variations about the type were unimportant and so even the arithmetic mean would appear to be irrelevant. This view was shared by the Thomists, in the new synthesis of knowledge due to a circle of theologians in the thirteenth century, of whom the leading figure was Thomas Aquinas* (*ca* 1225–1274). Indeed, biodiversity produces difficulties in the life sciences that are little felt in the classical history and philosophy of science.

After the revival of learning, many studies in biology were carried out in Europe, where biological diversity was less remarkable than in the tropics, a phenomenon in part due to climatic conditions and to the residual effects of the last Ice Age, in which many species were lost. A change in outlook took place after the long sea voyages commencing about 1500, revealing a world of creation far richer than any European had imagined. There are many examples in this book that show that little interest was shown in diversity; for example we may cite the weights of newborn infants in §14.1.

Galton (1889b) saw that the problem of diversity had been neglected and began collecting data sets mostly on or about human characteristics. Many jokes have been made about this work but it led to him making generalizations about methodology. He can be credited with realizing what an important measure of diversity the variance was (see his statement in §3.6). Others such as Quetelet failed to realize its importance. The biologist will find many examples where the normal curve with estimated mean and variance gives a good approximation to the frequency distribution of the original data (see the examples of Weldon given in §7.3). A great example of diversity comes from Galton's work on fingerprints, whereby many millions of patterns can be recognized and lexiconized.

Ultimately, biodiversity comes about by evolutionary processes but we can give examples of it in a single species. Within species, diversity has come about by

1. environmental causes such as climate;
2. genetic causes such as mutations, followed by polymorphisms; for example, there are several hundred variants of the abnormal hemoglobins, some of which protect members of the human species against malaria in hyperendemic regions (see §5.6; see also Thompson and Thoday, 1979);
3. the presence of other genes and so interactions between genes;
4. mimicry, which can be an important cause of diversity where the individual or the species is protected against predators that do not attack prospective victims because they have mimicked the appearance of a victim that the predator does not eat. Mimicry can take other forms, notably the classic melanism. (See §7.4 for the importance of Punnett and Norton for the numerical treatment of evolution.)

But there are natural causes acting to lessen biodiversity in limited regions; thus the latest Ice Age led to the destruction of many species in Europe. Of course, diversity can be reduced by artificial means in the laboratory, for example by selfing in plants with the production of pure lines (see §4.6).

Fisher (1953) gives an example showing that the eels did not develop a secondary species, because their life-style included mating in the region around the Sargasso Sea, whereas a period of isolation is needed for the development of new species. A gradual enlargement of the Atlantic Ocean locked the eels into a remarkable voyage.

As the related sciences progress, more examples of diversity appear. Special forms of diversity were discovered by Louis Pasteur* (1822–1895), namely differing forms of symmetry or asymmetry, especially in organic bodies. An interesting variation came from George Henry Falkiner Nuttall (1862–1937), who made precipitin tests on the bloods of 600 species of vertebrates, finding a distinct similarity between his immunological findings and the standard classifications by phylogenetical methods (see Nuttall, 1904).

Mayr (1982, p. 99) mentions the development of instruments, especially the microscope, as introducing new varieties of living things (see §§3.2 and 3.3).

An example of diversity relevant to the themes in this book is that infective disease is caused by many different living agents, with a great diversity of modes of spread from case to case; Hopwood and Chater (1989) write on the *Genetics of Bacterial Diversity* and diversity now enters into the titles of many books.

[For the mathematics of diversity see Kingman (1980).]

## 3.2 From Microbes to Cells

Now we turn to a converse question: Is there any general theme on the structure of living agents? We note that the microscope had revealed a new world of the minute and an extension of the biodiversity. Biochemical methods revealed the microbes as autonomous, living organisms; for example, yeasts could cause fermentation of sugar and perhaps also the body cells.

An example of the importance of instrumentation in biology is provided by the works of Antoni van Leeuwenhoek* (1632–1723) and Marcello Malpighi* (1628–1694), who saw for the first time the microscopic structure of both animals and plants. Very little progress had been made after the account of Leeuwenhoek sent to the Royal Society (of London) in 1680, in which he described the globules of yeast in beer. Thus a new world for investigation was created but it was some time, almost two centuries, before the two new forms of life were properly analyzed; perhaps the difficult physical conditions of the microscopist were a cause of the apparent lack of interest. However, around 1800 there was increasing interest in the minute organisms and microscopes were improving.

We note that in 1832 Robert Brown observed the nucleus in cells, describing it as more opaque than the membrane of the cell. Little precise information on microbes was available until the publications of three authors in 1837, namely Charles Cagniard de Latour* (1777–1859), Theodor Ambrose Hubert Schwann* (1810–1882), and Friedrich Traugott Kuetzing* (1807–1893), independently reported that yeasts were the agents of fermentations. The simultaneity of their discoveries has been attributed to the development of achromatic compound microscopes (see Fruton, 1972, p. 44). Kuetzing in his report remarked that yeast must be struck off the list of chemical compounds, thus denying current chemical doctrine. Pierre-Jean-François Turpin* (1775–1840), writing on behalf of a committee appointed by the French Academy, reported favorably on the discovery and agreed that the three authors had arrived at the result almost simultaneously. Liebig would have none of these interpretations; his ideas delayed the full acceptance of the work of the three authors just mentioned and the acceptance of microbes as a cause of many infectious diseases up even to the present century (see §8.5).

Fortunately, these revolutionary findings were of great interest to Louis Pasteur, who early in his career had been interested in the production of asymmetric organic compounds by living agents. His published work on fermentations dates from about 1857; he was anxious to support the notion that yeasts were the active agents in fermenting cane sugar and found that living agents brought about the fermentation of lactose in milk. He set up a critical experiment whereby only an ammonium salt and the minerals obtained by an incineration of yeast were added to a pure solution of cane sugar and then a trace of pure brewer's yeast. This experiment usually succeeded if all the components were brought together but failed if any one of them was omitted. Pasteur then concluded that the yeast assimilated carbon from the sugar, decomposing it into carbonic acid and alcohols and forming organic nitrogen compounds. [For the three recognized discoverers, see Bulloch (1938/1960, pp. 47–57) and Fruton (1972, pp. 22–86).]

The solution to the fermentation problem suggested that there might exist other microbes that could cause fermentation in other substrates or even within animal bodies and perhaps disease. Furthermore, it suggested that the cell might be the important element in multicellular organisms, to which we now turn.

In the improvement of the microscope, a leading role was played by Joseph Jackson Lister* (1786–1869) with his most active years 1824 through 1843, when he was skilfully designing microscopes with the use of mathematical and physical methods. G.L'E. Turner (*DSB*, VIII, pp. 413–415) remarks that he changed the microscope from a toy to an important scientific instrument [see Bracegirdle (1978), Bradbury (1967), Bradbury and Turner (1967), and Hughes (1959)].

## 3.3  Size and Structure of Living Organisms

The size of living organisms is bounded below by the finite divisibility of matter (see §2.1). Upper limits to size are brought about by similitude (see §14.2). According to Ackerknecht (1953, p. 71) there had been several common principles called upon in the eighteenth century as the basis of life: (a) fibres that began as globules, (b) globules and particles (with inevitable optical illusions because of the poor powers of resolution of the microscopes of the time), (c) identity of cell forms in plants and animals, and (d) continuity of life. Biologists were still looking for a common principle for both plants and animals; among them were Theodor Schwann (already mentioned in §3.2) and Jacob Mathias Schleiden* (1804–1881). The latter had qualified as a lawyer but turned to biology; he was in Berlin for some years and was able to meet among others Friedrich Wilhelm Heinrich Alexander von Humboldt* (1769–1859), Robert Brown, and Johannes Peter Mueller* (1801–1858). There he engaged in an enquiry into the elementary units common to the animal and plant kingdoms. A statement of his general conclusions is given in his *Grundzüge der wissenschaftlichen Botanik* (1842–1843), which was the foundation of his teaching and literature and which nevertheless was incorrect in detail.

It is difficult with so many workers in the field to know which worker elaborated the cell theory as it became known. François-Vincent Raspail* (1794–1878) is mentioned by his biographer Marc Klein (*DSB*, XI, pp. 300–302) as making a prophetic statement on the cell theory: "The plant cell, like the animal cell, is a type of laboratory of cellular tissues that organize themselves and develop within its innermost substance." Hugo von Mohl* (1805–1872) is said not to have been aware of his work and Schwann and Schleiden held it in deep contempt.

Tradition has it that the final steps in the cell theory resulted from a meeting of Schleiden and Schwann. We may follow the account given by Florkin (*DSB*, XII, pp. 240–245) of Schwann's *Mikroskopische Untersuchun-*

*gen* (1839). Schwann found that the *chorda dorsalis* of the frog larva contained polyhedral cells "which have in or on the internal surface of their wall a structure corresponding to the nucleus of plant cells. New cells are formed within parent cells." Similarly, the structure of cartilage resembled the tissues of plants. He believed that the cartilage cell contains a nucleus and that new cells develop within formed cells first by division of the nucleus and the construction of a new wall between the two daughter cells, each containing a nucleus and vacuole and surrounded by a membrane. Schwann went on to state that all the "elementary" parts of tissues developed from a specialization of cells.

So finally it could be stated that living bodies consist of cells, some specialized, and structures formed by the cells, for example, cerebrospinal fluid and cartilage.

It is necessary to clear away misconceptions about the term *spontaneous generation* in that it is now believed that the beginning of life was due to developments in nonbiotic matter over a long period of time, whereas the opinion to be combatted was that fully developed organisms could appear spontaneously. For example, Aristotle believed that flies and maggots appeared from rotting meat, frogs, and mice, from moist earth, and so on.

Francesco Redi* (1626–1698) demonstrated that, in examples of so-called spontaneous generation, contagion, with insects and other animals that were not too small, could be proved. His survey revealed to him the complex mechanisms of insects and he was able to examine the morphological characteristics of the animal (Redi, 1684). Nevertheless, Redi still kept an open mind on smaller organisms. A reading of Farley (1977) shows that the case against the existence of spontaneous generation cannot be proved; perhaps this can be accepted as a working hypothesis in any specific case. (See §8.7 for the experimental view.)

Rudolf Carl Virchow* (1821–1902), a pathologist, stated the main thesis of the cellular theory, *omnis cellula e cellula*, applied to demolish the humoralist or blastema theory, revived by Karl von Rokitansky (1804–1878) and given in the first volume of his *Handbuch* (1842–1846), by which all constituents of the body are said to be developed in coagulated lymph. Such lymph derives substance and nourishment from the original structures but is capable of effecting its organization by its own inherent life power (see Mettler, 1947, p. 259, and Ackerknecht, 1953, p. 82). We cite a translation of Virchow (1863):

At the present time, neither fibres, nor globules [actually microscopical artefacts], nor elementary granules, can be looked upon as histological starting-points. As long as living elements were conceived to be produced out of parts previously destitute of shape, such as formative fluids, or matters (*plastic matter, blastema, cytoblastema*), any one of the above views could of course be entertained, but it is in this very particular that the revolution which the last few years have brought with them has been the most marked. Even in pathology we can now go so far as to establish, as a general principle, *that no development of any kind begins* de novo, *and consequently as to reject the theory of equivocal* [spontaneous] *generation just as much in the history of the development of individual parts as we do in that of entire organisms....* Where a cell arises, there a cell must have previously existed (*omnis cellula e cellula*), just as an animal can spring only from an animal, a plant only from a plant.

The cell theory carries with it the denial of the possibility of spontaneous generation. (For a notion, cognate with Soemmerring's view given in §2.2, see §8.7, where Henle is combatting the view that an infective disease could be a living entity reproducing itself.)

Florkin, in his biography of Schwann (*DSB*, XII, p. 241*b*), notes that the cell theory prolonged in biology the old debate on continuity and discontinuity.

Pasteur went on to make many more ingenious experiments to deny the doctrine of spontaneous generation. Notable work was done also by F.J. Cohn and J. Tyndall (see §8.7).

(For spontaneous generation, see Mayr, 1982, Chapter 3, especially pp. 105–112 and p. 582.) (For the stereochemistry of Pasteur, see Palladino, 1990.)

Vitalism in some of its forms assumed that atoms or molecules in living beings differed in some ways from those in dead bodies or inorganic material. As chemistry developed, more and more organic substances were synthesized without live substances being used. [See §15.10 for comparisons of energy transformed in metabolism in animals or in the laboratory, for which see Florkin (1972, pp. 81–96).] In either case the experiments were evidence against the theory of vitalism.

(See §7.5 for Četverikov's remarks on size and insect evolution.)


## 3.4  Applied Mathematics

Many mathematicians would have realized that for progress to be made in applied statistics a deeper mathematical theory must be created, summed up in the following remark by Jean Baptiste Joseph Fourier* (1768–1830), cited by Quetelet (1826, vol. 2, p. 177) and by Sarton (1935, p. 13): "*Les sciences statistiques ne feront de véritables progrès que lorsqu'elles seront confiées à ceux qui ont approfondi les théories mathématiques.*" [The statistical sciences will only make real progress when they are entrusted to those who have deepened mathematical theory.]

Some of the mathematics had already been developed at the time Fourier wrote these words. It remained to interpret the meaning of variance and correlation for applications to biology and, indeed, sciences other than astronomy and related sciences in which the normal distribution appears as that of the errors. Galton realized the importance of the variance in describing biological data (see §3.6). He also gave a meaning to the variance-covariance matrix of two random variables and proved that regression was often linear and homoscedastic. Karl Pearson extended the theory for taking several variables and then standardizing them to have unit variance, specializing the variance-covariance matrix to have a special form, units down the principal diagonal, and coefficients of correlation in the off-diagonal positions; this defines the correlation matrix. (See §19.3 for a discussion of the normal distributions.)

## 3.5 Biologists' Views on the Use of Mathematics

The views of practising zoologists could be sought. Simpson, Roe, and Lewontin (1960), with their book of 440 pages, devote 66 pages to a chapter on "growth," whereas the rest of the book is devoted to what is effectively an elementary text on probability and statistics with biological examples. From this text it could be concluded that mathematics enters into biology as an adjunct to measuring and counting problems, that is, in most cases, as an aid to the technology of, rather than the science of, biology.

In modern times, mathematics has taken a place in *biological science*. A leader here was D'Arcy Wentworth Thompson* (1860–1948). Thompson (1917/1942, p. 2) points out that the morphologist has been slow to give mathematical explanations of the forms of living structures. "To treat the living body as a mechanism was repugnant, and seemed even ludicrous, to Pascal; and Goethe, lover of nature as he was, ruled mathematics out of place in natural history. Even now the zoologist has scarce begun to dream of defining in mathematical language even the simplest organic forms."[1]

Teleological concepts such as end, "design," and vitalism were and still are used. Mayr (1982) says that a development of the mechanistic world view may be credited to Galileo Galilei* (1564–1642). This led to the construction of models and the use of the deductive method by René du Perron Descartes* (1596–1650).

For biologists, who have found it necessary to call on mathematicians or mathematics, we refer the reader to the following: Allison, anemia, §5.6; Galton, genetics, normal distribution, §3.6; Garrod, human disease, §5.2; Holmes, contagion, §9.1; Johannsen, genetics, §4.6; Medawar, growth, §14.2; Mendel, genetics, §4.2; Penrose, genetics, §5.5; Punnett, genetics, §4.3; Škoda, clinical signs, §15.5; Weinberg, human characters, §4.8; Weldon, correlation, §7.3.

## 3.6 Francis Galton

Francis Galton* (1822–1911) was born into a well-to-do family, closely related to the Darwin family, and like it known for its intellectual abilities and interests. His earliest scientific researches were in the exploration of south-

---

[1] From Thompson, D.W. *On Growth and Form.* Copyright © Cambridge University Press, New York, 1942. Reprinted with permission.

west Africa during 1850–1852, which led to his election to the Royal Society in 1856. He took an active part in the reform of the Society and acted on numerous committees and commissions of enquiry on a great variety of topics; he was awarded their most prestigious Copley Medal by the Society in 1910.

When Galton became motivated by an interest in genetics and eugenics, he noted the lack of quantitative method and of sufficiently ample material in anthropometry. Galton (1874, p. 143) states, "The topics suitable to statistics are too numerous to specify; they include everything to which such phrases as 'usually,' 'seldom,' 'very often' and the like are applicable, which vex the intelligent reader by their vagueness and make him impatient at the absence of more precise data," which is reminiscent of Gavarret's remarks in §17.2. Galton refers to the necessity of homogeneity in defining groups for analysis and the need to break up groups where there is variation governed by an important factor, for example, age; further, if the population is too large to permit all its elements to be measured, some random sampling method will be needed to choose the elements to be measured. Thus, before 1883 we find him devising instruments such as whistles for determining the upper limits of audible sounds in different persons, graded weights in boxes for testing muscle sense, graded distance recorders for testing acuity of vision, and artificial glass eyes and graded samples of hair to test eye and hair colours. His observations might be qualitative or quantitative.

Quantitative observations would usually be made under laboratory conditions. Galton could make generalizations on the distributions of quantitative biological variables. Sometimes it appeared that the logarithm of the variable observed would be the appropriate variable for analysis. He found that many distributions of sums of such variables were symmetrical; further that the graph of the cumulated sum of the frequencies below any point, $x$ say, took a characteristic S-shaped form. This is the distribution function, $F(x)$, of modern statistics corresponding to a density function, $f(x)$ say. He had thus rediscovered the cumulative normal distribution of André Michel Guerry (1802–1866) and graphed it as an "ogive"; with its aid, the median, quartiles, or any particular percentile could be determined. He pointed out that the use of the ogive method permits the generalization and that many empirical distributions are symmetrical, and, further, often well approximated by the normal, sometimes surprisingly well. Galton (1875) argued that "a medley of small and minute causes may, as a first approximation to the truth, be looked upon as an aggregate of a moderate number of 'small' and *equal* influences," and so yield approximately a binomial distribution, which is itself approximable by a normal distribution. (See later in §3.6 and in §19.3 for the nonnecessity of "equal" and notice that Russian mathematicians have weakened the restrictions on the summands, especially the symmetry.)

Although scorn and ridicule have often been poured on Galton's observational material, he was aware of the bearing of some variables on biological and specifically hygienic problems. Thus he thought that anthropometrical

surveys should be undertaken in all classes of schools in order to measure
fitness. He hoped that longitudinal studies would be made on anthropometrical variables, first at school and then at intervals of four years into adult
life (see Pearson, 1914–1930, II, pp. 347–348). In the following pages of that
work, Galton's design and issuing of schedules for the recording of anthropometrical readings are described. [See also Galton (1889a) in his reply to H.M.
Civil Service Commissioners, xxxi, p. 15.]

Galton's contributions to statistics are far wider than is usually thought;
they are discussed in many later sections. Here it is appropriate to mention
his attitude to the univariate normal distribution. A modern observer may
note that Quetelet never used the variance of a one-dimensional distribution
and that his biographers other than Karl Pearson have failed to note this
defect. As Galton (1889b) remarks, reprinted in Pearson:

It seems to be a great loss of opportunity when, after observations have been labori-
ously collected, and been subsequently discussed in order to obtain mean values from
them, that the small amount of extra trouble is not taken, which would determine
other values whereby to express the variety of all the individuals in those groups.
Much experience some years back, and much new experience during the past year,
proved to me the ease with which variety may be adequately expressed, and the high
importance of taking it into account. There are numerous problems of special interest
to anthropologists that deal solely with variety.

There can be little doubt that most persons fail to have adequate conceptions of the
orderliness of variability, and think it is useless to pay scientific attention to variety,
as being, in their view, a subject wholly beyond the powers of definition. They forget
that what is confessedly undefined in the individual may be definite in the group, and
that uncertainty as regards the one is in no way incompatible with statistical assurance as regards the other. Almost everybody is familiar nowadays with the constancy
of the average in different samples of the same large group, but they do not often
realise the completeness with which a similar statistical constancy permeates the
whole of the group. The mean or the average is practically nothing more than the
middlemost value in a marshalled series. A constancy analogous to that of the mean
characterises the values that occupy any other fractional position that we please to
name such as the 10th per cent, or the 20th per cent; it is not peculiar to the 50th per
cent, or middlemost. Still less do they realise the fact that all Variety has a strong
family likeness, by approximating more or less closely to the normal type, which is
that which mathematicians prove must be the consequence of Variety being due to
the aggregate effect of a very large number of small and independent influences. (II,
pp. 384–385)

We can see now the point of Fourier's remark in §3.4; the use of the
variance to give a measure of the spread of distributions is obvious to us now
but it needed a genius to observe the value of such use. Quetelet's method of
fitting the normal curve does not introduce the variance parameter, $\sigma^2$, and
hence he does not realize that the normal curve is completely specified by its
mean and variance. It follows easily from the $\alpha$, $\sigma^2$ parametrization of equation (19.3.1) that an appropriate linear transformation of the variable leads to
a standard distribution. Galton devised the *quincunx* with which to obtain a

normal distribution approximately and was able to note that normal distributions obey a stable law, by which the sum of independent normal variables is again normal. Galton (1889b) ends his discussion with the statement that the properties of the law of frequency (that is, the normal distribution) are

largely available in anthropometric inquiry. They enable us to define the trustworthiness of our results, and to deal with such interesting problems as those of correlation and family resemblance, which cannot be solved without its help. Anthropologists seem to have little idea of the wide fields of inquiry open to them as soon as they are prepared to deal with individual variety and cease to narrow their view to the consideration of the Average. (p. 21)

In view of the bitterness in the later struggle between the Galton-Pearson school and the Mendelians, we should note Galton's opinions. First, in §4.4, we have Galton's letter to Charles Darwin, his possible models, and K. Pearson's comments that Galton's thought ran very closely to the Mendelian views. A paragraph of Galton (1908) gave praise to Mendel for the cogency of his arguments and appreciation of the fact that Mendel worked alone but continuously on his problems. Galton wrote: "Mendel clearly showed that there were such things as alternative atomic characters of equal potency in descent. How far characters generally may be due to simple, or to molecular characters more or less correlated together, has yet to be discovered." It is evident from the context that Galton used "atomic" in the sense of "discrete." K. Pearson (1914–1930) as his biographer makes no connection between the two uses of "atomic" and no mention of the word in his index, a notion that might have destroyed his belief in the Galton "law of ancestral heredity" and allowed him to keep in touch with the Mendelians (see also §§1.1 and 4.11).

# 4
# Genetics

## 4.1 Genetics Before Mendel and Galton

Two remarkable pedigrees by René-Antoine Ferchault de Réaumur* (1683–1757) and Pierre Louis Moreau de Maupertuis* (1698–1759) on polydactyly may be taken as the dawn of modern scientific genetics, at a time when the theory of heredity was dominated by the preformationists.

The preformationists believed that the embryo developed out of a body, formed before conception, which itself contained such a body and so on indefinitely. Conception to their minds consisted merely of activating such a body, whether in an ovum or a sperm. This might be called the *encasement* or *emboîtement* theory. Both Réaumur and Maupertuis had been uneasy about the absurdities of the theory but the alternative, originally due to Aristotle and often known as the epigenetic theory, offered even more absurdities [see J.B. Gough, Réaumur's biographer (*DSB*, XI, pp. 327–335).] It was evident from ordinary life experiences that traits passed down through the father or mother. A critical test of the hypothesis was indicated. Human polydactyly was sufficiently rare to give confidence that the appearance of the trait in the child was not due to a *mésalliance*. A pedigree established the point. Such observations, by Réaumur, Maupertuis, and Wolff, mentioned later in this section, led many biologists to abandon the preformation theory and take up pangenetic theories.

Maupertuis had heard of hereditary albinism in Negroes and sought some other character to test; here he was fortunate in discovering the polydactylous Ruhe family, of whose members some had six fingers and six toes, others had no defects, and others had some defects intermediate between the fully blown syndrome and normality. From Maupertuis's description, a family tree can be constructed showing that E.H. of Rostock had the defect, passing it on to her daughter who became Elisabeth Ruhe; of Elisabeth's eight children, four (including surgeon Jacob Ruhe) showed the defect. Jacob married, like those mentioned previously, a spouse free of the defect; of their offspring, two males of the sibship of six had the defect (see Maupertuis, 1752). These findings enabled Glass (1959) to construct a genealogical table

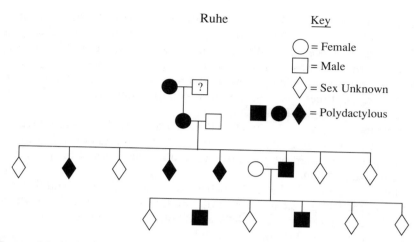

FIGURE 4.1.1. Polydactyly in the Ruhe family, according to Maupertuis. [From Glass, B., Temkin, O., and Straus, W.L., Jr. (Eds.), *Forerunners of Darwin: 1745–1859*, p. 65. Copyright © Johns Hopkins University Press, Baltimore, 1959. Reprinted with permission.]

(Figure 4.1.1). Critics would say that the observations were explicable by chance but Maupertuis gave a very satisfying defence, an early use of the "reference set" in applied probability (see §17.11).

Maupertuis (1752) can be regarded as the first great classic of probability theory applied to biology. In the preformation theory, it was necessary to have only one sex involved in the succession, so that opinion was divided between the "spermists" and "ovists." The Ruhe genealogy of Figure 4.1.1 showed a succession through a male and also one through a female, which is sufficient to reject the preformation theory. Glass (1959) concludes that Maupertuis may be added to the company of Lamarck, Karl Ernst von Baer, Blumenbach, Darwin, Wallace, and Hugo de Vries—"a man too far before his time, he was the most many-sided genius of them all."

Mayr (1982) gives an appreciation of Maupertuis ending with the passage:

Buffon saw clearly that both father and mother made a genetic contribution, but it was P. M. de Maupertuis, more than anyone else, who developed a theory of inheritance that can be considered as foreshadowing later developments (Glass, 1959; Stubbe, 1972). Maupertuis espoused a theory of pangenesis, based on the thoughts of Anaxagoras and Hippocrates, postulating particles ("elements") from both father and mother as responsible for the characters of the offspring. Most components of his theory can be found in the later theories of Naudin, Darwin, and Galton. (p. 646)

[For Maupertuis see also the opinion of Sturtevant (1965b, p. 6). For a more detailed account of the genetic work of Maupertuis and Réaumur, see Lancaster (1994a).]

Another successful attack was made on the preformation theory by Caspar Friedrich Wolff* (1734–1794) in his paper of 1759. Although the theory of preformation had been discredited, its influence was felt after 1900, for the notion that the chromosomes handed down a given character from generation to generation was held by some to be a mere revival of the preformation theory. In recent times, this objection has been overcome by the finding that the genes are self-replicating.

Experimental work was hindered in the early days by the expense. Mendel was able to get work done by his gardener, but others were not so fortunate, for example, Koelreuter, who was dismissed after the death of his protector (see *DSB*, VII, p. 440*b*).

Dobell (1914) and Stubbe (1972) believe that the first experimental treatment of a genetical problem was reported by Leeuwenhoek (1679), in which male wild rabbits were mated with "fancied" rabbits, the grayish (wild) color being dominant to the blue (fancied) color.

Other early workers preceding Mendel included Joseph Adams (1756–1818), a British medical practitioner, who studied human genetics as a background to genetic counselling and gave his conclusions in Adams (1814); a summary of his findings is available in Motulsky (1959) and Vogel and Motulsky (1982). Christian Friedrich Nasse (1778–1851) recognized the sex-linked mode of inheritance in hemophilia and gave a typical detailed pedigree of the disease in a family (Nasse, 1820).

Among other workers we may cite William Herbert* (1778–1847), who was interested in the classification of plants and later studied their hybridization on a grand scale. He noted the phenomena of dominance and recessiveness; his work was known to both Mendel and Darwin. Joseph Gottlieb Koelreuter* (1733–1806) studied the mechanism of fertilization in plants and established that there were contributions to the gamete from both parents, contrary to some prevailing theory. His work was confirmed and extended by Karl Friedrich von Gaertner* (1772–1850). From these workers, Mendel could see that he must study varieties in one species, as we would say now, if he wished to follow the generations through.

## 4.2  Mendel

Johann Gregor Mendel* (1822–1884) is one of the great figures of science. For our purposes, we note that he studied at the University of Olmütz (Olomouc) in the years 1840 to 1843, where he completed the two-year philosophy course (in the sense of natural science) studying philosophy, mathematics, and physics, which included a course in combinatorial mathematics given by J. Fux. He left the university with high commendations and entered the Augustinian monastery in Brno, taking the name Gregor, in 1843. The monastery was supported mainly by the income from its estates, and F.C. Napp, abbot since 1824, was attempting improvements in agriculture; a teacher of philosophy, Matthew Klácel (1808–1882), was directing an experimental garden and investigating variation, heredity, and evolution in plants. Klácel is said to have guided Mendel in his first studies in science and later put him in charge of the experimental garden (Kruta and Orel, *DSB*, IX, pp. 277–278). Klácel was later dismissed on a heresy charge and emigrated to America. During the early years at the monastery, Mendel also attended courses at the Brno Institute in agriculture, pomology, and viticulture, given by F. Diebl (1770–1859), an author of a two-volume work on plant production.

Mendel, after some experience as a pastor at Brno and as a teacher at Znojmo in southern Moravia, was sent up to the University of Vienna, where he studied under, among others, Franz Unger* (1800–1870) and Eduard Fenzl (1808–1879), botanists, and Andreas Ettingshausen (1796–1878), physicist and mathematician, who had moreover written a book on combinatorial analysis (Ettingshausen, 1826).

Mendel chose to work on the genus *Pisum*, principally of the species *sativum*, because its plants possessed constant characters, because the hybrids of such plants can, during the flowering period, be protected from the influence of all foreign pollen or be easily capable of such protection, and because the hybrids and their offspring should suffer no marked disturbance in their fertility in the successive generations. Mendel after preliminary tests chose an appropriate variety. It may be noted that Mendel was dealing with "major genes," as they are now called, the presence or absence of which can produce an easily observable effect (compare the choices of Réaumur and Maupertuis); such inheritance can be contrasted with multifactorial inheritance. Mendel defined characters as *dominant* or *recessive*; the dominant characters are transmitted entire or almost unchanged in the hybridization, whereas the recessive character withdraws or entirely disappears in the hybrids. Mendel wrote of hybrids between species where we would write of crosses between pure lines. Plants in the pure lines when selfed (self-pollinated) reproduce the same line.

Mendel's introductory remarks are of great interest. A translation was made for the Royal Horticultural Society and first published with permission by W. Bateson (1902/1909/1913) in his *Mendel's Principles of Heredity*. We shall refer to the English translation as Mendel (1865, transl.) so as to avoid difficulties as to the timing of Mendel's paper. Mendel (1865, transl.) begins with some observations:

The striking regularity with which the same hybrid forms always reappeared whenever fertilisation took place between the same species induced further experiments to be undertaken, the object of which was to follow up the developments of the hybrids in their progeny.

To this object numerous careful observers, such as Kölreuter, Gärtner, Herbert, Lecoq, Wichura, and others, have devoted a part of their life with inexhaustible perseverance ... Those who survey the work done in this department will arrive at the conviction that among all the numerous experiments made, not one has been carried out to such an extent and in such a way as to make it possible to determine the number of different forms under which the offspring of hybrids appear, or to arrange these forms with certainty according to their separate generations, or definitely to ascertain their statistical relations.

A new synthesis or paradigm, as we should now say, was needed. W. Bateson (1913) notes in a footnote that "it is to the clear conception of these three primary necessities that the whole success of Mendel's work is due. So far as I know this conception was absolutely new in his day."

Mendel crossed two pure lines, namely the dominant and the recessive, to obtain a generation of hybrids, and by selfing these he obtained the first hybrid generation; later in the literature, these two generations have been named, $F_1$ and $F_2$, respectively. We would say that the original cross gave $F_1$, the elements of which all have identical appearance. Mendel's great achievement was to show that by selfing the pure lines could be reestablished.

Numerous experiments showed that the ratio of dominant phenotypes to recessive phenotypes in $F_2$ was 3:1. Combinatorial theory enabled this to be expressed in symbols. For a cross in $F_1$, there are two possibilities for the first parent; either the dominant $A$ or the recessive $a$ will pass to the offspring; this was to be written

$$A + a \tag{4.2.1}$$

where there is a unit understood before $A$ and before $a$, stating that there is one way in which the symbol appears. The possibilities for the second parent were similar and could be written again in the form shown in equation (4.2.1). Now the offspring can obtain (a) $A$ from the first parent and $A$ from the second parent symbolized by $AA$, (b) $A$ from first and $a$ from second parent, (c) $a$ from first and $A$ from second, and (d) $a$ from each parent. These possibilities could be written $AA$, $Aa$, $aA$, and $aa$, but these are the terms of $(A + a)(A + a) = (A + a)^2$, for it was known that the parent from which $A$ was derived was immaterial so that $Aa = aA$. Mating in $F_1$ is therefore symbolized by

$$(A + a)(A + a) = (A + a)^2 = A^2 + 2Aa + a^2. \tag{4.2.2}$$

There are thus three types of hybrids (crosses) symbolized by $A^2$, $Aa$, and $a^2$, or as Mendel wrote $A$, $Aa$, and $a$, with relative frequencies $1:2:1$ since $Aa$ appears with coefficient 2 and $A^2$ and $a^2$ with coefficient unity, and so the ratio of the carriers of the dominant character to the carriers of the recessive character only is 3:1. (In modern language the genotypes appear with relative frequencies $1:2:1$ and the dominant phenotypes with relative frequencies 3:1.) Mendel gave a table of a mating system in which after $F_2$ all crossings occur between parents within the same genotype and showed that this will lead to the virtual disappearance of the $Aa$ type, and the establishment of pure lines.

Mendel (1865) then considered mating systems with two or more characters symbolized by $A$, $a$ and $B$, $b$. $F_1$ then consisted of crosses between a pure line with respect to $A$ and $B$ and a pure line with respect to $a$ and $b$. Parents can thus contribute $AB$, $Ab$, $aB$, or $ab$ to $F_2$. This could be symbolized by

$$(A + a)(B + b) \tag{4.2.3}$$

instead of (4.2.1). The possibilities of a mating were then to be symbolized by

$$(A + a)(B + b)(A + a)(B + b) = (A + a)^2(B + b)^2$$
$$= (A^2 + 2Aa + a^2)(B^2 + 2Bb + b^2)$$
$$= A^2B^2 + 2A^2Bb + A^2b^2 + 2AaB^2 + 4AaBb$$
$$+ 2Aab^2 + a^2B^2 + 2a^2Bb + a^2b^2, \tag{4.2.4}$$

an expression of nine terms, given by Mendel with all superscripts omitted. Equation (4.2.3) asserts the independent segregation of the genes (in modern terms, no linkage). Mendel then, in effect, introduced

$$(A + a)(B + b)(C + c) \tag{4.2.5}$$

and an expression of $3^3 = 27$ terms corresponding to equation (4.2.4) obtained by squaring equation (4.2.5).

Using the symbolism above, the formation of $F_1$, the crossing from pure lines, is given by

$$(A + A)(a + a) = 4Aa. \tag{4.2.6}$$

So in terms of the notation we can state the three laws of Mendel in modern terminology:

1. *Uniformity.* The progeny in the first filial generation are all identical and heterozygous.
2. *Segregation.* Intercrosses of heterozygotes yield offspring in the ratio 1:2:1 with heterozygotes having the proportion one-half; back crosses of the heterozygote with a pure line give progeny in a 1:1 ratio, heterozygote and homozygote being equally numerous.
3. *Independence.* With some choices of different characters, the segregation of the characters is independent. [This last law essentially asserts that the multiplications in equations (4.2.3) and (4.2.5) are valid for appropriate choices of $A$, $B$ and $C$, that is, in the absence of linkage.]

Mendel (1865) put all these laws to the test and the observed ratios of the various types of progeny were in reasonable agreement with the theoretical.

In Mendel's work, dominance and recessiveness played an important role, and it is often convenient to use the terms to distinguish two alleles. Sometimes it is inappropriate to use the terms as the heterozygote can be distinguished from both homozygotes; for example, sickle-cell trait can be regarded as due to a dominant gene, whereas sickle-cell anemia is a thalassemia due to a recessive gene. Lucien-Claude-Jules-Marie Cuénot* (1866–1951) first described a locus with multiple alleles, namely coat color in mice (Cuénot, 1905). F. Bernstein proved that there were at least three alleles at the locus for the ABO human blood groups (see §5.7).

## 4.3 Stability of the Gene Frequencies

We now prove the interesting and important (Hardy-)Weinberg law. It is assumed that the proportion of genotypes $AA$, $Aa$, and $aa$ in both males and females are $D$, $2H$, and $R$, $(D + 2H + R) = 1$ with no further restriction. After meiosis in the males (in the model it is supposed that this represents a random choice of gene from the particular male), $A$ will be in the proportion $D + H = p$, $a$ in the proportion $H + R = q$, so that $p + q = 1$; similarly in the females. This can be expressed by the probability generating function, $(pt + q)$; similarly for the females. Now the probability generating function for the offspring will be the product of these two, namely

$$(pt + q)(pt + q) = p^2t^2 + 2pqt + q^2. \tag{4.3.1}$$

The proportions of homozygous dominants, heterozygotes, and homozygous recessives are thus $p^2$, $2pq$, and $q^2$. The overall proportions of dominant and recessive genes are still $p$ and $q$, respectively. Under the same hypothesis, future generations will have the same proportions, so the population ratios are stable under panmixia, except for sampling variations. Mathematically, the procedure is the convolution of two binomial distributions, each with parameters, $p$, and unity.

This result is due first to Weinberg (1908); some six weeks after that publication, Reginald Crundall Punnett* (1875–1967) asked Godfrey Harold Hardy* (1877–1947) to consider the stability of the gene ratios in natural populations. Hardy (1908) was published in *Science* rather than in *Nature*, whose editorial policy was strongly anti-Mendelian at that time. Weinberg (1909) states that he had considered the problem for segregations at several loci simultaneously. There were some forerunners, Yule (1902), Castle (1903–1904), and Pearson (1904a), but they used no parameters other than $p = \frac{1}{2}$.

As B. Glass (1963) points out, the gene frequency at an individual locus has an objective meaning in an interbreeding population and its gene frequencies will tend to remain constant except for sampling errors from generation to generation; so it is proper to think of a collective genotype of the whole population comprising all the alleles at all the loci of its individual members. For this last observation, it can be noted that the Hardy-Weinberg law can be readily proved with several alleles at a locus by the same mathematical methods.

Mendel's pioneer experiments in genetics were celebrated in the *Proceedings of the American Philosophical Society* in 1965 by Dunn, Sturtevant, et al. The first pages and subject of their contributions were as follows: Dunn, p. 189, Mendel, his work and his place in history; Sturtevant, p. 199, the early

Mendelians; Dobzhansky, p. 205, Mendelism, Darwinism, and evolutionism; Stern, p. 216, Mendel and human genetics; Glass, p. 227, a century of biochemical genetics; Sonnenborn, p. 237, genetics and man's vision; Mangelsdorf, p. 242, genetics, agriculture, and the world food problem.

[See the 1965 edition of Mendel (1865) for a criticism of some of the results of Mendel's experimental data by R.A. Fisher, who believed that the differences between observed and expected were often too small to be considered as true results of an experiment.]

It is unfortunate that Mendel used the notation $(A, Aa, a)$ where modern authors would use the notation $(A^2, Aa, a^2)$ or $(AA, Aa, aa)$, clearly showing that there are two alleles with two places to be filled.

## 4.4  Evolution and Genetics

Mendel's work was done in the context of plant breeding and agriculture; it might be said that such studies related to the origin of the individual, the subject of one of the two great controversies in the eighteenth century. A theory of heredity was required also in the second great controversy of that century, namely that between the evolutionists and those believing in the fixity of species.

Charles Robert Darwin* (1809–1882) accepted the fusion or blending theory of inheritance but was uneasy about its truth; he realized that blending inheritance in bisexual reproduction would tend rapidly to produce uniformity and further that if variability persists, causes of new variation must be continually at work. His opinions were greatly affected by the criticism of Henry Charles Fleeming Jenkin* (1833–1885), who attacked the theory of evolution (Jenkin, 1867), especially emphasizing that the possible variations of an existing species must be limited, "contained within a sphere of variation," and that the probability of favorable variations (that is, mutations) in a single individual becoming incorporated in an existent population was small, because such variations were infrequent and the effect would be swamped by the blending inheritance. Darwin accepted in part Jenkin's objections (see Vorzimmer, 1963), but never satisfactorily solved them. The solution required, indeed, the rejection of the blending theory and the adoption of a particulate theory. Let us note R.A. Fisher's remarks on what might have been. Fisher (1930) has written:

It is a remarkable fact that had any thinker in the middle of the nineteenth century undertaken, as a piece of abstract and theoretical analysis, the task of constructing a particulate theory of inheritance, he would have been led, on the basis of a few very simple assumptions, to produce a system identical with the modern scheme of Mendelian or factorial inheritance. (p. 7)

There was indeed such a one, who tried to free Darwin of his difficulties. In a letter to Charles Darwin of 19 December 1875, given in Pearson (1914–1930,

II, p. 189), Galton considers a character (appearing as an intracellular unit) that is black or white with a hybrid that is "exactly intermediate, viz. gray." The character might be observed as consisting of cells with all white gemmules or all black gemmules, respectively, in the two parents, but the hybrid could take two forms, namely (a) the hybrid cell might have half its gemmules white and half black, or (b) have all its gemmules uniformly gray. The distinctive appearances of (a) and (b) would be visible only by microscopic examination. Galton notes that with the possibility of gemmules from each parent being present in a cell of the offspring, there might be two gemmules only, each of which might be black or white, with ratios $1:2:1$ for both black, precisely one black, and both white, respectively; with three gemmules there would be ratios $1:3:3:1$ and so on as in the successive lines of a Pascal triangle. Pearson comments in a footnote "this letter shows how very closely Galton's thought at this time ran on Mendelian lines." Pearson carries out a calculation using Galton's law of ancestral heredity, from which he concludes that "this simple application of his law would have led Galton to the fundamental equation of Mendelian hybridisation" (II, p. 84). Galton, however, had not pursued his early ideas on the notion of discontinuous variables.

We consider now the relevance of atomism to Galton's law of ancestral heredity. Galton (1865, p. 326) writes: "The share a man retains in the constitution of his remote descendants is inconceivably small. The father transmits, on an average, one-half of his nature, the grandfather one-fourth, the great-grandfather one-eighth; the share decreasing step by step in a geometrical ratio, with great rapidity." As Pearson (1914–1930, II, p. 84) points out, he leaves out the female components of and does not count the numbers in each donor generation, and obtains the formula, $1 = \dfrac{1}{2} + \dfrac{1}{2^2} + \cdots$, but we need not refer to his formula, for Galton has already admitted that his theory implies the infinitely divisible property of matter. Yet neither Galton nor his biographer notice that there has been a contradiction of an established physical theory. The same objection can be applied to Darwin's blending inheritance.

Fisher (1930) shows the inapplicability of the blending theory of heredity as follows. Let $X$ and $Y$ stand for the deviations from the mean of the heritable part of the character for the two parents; under panmixia $X$ is independent of $Y$. $EX^2$ and $EY^2$ are the variances equal to $\sigma^2$. Under blending inheritance, the offspring has a deviation $\frac{1}{2}E(X + Y)$ and this averaged out over the population is zero. The variance of $\frac{1}{2}(X + Y)$ under hypotheses made is $E[\frac{1}{2}(X + Y)]^2 = \frac{1}{4}[EX^2 + 2E(XY) + EY^2] = \frac{1}{4}(\sigma^2 + 0 + \sigma^2) = \frac{1}{2}\sigma^2$, $E(XY) = 0$ from the independence of $X$ and $Y$. Thus the variance is halved at each generation. Even if the panmixia condition is relaxed to $E(XY) = \rho$, the variance is $\frac{1}{2}(1 + \rho)\sigma^2$, so that with correlations between $X$ and $Y$ as large as $\frac{1}{2}$ there is still a diminution of one-quarter in the variance. There is no diminution if and only if the coefficient of correlation is unity; in this case to every parent with $X = x$ say, we have to find a second with $Y = x$. As Fisher (1930, p. 5) points out, an "inevitable inference of the blending theory is that the bulk of the heritable variance present at any moment is of extremely recent origin. One half is new in each generation."[1]

Particulate inheritance is an alternative to the blending hypothesis. Under it, there is no inherent tendency for the heritable variance to diminish, as follows from the Hardy-Weinberg law (see §4.3) or from similar computations, provided that there is no selection.

Galton (1865) writing on hereditary talent and character, commented on the small amount of real knowledge on heredity. He collected many data on the inheritance of scholarly, artistic, and athletic ablities from the records of notable families and concluded that there was strong evidence for the inheritance of such abilities. Galton's variables here were continuous, for example, weight, height, intelligence, and so, when he turned to experiments on the sweet pea, he chose to study the inheritance of continuous characters such as the weight of the seed; here he could find the analogy of the joint distribution of seed weights between parent and offspring and the joint distribution of heights of father and son. At this time, there was no general mathematical theory of joint distributions, that is, observations classified simultaneously by two variables whether continuous or discrete (discontinuous); even in the study of independent variables, notable mathematicians could fall into serious errors. Also at this time there was no general theory of association in science. The physicists or applied mathematicians were accustomed to work with strict forms of causation—given initial conditions of a planet in orbit at time 0, what will be its position at time $t$? The philosophers had only two statements, complete mutual dependence and (set theoretical) independence. Set theoretical independence means no cell is empty in the customary break-up of the joint distribution of the two variables, row and column, into cells. The biologist would say that the heights of father and son are associated, but he needs some measure of the degree of association. The same occurs if the actual measurements of heights are used. This type of problem had never been studied before.

Galton, according to Pearson (1914–1930, IIIA, p. 2), wished to solve a problem of Cuvier, who denoted by the word *correlation* an association between two organs or characters of a family, whereby, if correlation were present, from the discovery of an organ a prediction could be made as to the nature of others. Galton could not see as yet that the two problems were identical; it could be said that Galton's work was a mathematization and generalization of Cuvier's theory of correlation.

## 4.5 Rediscovery of Mendel's Work

Hugo de Vries* (1848–1935), a professor of botany at the University of Amsterdam, had worked on taxonomic problems and plant physiology but in the 1880s turned his attention to the problems of heredity. The result

---

[1] Fisher, R.A. *The Genetical Theory of Natural Selection.* Copyright © Oxford University Press, Oxford, 1930. Reprinted with permission.

was *Intracellular Pangenesis* (1889), in which the works of Spencer, Darwin, Naegeli, and Weismann were reviewed; he proposed the theory that "pangenes," so named in honor of Darwin, were the carriers of hereditary traits; each independent characteristic was associated with a material bearer, made up of numerous molecules and having the ability to take up nourishment and to divide to yield two new pangenes. The pangenes were located within cells and after cell division, a set of pangenes went to each daughter cell. In each reproductive cell, there was present at least one of the pangenes, representative of the particular trait. De Vries certainly knew the segregation laws in 1896, deduced from his own experimental work and not suggested to him by any means, direct or indirect, from Mendel.

Carl Franz Joseph Erich Correns* (1864–1933), according to Olby (*DSB*, III, pp. 421–423), arrived at a conclusion resembling Mendel's on heredity in peas in October 1899; after he had reread W.O. Focke's *Die Pflanzenmischlinge*, he was led to read Mendel's paper of 1865 and realized the identity of Mendel's general theory with his own. A few weeks later he received a copy of de Vries (1900), which had come to similar conclusions; he immediately sent a copy of his work to the German Botanical Society. Correns was the first to correlate Mendelian segregation with the reduction division of the nucleus, namely meiosis (see §4.9); he showed that the Mendelian segregation does not require the dominant-recessive relationship; he also showed that Mendel's second law on independent segregation was not universally true, for the genes may be coupled (conjugated or linked). In 1902 he produced a chromosome explanation of heredity. In 1903 he predicted that sex would be found to be inherited in the Mendelian manner. He later gave a proof of the existence of non-Mendelian inheritance, denied by modern writers.

Erich Tschermak von Seysenegg* (1871–1962), a successful breeder of new agricultural varieties and a versatile scientist, was led to a rediscovery of Mendel's laws by his hybridization experiments with peas at the botanical garden in Ghent, Belgium (see Tschermak von Seysenegg, 1900).

In May 1900, William Bateson, travelling by train from Cambridge to London to deliver a lecture, read the account of Mendel's work by de Vries (1900) and was so deeply impressed with it that he at once incorporated it into his lecture (Bateson, 1900).

It is often asked why Mendelism, neglected in 1865, was accepted so generally after 1902. It is evident that there had been great advances in general biology in the intervening time and that much experimental work had been done. In Mendel's time, although a mathematical model could be constructed to account for his ratios, there was no physical (that is, anatomical) model. There was a deepening of the understanding of the events of fertilization of the egg, due to progress in microscopic techniques and the application of much talent to elucidate the various steps (see §4.9), so that, in 1902, there were scientists who could identify Mendelian segregation with visible effects in fertilization of the ovum. In particular, the reducing division (meiosis) could be identified with the choosing of one of the allelomorphs and the union of the paternal and maternal chromosome could be identified with the restoration of the number of genes at each locus. It may well be that the botanists of 1865, for example Naegeli, disliked the introduction of mathematics into the study of a botanical topic.

The reader may prefer the summing up of Stern (1965):

The reason that [Mendel's paper] remained without influence for thirty-five years and then, on its "rediscovery" was immediately recognized in its importance, depended primarily on what had happened during the intervening period. This period had included the discovery of chromosomes and their behavior in cell division and gametogenesis, an intensive study of biological variation, and the formulation, by Weismann, of a conceptual framework for a theory of heredity, development and evolution. The time was ripe for Mendelism.[2] (p. 216)

[See Coleman (1965) for an account of the development of the theory by the anatomists, especially pp. 156 and 157. See also Wilson (1914).]

We may add that in human genetics, Mendelism was not at first associated with common traits (but see Maupertuis in §4.1 and Garrod in §5.2). [See Allen (1979) for the differences in outlook between naturalists and experimentalists in their treatments of genotype and phenotype.]

Many abstracts and translations of classic works on genetics are available in Moore (1972) and Peters (1959).

## 4.6 Pure Lines

In the early part of the nineteenth century, research in the heredity of plants was carried on chiefly by the plant hybridizers, hoping to find new breeds by making crosses within species, between species, or even between genera; they necessarily became interested in the topic of hybrid sterility, although in some of their crosses between species or genera the hybrids were fertile and bred true. Mendel, as we have seen in §4.2, made crosses within species; in the table of §7 of Mendel (1865), it is shown that, for a single character, it is possible to obtain pure lines, in which every mating within the line yields offspring again belonging to the line. Quite independently of Mendel, Johannsen (1903/1959) investigated such pure lines.

Wilhelm Ludvig Johannsen* (1857–1927) was born into a cultured family; since they could not support him to enter a university, he was apprenticed to a pharmacist, studying especially chemistry and botany (Winge, 1958). In 1881, he became assistant to the chemist Johann Gustav Christoffer Kjeldahl* (1849–1900) in the newly established Carlsberg Laboratory at Copenhagen. His first researches were in the metabolic processes of ripening, dormancy, and germination in plants. His chief interest to us here was that he aimed to elucidate Galton's law of ancestral heredity.

---

[2] Copyright © *Proceedings of the American Philosophical Society* 1965; 109:216–226. Reprinted with permission.

Johannsen established "pure" lines by repeated selfing in peas and beans. (We can interpret this now as meaning that an individual in a "pure" line has homozygosity at every relevant locus, that is, the two elements at each such locus are identical.) He was able to demonstrate that in a pure line, the mean and variance of a metrical character in one generation were equal to the mean and variance of the previous generation, excepting changes brought about by the environment and by sampling errors. Selection in such pure lines was ineffective, for the offspring of an extreme individual had the same distribution of genes as the parent. There is some permanent feature in the pure line. Johannsen (1905) was later to adopt the term *gene* from de Vries (see §4.5). Johannsen later used the term *genotype* to mean also the totality of the genes in a population. With the use of these terms, we can say that the genotype of a pure line is fixed and remains unaltered in further generations, with exceptions as noted below. The physical expression of the genotype in metrical or qualitative properties is called the *phenotype*. In pure lines, the phenotype will be determined partly by environmental conditions. Variations in the phenotype are measured by variances; Johannsen (1903/1959) was the first to define the nonheritable variation. He notes that within pure lines the regression to the mean is complete, that is, the regression of offspring heights on parental heights, say, is zero. He then interprets Galton's law to mean that the average character of the offspring differs less from the mean of the population than their parents but in the same direction as their parents. He concludes, in the translation of Peters (1959) that "selection within a population causes a greater or lesser shift in the direction of the selection of the average for characteristics around which the individuals concerned are fluctuating," but this can only be so because the population before selection was heterogeneous, consisting in fact of a mixture of lines. He says that the typical stepwise progression in the direction of selection in the course of each generation depends on a stepwise progression in the relative frequencies of the differing lines concerned. Selection, therefore, must stop when the isolation of the most strongly divergent pure line is complete. Here his experimental work has led him to a confirmation of the objections of Jenkin (1867) to Darwin's hypothesis of blending heredity (see §4.4).

Johannsen (1903/1959) states that the Galton-Pearson law of ancestral inheritance in his pure lines is other than what has usually been taken for granted, "the individual peculiarities of the parents, grandparents, or any other ancestor, have—insofar as my researches are concerned—no influence on the average characteristics of the offspring." Indeed, the genotype has been assumed to be constant and the phenotype will depend on the environmental characteristics. Johannsen (1903/1959) gives the following exceptions: (a) there may be fluctuating variants in the genotype itself, although this phenomenon has not yet been demonstrated, or the lines may not be completely pure; (b) cross-breeding must be considered, but this is not part of the current experimental program; and (c) there may be mutations.

Finally, we may cite some interesting historical comments by Johannsen (1903/1959); p. 26 of Peters (1959):

Although Weismann very recently regarded Galton as the "voice" of cellular limitation through "Determinants"—or however one might name these theoretical hereditary corpuscles, de Vries deserves the great credit for having recognized the unitary nature of hereditary particles, which he called "pangenes"—a concept he first published in 1889 and further advanced in the "Mutationstheorie." It seems to me that the Galton-de Vries theory is the only truly useful theory of heredity.[3]

The zoologist Weldon should have realized that in spite of defects there was important matter in the Johannsen paper. Johannsen had referred respectfully to the *Biometriker*, meaning the Pearson school with special reference to Galton, as we have seen. These considerations did not count with W(eldon) and P(earson) (1902–1903), who maintained that Johannsen's correlations did not appear to be zero but rather to approximate to 0.5. Nor did these industrious reviewers note that the existence of nonheritable variance would necessarily bias downward all calculations of correlations.

With so many persons working on a new subject, there must always be doubts about absolute priorities; thus William Ernest Castle* (1867–1962) could say, "Mendelian predictions are based, not on the *somatic character of the parents*, but on the character of the *germ-cells* formed by the parents" (1903–1904, p. 228). Later, Sergeĭ Natanovič Bernšteĭn* (1880–1968) could say (Bernstein, 1932) that genetics would be a fertile field for the applications of mathematics and that the processes of genetics were Markovian; because, we might add, the genetic makeup of the offspring, given those of the parents, is independent of earlier ancestors.

## 4.7 Major Genes and Polygenes

In the English-language histories, there is usually a stress on the fact that Mendel (1865) was dealing with genes that have a major effect on the production of a character, and polygenes are brought in later in the discussion. Stern (1965) believes that this is unhistorical for Mendel (1865 transl.) had used the notion of polygenes as in the following:

Despite the many disturbing factors with which the observations had to contend, it is nevertheless seen by this experiment that the development of the hybrids, with regard to those characters which concern the form of the plants, follows the same laws as in *Pisum*. With regard to the colour characters, it certainly appears difficult to perceive a substantial agreement. Apart from the fact that from the union of a white and a purple-red colouring a whole series of colours results [in $F_2$], from purple to

---

[3] Citations from the translation of Johannsen (1903). In Peters, J.A. (Ed.) *Classic Papers in Genetics*. Copyright © Prentice Hall, Englewood Cliffs, NJ, 1959. Reprinted with permission.

pale violet and white, the circumstance is a striking one that among thirty-one flowering plants only one received the recessive character of the white colour, while in *Pisum* this occurs on the average in every fourth plant. Even these enigmatical results, however, might probably be explained by the law governing *Pisum* if we might assume that the colour of the flowers and seeds of *Ph. multiflorus* is a combination of two or more entirely independent colours, which individually act like any other constant character in the plant. If the flower-colour $A$ were a combination of the individual characters $A_1 + A_2 + \cdots$ which produce the total impression of a purple coloration, then by fertilisation with the differentiating character, white colour, $a$, there would be produced the hybrid unions $A_1a + A_2a + \cdots$ and so would it be with the corresponding colouring of the seed-coats. According to the above assumption, each of these hybrid colour unions would be independent, and would consequently develop quite independently from the others. It is then easily seen that from the combination of the separate developmental series a complete colour-series must result. If, for instance, $A = A_1 + A_2$, then the hybrids $A_1a$ and $A_2a$ form the developmental series—

$$A_1 + 2A_1a + a$$
$$A_2 + 2A_2a + a. \qquad\qquad [4.7.1]$$

(pp. 367–368)

After the rediscovery of the Mendel theory, several continental biologists were working at related themes, for example, Johannsen (see §4.6). Herman Nilsson-Ehle* (1873–1949) was working on the genetics of wheat and found that genes at different loci could be responsible for the same phenotypic character. He divided such genes into two classes, *homerous* (presumably shortened from homomerous) and *polymerous*. In the first class, the presence of a dominant at any locus brought about the character, but the presence of dominants at other loci brought no further intensity in the character (such as colour of grain). In the second class, the presence of dominant genes at further loci brought about further intensity; the effects in this polymerous class were additive.

From Johannsen's work on pure lines, there were clear implications that the variance of a metrical character in the phenotype could be split into genetic and environmental parts; from the work of Nilsson-Ehle, the character might be dependent on several genes and so there would be a sum of such genetic components and of the independent environmental effects. Weinberg (1908, 1910) was able to partition the total variance of a phenotypic character and the correlations between relatives. This work appears to have been accepted on the European continent. In England and America, G.U. Yule, F. Galton, W. Bateson, and T.H. Morgan had seen that the apparently continuous variation could be brought about by the combined action of many genes, whose effects were additive (see Sturtevant, 1965b); indeed, there had been calculations to show that such additive genes could bring about correlations between relatives, although the results were disputed by the biometricians because they would contradict Galton's law of ancestral inheritance based on a blending theory of inheritance.

We cannot trace the line of thought that led R.A. Fisher to his first genetical paper (1918), in which he showed that Mendelian theory implied the kind of correlations observed and that the sources of the correlations could be assigned to heritable and nonheritable fractions; the heritable fraction could further be partitioned into parts due to additive gene action, to dominance, and to interaction between genes; he showed further that allowance could be made for the correlation between spouses. Moreover, the excess of correlations between sibs over those between parent and offspring could be explained by the Mendelian theory. [See Yates and Mather (1963) and Moran and Smith (1966) for discussions in detail of these points.] The inheritance of continuous variation was thus shown to be explicable in Mendelian terms and this conclusion is the basis of modern genetical thought.

Let us consider the multivariate case where for simplicity we suppose that the variables, $z_i$, are centered and that

$$X = \sum_1^n a_i z_i, \qquad Y = \sum_1^n b_i z_i \qquad (4.7.2)$$

where $X$ and $Y$ each are approximately normal, then $X$ and $Y$ are jointly approximately normal, $\operatorname{var} X = \sum_1^n a_i^2 \sigma_i^2$, $\operatorname{var} Y = \sum_1^n b_i^2 \sigma_i^2$:

$$\text{correlation } (X, Y) = \sum_1^n a_i b_i \sigma_i^2 \bigg/ \left( \sum_1^n a_i^2 \sigma_i^2 \sum_1^n b_i^2 \sigma_i^2 \right)^{1/2} \qquad (4.7.3)$$

and a similar finding holds for $X, Y, \ldots, T$, any finite number of variables. If the variables, $z_i$, are normal, equation (4.7.3) is exact.

Yule (1902, p. 235) showed that if a character such as height depended on a large number of genes each contributing a summand, the distribution of the sum would appear to be continuous. He did not call on the central limit theorem. Fisher (1918), using ideas related to this theorem, suggested that the sum would be approximately normal also.

## 4.8 Wilhelm Weinberg as a Medical Statistician

Wilhelm Weinberg* (1862–1937), the son of a Stuttgart merchant, is one of the great amateurs of statistics. He studied medicine at Tübingen and Munich, graduating with an M.D. degree in 1886. After clinical experience in Berlin, Vienna, and Frankfurt, he established himself as a general practitioner and obstetrician in Stuttgart in 1889.

It is reported by Stern (1962) that Weinberg published, in the period 1886 to 1935, more than 160 papers on a variety of genetical, demographic, and medical statistical topics. Here are some examples:

*An interpretation of the demographic data on twins.* Although Francis Galton (1876) had considered that twin studies might well be carried out to determine the relative importance of nature and nurture in individual development, he possibly did not know that twins are of two types, monovular and binovular. This distinction was first made clear by the embryologist Gabriel Madeleine Camille Dareste (1822–1899) in Dareste (1874). Weinberg (1901) noted that binovular twinning would follow the binomial law with relative frequencies, $p^2$, $2pq$, and $q^2$ for 2 males, 1 male, or 0 male in the twin pair, where $p$ was the probability of a male gene and where the contributions of the genes of the two parents could be regarded as independent events; further, $p + q = 1$. Now $p = 0.5 + \delta$, where $\delta$ is a small quantity and so the pair would be of probability $2pq = 2(0.5 + \delta)(0.5 - \delta) = 0.5 - 2\delta^2$. Of course, all such pairs of mixed sexes would be dizygotic (that is, binovular). In any series, the number of binovular twins could be estimated to be twice the number of mixed-sex pairs. This is known as the *difference method*, for the estimate of the number of monovular twinnings is just the difference between the total and binovular numbers. Weinberg found from official birth statistics that the frequency of monovular twinning was independent of maternal age but that the frequency of binovular twinning varied with maternal age; furthermore, binovular twinning had some genetic component and the law he determined was incompatible with blending inheritance. Weinberg cites other authors, Bertillon, Hensen, and Moser, as being aware of the difference method.

*Population genetics.* In §4.3 we have mentioned Weinberg's part in the formulation of the Hardy-Weinberg law. Weinberg (1909) generalized the theorem in terms valid for multiple alleles and investigated polyhybrid populations, in which he recognized their essentially different methods of obtaining equilibrium. These studies were part of the foundation of population genetics.

*Ascertainment and sampling from populations.* Weinberg (1912) first drew attention to problems of ascertainments. Bateson (1902/1909/1913) had found that in sibships, resulting from crosses of the form $Aa \times Aa$, the ratio of the number of doubly recessives to sibship size observed was greater than $\frac{1}{4}$. Weinberg realized that this false inference had followed from observing sibships that always contained one doubly recessive, the proband as would be said now. His solution was to consider only the sibs of the proband and calculate the ratio in this subset, or alternatively to give the observed distribution for sibships of $m$ (excluding the proband). He saw that this problem was met in other investigations; for example, it had been considered that parents had larger mean fertility than their children. He proposed to compare the fertility of the sibs of the parents with that of the children (that is, of the next generation).

Besides his fundamental contributions to genetics, Weinberg made detailed studies of mortality statistics and the genetics of specific diseases. He also constructed morbidity tables analogous to life tables. A detailed bibliography of his works has been given by Sherwood (1964). Weinberg was Chairman of the Stuttgart Chapter of the Deutsche Gesellschaft für Rassenhygiene. (See Stern, 1962, and *DSB*, XIV, pp. 230–231.)

## 4.9  Anatomical Basis of Heredity

Statistical authors, as we shall see, have often concentrated too much on the mathematical findings without attempting to relate their models to the biological realities. Among the biologists, too, there had been confusion as to the functions and importance to heredity of the ovum and spermatozoon. We may begin with Wilhelm August Oscar Hertwig* (1849–1922), accompanying E.H.P.A. Haeckel* (1834–1919) on a research expedition to the Mediterranean in 1875. Hertwig found the sea urchin *Toxopneustes lividus* to be an admirable experimental animal for his purposes, for it was small and remarkably transparent, and the ovum was surrounded by a finely divided yolk without any noticeable membrane. He was able to detect the egg nucleus before and after entry of the spermatozoon and the fusion of the two nuclei five to ten minutes later. He observed that there was no breakup of the new nucleus but there was morphological continuity between it and the cleavage nuclei of the developing embryo. Only one spermatozoon was required to fertilize one egg and indeed the entry of further spermatozoa was prevented by the development of a vitelline membrane. Hermann Fol* (1845–1892) made these observations more precise; he showed that "the egg nucleus is not a separate cell but rather an important structural and functional part of the ovum; [he showed also] that during maturation of the egg three daughter nuclei are cast off as polar bodies so that the mature ovum contains only one nucleus."[4]

Walther Flemming* (1843–1905), with novel precise staining methods, demonstrated in 1879 the phases of cell division and perceived that the longitudinal division of the chromatin loops was the basis of nuclear multiplication; formerly it had been thought that the division of these loops was transverse.

The next advance was to identify the substance(s) responsible for carrying the genetic information, often spoken of as the *idioplasm*, about which there was much literature. Hertwig in 1885 believed that egg and sperm nuclei had equal power and so for him the nuclein or chromatin (staining substances) must be the material basis of heredity and responsible for both fertilization and the transmission of the hereditary properties. In 1884, Eduard Adolf Strasburger* (1844–1912) considered that the male and female germ cells, spermatozoon and ovum, were equivalent in the bearing of heredity. He also maintained that the longitudinal splitting in the chromosomes assured the proportionate distribution in the nuclear threads. August Friedrich Leopold Weismann* (1834–1914) had reached similar views, it is said, before he read Hertwig's paper. Carl Rabl* (1853–1917) published in 1885 a detailed description of cell division and maintained that the chromatin staining fila-

---

[4] G.E. Allen on Fol, *Dictionary of Scientific Biography*, vol. V, p. 51*b*. Charles Scribner's Sons, New York. Copyright © American Council of Learned Societies, 1972. Reprinted with permission.

ments in the nucleus, later called the chromosomes, must actually persist through the interphase, even though for a time they seemed to disappear, according to G. Robinson (*DSB*, XI, pp. 254–256).

The morphologists had thus concluded that important events in fertilization took place in the nucleus of the germ cells. In particular, the theory of Haeckel (1866) was founded upon the idea that heredity is brought about by the transmission from one generation to the next of a substance with a definite chemical and molecular constitution. Weismann, Strasburger, Hertwig, and Koelliker in 1884 were agreed that the nucleus was an important factor in inheritance. Weismann (1885/1891) located the germ plasm within the nucleus. Once a chemical theory was established and Weismann had brought forward the theory of the continuity of the germ plasm, it was difficult to believe in the inheritance of acquired characters.

The changes in morphological understanding of heredity over the period 1840 to 1885 made no impression on the work of Galton and later Weldon and Karl Pearson, but see the comment from Galton (1908) cited in §3.6. However, they made the work of the Mendelians easier for, after the rediscovery of Mendel's work, Theodor Boveri* (1862–1915) in Boveri (1902, 1904) and Walter Stanborough Sutton* (1877–1916) in Sutton (1902, 1903) realized that the behavior of the Mendelian factors exactly parallels the behavior of the chromosomes in sexual reproduction. Sutton (1903) in the *Biological Bulletin* makes six points, of which the last three are uniquely his own:

1. The somatic chromosomes comprise two equivalent groups, one of maternal derivation and one of paternal derivation.
2. The synapsis consists of pairing of corresponding (homologous) maternal and paternal chromosomes.
3. The chromosomes retain their morphologic and functional individuality throughout the life cycle.
4. The synaptic mates contain the physical units that correspond to the Mendelian allelomorphs; that is, the chromosomes contain the genes.
5. The maternal and paternal chromosomes of different pairs separate independently from each other—"The number of possible combinations in the germ-products of a single individual of any species is represented by the simple formula $2^n$ in which $n$ represents the number of chromosomes in the reduced series."
6. "Some chromosomes at least are related to a number of different allelomorphs ... all the allelomorphs represented by any one chromosome must be inherited together." (Peters, 1959, pp. 28–34).

Some modifications must be made in Sutton's statements. In (5) he is thinking of a dominant or recessive at each locus but there can be allelomorphs so that the number of possibilities (possible combinations) is not $2^n$ but $\prod_{k=1}^{n} a_k$, $k = 1, 2, \ldots$ where $a_k$ is the number of allelomorphs at locus $k$. In (6) it may be noted that every gene on the chromosome will contribute to the zygote.

Theodor Boveri made discoveries overlapping those of Sutton noted above. In the years 1887 to 1890, he described some aspects of the maturation of the egg and the formation of its polar bodies; he demonstrated the individuality of the chromosomes and their continuous organized existence in the cell nucleus; he also proved that at fertilization, egg and spermatozoon contributed equivalent sets of chromosomes to the new individual.

Although workers, some of whom are mentioned above, had suggested a relationship between chromosome movement and the segregation of Mendelian factors, Thomas Hunt Morgan* (1866–1945) in 1902 still held objections that are detailed in Allen's article in the *DSB* (IX, p. 520) and in Allen (1978). Morgan had been trained as a morphologist but became more interested in experimental work, embryology (1895–1902), evolution (1903–1910), heredity (1910–1925), and embryology and its relations to heredity and evolution (1925–1945), according to Allen in the same entries. Morgan considered that, as no large group of characteristics had been found, the identification of Mendel's factors and chromosomes was unlikely; many characteristics seemed to be mixtures of the parental types; Mendelism offered no easy solution to the inheritance of sex; Mendel's laws were defined too narrowly, as indeed turned out to be the case; Mendelism seemed to imply a preformationism; not enough experimentation had been carried out. De Vries had been carrying out much experimental work on plants and Morgan was anxious to study similar problems in animals. He decided to use *Drosophila melanogaster* as his subject; it had the advantages of a short breeding time and only four chromosomes. In 1910, he observed the first mutation, a white-eyed fly, white being recessive to red. An analysis of his findings led him to see that Mendelian and chromosome theories were but two sides of the same theory.

Many new discoveries were made in Morgan's laboratory. Thus Alfred Henry Sturtevant* (1891–1970) showed that the degree of linkage could give a measure of nearness (later named after Morgan) on the chromosome of gene loci. "Lethal" genes and multiple (at the same locus) alleles were noted. These and other findings were summarized as *The Mechanism of Mendelian Heredity* by Morgan, Sturtevant, et al. (1915).

So Mendelism was extended and its physical counterpart made clear; further, new concepts were developed. As Allen (1978) suggests, Morgan showed that heredity could be treated quantitatively and rigorously.

## 4.10 Genetic Linkage

It is now known that genes are located (have loci) along the chromosomes and that genes on the same chromosome tend to be transmitted to the offspring together, more often if they are close together than if they are widely separated; confirmation for this comes from cytological studies by which the chiasmata are observed to form and chromosome segments are exchanged. This is the phenomenon of "crossing over." It may even appear that genes on the same chromosome are segregating independently. Such pairs of genes are said to be *syntenic* but not linked. Morgan and his school, in the first two decades of this century, were able to form maps giving the positions of genes on the chromosomes of *Drosophila melanogaster* by appropriate breeding experiments. Mendel had been fortunate in his choice of characters, for they were not linked. Mendel was able to state his third law as mentioned in §4.2, whereby the genes are stated to segregate independently. A table may be formed of matings of the various paternal gametic possibilities of a double heterozygote against the double recessive maternal homozygote:

| Paternal gametes | $AB$ | $Ab$ | $aB$ | $ab$ |
|---|---|---|---|---|
| Maternal gametes | $ab$ | $ab$ | $ab$ | $ab$ |
| Offspring genotype | $AaBb$ | $Aabb$ | $aaBb$ | $aabb$ |
| Frequency | 1/4 | 1/4 | 1/4 | 1/4 |

No linkage is present.

Such a simple result does not follow if there is linkage. We then have the following possibilities:

| Paternal gametes | $AB$ | $Ab$ | $aB$ | $ab$ |
|---|---|---|---|---|
| Maternal gametes | $ab$ | $ab$ | $ab$ | $ab$ |
| Offspring genotype | $AaBb$ | $Aabb$ | $aaBb$ | $aabb$ |
| Frequency (coupling) | $\frac{1}{2} - \theta$ | $\theta$ | $\theta$ | $\frac{1}{2} - \theta$ |
| Frequency (repulsion) | $\theta$ | $\frac{1}{2} - \theta$ | $\frac{1}{2} - \theta$ | $\theta$ |

where $\theta$ is the recombination fraction, $0 \leqslant \theta \leqslant \frac{1}{2}$. Suppose that $\theta = 0$, then $A$ and $B$ or $a$ and $b$ pass over together to the gamete; if $\theta = \frac{1}{2}$, $A$ and $B$ are not on the same paternal chromosome but on the homologous chromosome and there has been a crossing over. *Coupling* is said to be present when $A$ and $B$ are on the same chromosome with formula $\dfrac{AB}{ab}$ and *repulsion* with formula $\dfrac{Ab}{aB}$ or equivalently they are said to be in the *cis* and *trans* positions, respectively. If $\theta$ is small the $A$ and $B$ loci are close together; if $\theta$ approaches $\frac{1}{2}$ the two loci are separated by a long distance along the chromosome. Indeed, the distance between two loci is defined as a function of $\theta$, the Morgan distance (see §4.9).

(We discuss the inheritance of sex in §5.4, together with sex limited or controlled traits, as in some species the sex ratios may be very high.)

## 4.11  Biometrical Genetics

Galton looked to some biological model, as we should say, where large numbers of observations could be readily accumulated. He chose to study the sweet pea *Lathyrus odoratus* rather than the edible pea *Pisum sativum* of Mendel's choice, because he would not be troubled by too much variation between the sizes of peas within the same pod. Galton had already begun to accumulate observations in 1875.

Galton assumes that the normal curve would adequately describe the deviations of his populations from their mean value. He takes seeds of equal weight (which may be the ordinary produce of large-seeded plants or the exceptional produce of small-seeded plants) and then determines the distribution of the daughter seeds conditional on the parental seed weight. A linear relation is found to hold approximately but the slope is less steep than it would be if the conditional mean of the daughters was equal to the value of the parental mean. The ratio between the two slopes is called the *coefficient of reversion*. Galton finds also that the variance of the daughter seeds is a constant, that is, independent of the size of the parental seed. There are some criticisms of the experiment by Pearson (1914–1930) but the joint normal distribution was to have many applications in biology, psychology, and other fields.

Galton implies that each parent contributes a proportion, $\frac{1}{4}$, to the heritage, each grandparent $\frac{1}{16}, \ldots$ and $n$ generations back, each individual $2^{-2n}$. He takes stature, a continuous quantity, and eye colour, a discontinuous quality. Galton (1886) states that

The blending in stature is due to its being the aggregate of the quasi-independent inheritances of many separate parts, while eye-colour appears to be much less various in its origin. If then it can be shown, as I shall be able to do, that notwithstanding this two-fold difference between the qualities of stature and eye-colour, the shares of hereditary contribution from the various ancestors are in each case alike, we may with some confidence expect that the law by which these hereditary contributions are governed will be widely, and perhaps even universally, applicable. (pp. 402–403)

Here Galton has forgotten or given up the notions of heredity of discontinuous variables such as we have mentioned in §4.4, and eye colour is given the same treatment as stature.

Galton (1897) returns to the estimation of the contributions of the ancestors to the heredity of an individual as follows:

The two parents contribute between them, on the average one-half, or (0.5) of the total heritage of the offspring; the four grandparents, one-quarter, or (0.5)$^2$; the eight great grandparents, one-eighth, or (0.5)$^3$ and so on. Thus the sum of the contributions is expressed by the series $\{(0.5) + (0.5)^2 + (0.5)^3 +, \text{etc.}\}$, which, being equal to 1, accounts for the whole heritage. (p. 402)

It was not apparent until Johannsen (1903/1959) studied his pure lines that, conditional on the parents belonging to the same pure line, the mean value (or even the distribution) of a measurable quantity in the offspring was independent of the parental values. We might summarize Johannsen's view by saying that in a pure line the phenotype of the offspring is independent of the phenotypes of the parents. Weinberg (1908, 1910) was the first to partition the total variance of phenotypes into genetic and environmental portions; he calculated the correlations to be expected between relatives, taking account of both genetic and environmental conditions.

It was not until Fisher (1918) that the importance of this point was fully realized. (See §4.6 for a note on the review of the Johannsen paper by Weldon and Pearson.)

Yule (1902), without committing himself to it, believed that the law of ancestral heredity should be stated in the form, "the mean character of the offspring can be calculated [that is, estimated] with the more exactness, the more extensive our knowledge of the corresponding characters of the ancestry."

## 4.12 Karl Pearson as a Geneticist

Karl Pearson* (1857–1936) graduated from the University of Cambridge, Third Wrangler in the mathematical tripos in 1879 and so third in the first class in mathematics, and received his LL.B. degree in 1881, although he never practised law. He was Goldsmid Professor of Applied Mathematics and Mechanics at University College, London, from 1884 to 1911, when he resigned to become the first Galton Professor of Eugenics, a post offered to him by University College in keeping with the will of Francis Galton. He retired in 1933. He was elected Fellow of the Royal Society in 1896 and was awarded the Darwin Medal of the Royal Society in 1898. As part of his duties at Gresham College (1891–1894), he published *The Grammar of Science* (1892), an influential book.

His steady progress as an applied mathematician continued until he, like Weldon, read Galton (1889c) and Weldon was appointed to the chair of zoology at University College. Pearson saw that Galton's correlation theory yielded a category, broader than causation but of which causation was only the limit, and that the theory would be applicable to psychology, anthropology, medicine, and sociology. In the next fifteen years, 1890 to 1904, Pearson firmly established mathematical statistics as a body of knowledge. After it had been laid down by the Royal Society that biology and mathematics were not to be discussed together, Pearson with the help of Weldon and others founded the journal *Biometrika*, with aims stated by Weldon in its first editorial (cited in §19.2). Pearson was an inspiring head of department and his new interest, biometrics and mathematical statistics, soon attracted colleagues and students, for example, George Udny Yule, Louis Napoleon George Filon (1875–1937), and later Major Greenwood (1880–1949) and Joseph Oscar Irwin (1898–1982); he also had such assistants as Alice Lee, Ethel M. Elderton, and Julia Bell, who enabled him to develop tables of the various test functions and to analyze great masses of data on human heredity (see Haldane, 1957). For a discussion of his personality, see the two tributes by Major Greenwood in the *Dictionary of National Biography* and in the *Journal of the Royal Statistical Society*.

Pearson's great series, *Contributions to the Mathematical Theory of Evolution*, I and II, and *Mathematical Contributions to the Theory of Evolution*, III to XVI, served to bring together and to give mathematical form to the problems of genetics and evolution. However, Darwin's ideas on evolution by continuous change colored his views and Pearson never accepted Mendelism, in particular, the possibility that evolution might have come about by major mutations followed by selection processes. His views seem to have been unaffected by the cytological discoveries detailed in §4.9. He accepted Galton's law of ancestral heredity but did not see that Johannsen's work on pure lines gave an insight into the law different from Galton's (see §4.6); nor did he realize that the distributions observed were of phenotypes and not genotypes and hence that the estimated correlations would be too small because of the nonheritable variances. These errors of judgment, including his rejection of Mendelism, led to a feud between the Mendelians and the biometricians. In this feud, Weldon, the biologist, must be rated a greater sinner than Pearson. Johannsen (1929) fiercely criticized Pearson's genetic views and his dogmatism on Galton's law of ancestral heredity. R.A. Fisher's views on K. Pearson are well known. [See also Weldon's remarks after meeting Pearson in 1895 in Stigler (1986, p. 337).]

## 4.13 R.A. Fisher as a Geneticist

Ronald Aylmer Fisher* was born in East Finchley, London in 1890 and died in Adelaide, South Australia in 1962. He has been so well served by his biographers, F. Yates and K. Mather (1963), his daughter Joan (Fisher) Box (1978), and many others that it is necessary for us here only to discuss his great contributions to the science of human genetics, evolution, and anthropometry. His influence on the development of clinical trials is discussed in Chapter 18.

We are told that from an early age, Fisher had strong biological interests. At Cambridge University, he became a wrangler in 1912 and studied further physical and applied mathematical topics under James Jeans and the theory of errors under F.J.M. Stratton. While at Cambridge he read Karl Pearson's *Mathematical Contributions to the Theory of Evolution*, which directed his attention to evolutionary and genetical problems. In 1919 Fisher accepted the newly created post of statistician at Rothamsted Experimental Station, in preference to the post of chief statistician under Karl Pearson at University College, London. Under the enlightened directorship of Sir John Russell, FRS, Fisher was able to meet a great variety of researchers in biological and other fields and to act as consultant in their work. Those who met him in later life were surprised at the wide range of his interests and his ability to call on an extensive knowledge in many fields. At Rothamsted, he recast the theoretical basis of mathematical statistics, making many advances in distribution theory, in estimation theory, in the testing of hypotheses and in the design of experiment. His elucidation of the problem of the *Rhesus* factor is discussed in §5.7.

There are many biographies of Fisher besides those mentioned above. There is also Fisher (1971–1974), and the collected works edited by J.H. Bennett (1983), which gives extracts of the correspondence between Fisher and Leonard Darwin (1850–1943) and provides many interesting details, for example, about the difficulties in finding a publisher for Fisher (1918).

Fisher in 1943, according to Box (1978, p. 416), decided to put into practice what he had long felt needed doing, namely "the creation of permanent inbred lines in all ... of the genes recognizable in mice." Further, Fisher thought there should be cooperation between laboratories worldwide. He should be credited with attempting to create a genome of the house mouse species. The human species did not seem susceptible, owing to huge amounts of money required for a human genome project.

# 5
# Human Genetics

## 5.1 Autosomal Dominant Inheritance

Some examples of human inheritance are given in this chapter; in this section they are of the autosomal dominant type. Some further human examples are given in §4.8.

1. *Polydactyly.* This anomaly has already been discussed in §4.1 for the Ruhe family. R.A.F. de Réaumur (1751) had given another such family.
2. *Brachydactyly.* The inheritance of this anomaly was described by Farabee (1905) as a shortening of the phalanges (digits) of the hands and feet and sometimes their numbers were reduced. A family tree is given by Vogel and Motulsky (1982, p. 84, Figure 3.2).
3. *Huntington's chorea.* This anomaly is, perhaps, the classic among the autosomal dominant types. Reed and Neel (1959) and Wendt and Drohm (1972) gave surveys of cases of the anomaly in Michigan and West Germany, respectively.
4. *Porphyria variegata.* Porphyria is the excretion of bile pigments in the urine; the chemical substances known as porphyrins are present in normal bile. This disease, named by Geoffrey Dean (b. 1918), appeared as an epidemic disease in South Africa, or the disease was correctly diagnosed there, just after the end of World War II. Indeed, soon after his migration to South Africa in 1947, Dean was called into consultation to examine a series of patients, who seemed to be hysterical and then developed generalized paralysis (Dean, 1971). After seeing several patients, Dean recognized that acute porphyria was present.

   The first patient so recognized was an intelligent, middle-aged businesswoman, complaining of abdominal pain for many years; repeated visits to doctors with physical and radiographic examinations had failed to reveal any cause. Barbiturates had worsened her symptoms; she was emotional about her condition and her medical advisors had considered her to be rather unstable and neurotic. After her first exploratory operation, she had severe abdominal pains and became mentally disoriented and remained in hospital for five months. After a second operation, when she was given a barbiturate anesthetic, she again suffered severe pains. On consultation, Dean found her muscles weak and noticed her reddish-brown urine, due to porphyrin. She became paralysed and died despite all therapeutic care. The skin on the back of her hands had blistered easily, characteristic of the porphyrias. Three of her sisters had already died with similar histories of emotional symptoms, the taking of sleeping tablets, paralysis, and the passing of reddish urine.

Three weeks later, Dean was to see a nurse, 19 years old, who also died. Dean succeeded in tracing 478 descendants of her great-grandfather (1814–1884); of these descendants 434 were still living in 1951; 60 descendants over the age of 18 years had inherited porphyria and in later years a further 46 descendants were found to have porphyria. Dean (1971, p. 5) has a table showing that about half the children of an affected parent were again affected, and there is a detailed family tree of the family in question (opposite p. 18). All cases are traced back in a larger family tree to a Dutch couple in South Africa who married in 1688.

Dean later found that patients with variegata type porphyria had been noted in Holland, but apparently no family relationship has so far been established with the South African founder of 1688. From the table (Dean, 1971, p. 5), the proportion of children who develop the disease in adult life is given as a percentage of the total children in each generation: 5 out of 10 in the second generation, or 50%; 16 out of 37 in the third generation, or 43.2%; 32 out of 59 in the fourth generation, or 54.2%; and 7 out of 19 in the fifth generation, or 46.8%. The proportions in the generations are thus what is expected of a disease due to a dominant gene.

## Founder Effects

Some populations of large size have in recent centuries developed from an originally small number of persons, the *founders*; good examples are the Maoris in New Zealand and various islands in the Pacific Ocean, the Amerindian population of the Americas migrating across Beringia, and the 19 original Dutch families settling in South Africa. This is now a well-known problem in anthropological studies and it occurs in animal studies in the wild. Mathematical models have been set up, including a very special form in Vogel and Motulsky (1982), which are supported by some laboratory work. One consequence is that some alleles may be lost from the daughter population; another is that gene frequencies in the daughter populations from the same mother population may differ considerably, even at the foundation of the daughter populations. The incidence of porphyria variegata in South Africa is an example of such a founder effect. The theory shows that it may be quite illusory to draw conclusions from gradients in blood group frequencies about origins of populations, for example, subpopulations of the Amerinds [see also Giddings, Kaneshiro, and Anderson (1989)].

Perhaps, the first example of such autosomal dominant inheritance was given by Maupertuis (1744–1745), for which Glass (1959) gives the following citation:

He had already, prior to 1745, conducted a careful study in human genetics that is matched by only one other of its time. It seems that a young albino Negro had been brought to Paris and naturally enough created a great stir. This started Maupertuis to thinking about heredity. Surely the albino condition must be hereditary, for although it seemed to appear sporadically among Negroes, it was reportedly not rare in Senegal, where whole families were said to be "white." Of the famous albino Indians of Panama he had also heard, and discoursed about them with no little sentimentality. "The black color," he said, "is just as hereditary in crows and blackbirds, as it is in Negroes: I have nevertheless seen white blackbirds and white crows a number of times. (p. 63)"[1]

## 5.2 Autosomal Recessive Inheritance

In this mode of inheritance, the heterozygote does not differ phenotypically from the normal homozygote, although in some cases special methods, perhaps biochemical, will be able to detect a difference. If the parents are both heterozygotes, the ratio of the three genotypes $AA$, $Aa$, $aa$ will be $1:2:1$; that is, one-quarter of offspring will be doubly recessive for the defect. Recessive conditions are more readily studied in large families, as there is a chance of $(\frac{3}{4})^n$ of the condition failing to appear in a sibship of $n$, and the chance of 1, $2, 3, \ldots, n$ of the children in the family being doubly recessive is given by the terms of the binomial $(0.75 + 0.25t)^n$, $t$ being just an indeterminate (or marker). The average number of double recessives in a family with heterozygous parents is thus $0.25n = \frac{1}{4}n$. It is nescessary to look to a careful method of assignment if the family is obtained by following up a proband; then the average is taken over the other members of the family and the expected number is $0.25(n - 1) = \frac{1}{4}(n - 1)$.

Archibald Edward Garrod* (1857–1936) in Garrod (1902) applied Mendel's theory to the human disease alkaptonuria, in which homogentisic acid is excreted in the urine, causing it to darken on exposure to light; he remarked that this excretion is an all-or-nothing symptom, it is usually congenital, it occurs in sibships, the parents are free of the disease, and in six of ten reported families the parents were first cousins, whereas in contemporary England the rate of first cousin marriages was at most only 3%. Further, the disease was a rarity even in first-cousin marriages. After discussion with William Bateson, it appeared that alkaptonuria is evidently due to a recessive gene. Garrod (1902) remarks that other such diseases may exist, for example,

albinism and cystinuria. Such diseases may be extreme cases of a class
of diseases with a "much more widespread applicability." [See Biochemical
Society Symposium No. 4 (1950) and Penrose (1953). See also the long discussion of inheritance of this type in Vogel and Motulsky (1982), especially
at p. 9; see also Dahlberg (1929).]

## 5.3 Sex-Linked Inheritance

Christian Friedrich Nasse (1820) recognized the sex-linked mode of inheritance of hemophilia and gave a typical detailed pedigree of the disease. Vogel
and Motulsky (1982, p. 91) give the family tree of Victoria, Queen of England,
and a pedigree with two female homozygotes for $X$-linked hemophilia.

Much work has been carried out on the $X$-chromosome by J.B.S. Haldane
and Julia Bell (1879–1979), especially examining the pedigrees in the *Treasury of Human Inheritance* in London (see §5.9) and by Victor Almon
McKusick (b. 1921) in the United States.

## 5.4 Inheritance of the Sex Ratio

Clarence Erwin McClung* (1870–1946) showed that the inheritance of sex
was Mendelian and corresponded to a cross ($YX \times XX$), where $Y$ is dominant to $X$. [See also Edmund Beecher Wilson* (1856–1939) (*DSB*, XIV, pp.
423–436) and his student, Nettie Maria Stevens* (1861–1912) (*DSB*, XVIII,
pp. 867–869).] It might be thought that the primary sex ratio would be 1:1,
but in most human populations observed it has been in the neighborhood of
1.06:1 at birth. Let us write the distribution of a single birth, ($pt + q$), meaning that $p$ is the probability of the birth being a male.

### Example

Laplace (1814/1951) noted that the ratio of male births to female births had
been approximately equal in all the countries of Europe. In the small community of Carcelle-le-Grignon there had been born 983 boys and 1,026 girls
out of 2,009 births; Laplace considered this to be a deviation from the expected mean of $\frac{1}{2}$ due to chance. A special investigation, initiated by him,
in chosen districts throughout France, yielded a count of 110,312 boys and
105,287 girls born in 1800, 1801, and 1802, 46,037 marriages, and the deaths
of 103,659 males and 99,443 females. He estimated a ratio of population to
annual births as 28.35 from these statistics and a knowledge of the living
populations. He then estimated the annual births for all France to be a
million and, using the ratio, the population of all France to be 28,352,845. He
was able to compute that the error was not more than half a million, with a
probability of less than 1 in 300,000.

Has $p$ a non-degenerate distribution? Does it vary between sibships? In other words, is $p$ a universal constant for families?

The null hypothesis will specify that $p$ is a constant. A simplifying assumption is that $p$ is a constant throughout the time of generation of a sibship. We can examine sibships of $N = 2, 3, 4, \ldots$. The distribution of sex in a given sibship is $(pt + q)^n$, $n = 2, 3, \ldots$, the probability generating function of a binomial variable with parameters $p$ and $n$.

Geissler (1889) published tables of the frequency distribution of the sex ratio of sibships of a given size. His findings for sibships of eight are well known, being used as an example of the binomial distribution by R.A. Fisher (1925). There is a notorious bias in demographic data toward even numbers, so that we must regard such findings as suspect. Fisher found that in the 53,680 sibships of eight, observed numbers exceeded the numbers expected on the hypothesis of independence of the sex of successive sibs, namely the terms of $53,680(pt + q)^8$, $p = 1 - q = 0.5147676$, in every case except for three boys or five boys where there are large deficits. He did not conclude that the parameter $p$, the probability of a birth being male, varied between families. His comments should be read to see how many ideas can be mentioned in good concise English.

Lancaster (1950b) carried out the same test but on the collection of 38,495 sibships of size nine observed before the birth of the ninth child. In these sibships of eight there were differences (observed minus expected) of $-404$, $-226$, and $-307$ against expected values of 8,902, 10,489, and 7,910 for five, four, and three sons, respectively. It appeared from this and other analyses that there were inconsistencies in Geissler's data, most probably because of faulty answers by the parents. Corrado Gini (1884–1965), who had written *Il Sesso dal Punto di Vista Statistico* in 1908, offered various criticisms of the views of Lancaster (Gini, 1951), but most demographers would now agree with Lancaster that "definite conclusions cannot be drawn from them [Geissler's data] to support the theory that there is heterogeneity in $p$, the probability of a birth being a male, between sibships." Nor have any other data been published that would upset the agreement.

## Sex-Limited Traits

Genes present in both sexes may only act in one sex; beard growth, hair distribution, and possibly male-pattern baldness represent examples of sex-limited traits in humans.

## Sex-Controlled Traits

Nora and Fraser (1974) state that this term is used in the context of multifactorial inheritance; it is believed that the penetrance of some multifactorial systems is affected by "environmental" causes and sex may be regarded as such; such an effect leads to the females having a higher incidence of congenital dislocation of the hip, patent *ductus arteriosus*, and atrial septal defect, and conversely to the males having higher incidence of pyloric stenosis, coarctation of the aorta, and transposition of the great vessels.

There appear to be doubts as to what genetic characters are passed on by the Y-chromosome other than sex and hairy ears; so sex-linked inheritance according to the present theory is usually X-linked. Hemophilia is the best known of these defects (see §5.3). The reader may refer to the various texts. [See Drayna and White (1985) for a survey of recent work on the mapping of genes on the X-chromosome.]

## 5.5  Human Mutation Rates: Penrose

According to Vogel and Motulsky (1982), mutations are of three types:

1. In genome mutations, there are changes in the number of chromosomes; whole sets of chromosomes may be multiplied (multiploidy), the number of single chromosomes may be increased (trisomy) or decreased (monosomy).
2. In chromosome mutations, microscopically detectable changes occur in the structure of chromosomes.
3. In gene mutations, changes cannot be detected microscopically but mutations are inferred from changes in the phenotype.

Mutations occurring in the germ cells may be transmitted to the offspring; mutations in the somatic cells may be detected and the individual will be seen to be a mosaic.

The first study of Lionel Sharples Penrose (1898–1972) at the Eastern Counties Hospital, Colchester, England, was a detailed clinical and genetic investigation of 1,280 patients in this mental institution; it involved collecting appropriate data on their parents, relatives, and 6,629 sibs. The patients were graded by intelligence tests, then by type of mental illness. Twenty cases of epiloia were found; in this disease there is a characteristic facial rash known as sebaceous adenoma, multiple nodules in the brain, and epilepsy; the mental condition ranges from idiocy into the normal range. The syndrome could be explained as due to a single dominant autosomal gene. Penrose was able to calculate the frequency of mentally defective epiloia as about 1 in 30,000 of the general community; the family data suggested that between one-half and one-quarter of clinical cases were due to fresh mutations and thus that the mutation rate was between 1 in 120,000 and 1 in 60,000; this was the first estimate by the "direct" method of a mutation rate in man. J.B.S. Haldane obtained by the "indirect" method an estimate of the mutation rate of the gene for hemophilia; the two estimates were published as a joint paper (Penrose and Haldane, 1935).

Penrose studied another genetic disease, now known after Down, over many years. By 1933 he had collected accurate data on parental ages in 150 families. By statistical reasoning, he was able to show that maternal age was an important factor and that with fixed maternal age, paternal age was irrelevant; he also showed that the distribution of maternal age was bimodal. Penrose demonstrated that palm printing or dermatoglyphics could be an aid in the family studies, because certain angles in the palm prints were continuous and measurable variables that were amenable to multivariate analysis. With advances in microscopic technique that allowed the chromosomes to be counted, many of the problems of Down syndrome were solved. Some cases are due to a trisomy at chromosome 21; some are due to a mosaic mother, and Penrose explained the effect this would have on the incidence in relatives. It is clear that he did much by clinical observation and statistical methods to clarify the etiology of this important medical disease, and of a form of mental defect, phenylketonuria, associated with the excretion of phenylpyruvic acid in the urine; he found such a patient in his series after testing over 500 urines, and he was able to show that this disease was transmitted as an autosomal recessive.

[For further references see Bora, Douglas, and Nestleman (1982), Denniston (1982), and Crow and Denniston (1985). See also Harris (1973) for an obituary of Penrose.]

## 5.6 Polymorphisms and Human Hemoglobinopathies

A polymorphism is a Mendelian or monogenic trait that exists in the population in at least two phenotypes (and also in at least two genotypes), neither of which is rare, that is, with a frequency greater than, say, 1% to 2% in some group of the species. The best known genetic polymorphism is the red blood cell surface antigens with alleles represented by $A_1$, $A_2$, B, and O. Other polymorphisms are known in the blood group or other human systems, and, of course, in racial differences.

It was known from mathematical theory that a mutation at a locus would very rarely result in a substantial relative frequency of the mutated gene in a large population unless it gave a selective advantage to its possessors. So we find in Bennett (1983, p. 37) that R.A. Fisher had written in 1930, in a reply to a remark that blood group serological differences were apparently without selective advantage, "There are a good many climatically limited blood diseases, such as malaria and yellow fever, so I would not be too sure of the absence of selection." In Bennett (1983, p. 184) Fisher in 1934 writes to W.C. Boyd that he "cannot see any escape from the view that the frequencies have been determined by more or less favourable selection in different regions, governed not improbably by the varying incidence of different endemic diseases in which the reaction of the blood may well be of slight but appreciable importance."[2]

---

[2] From Bennett, J.H. (Ed.) *Natural Selection, Heredity and Eugenics.* Copyright © The Clarendon Press, New York, 1983. Reprinted by permission of Oxford University Press.

Beet (1946, 1947), working in Northern Rhodesia (now Zambia), found that children, heterozygous at the locus for the sickle-cell anemia gene, tended to have lower percentages of malaria in blood smears and to have less splenomegaly than the normal homozygote.

The syndrome of a severe anemia with strenuous attempts of the body to form new red cells by extending the bone marrow had long been known under various names, including thalassemia, which means a Mediterranean blood disease. Anthony Clifford Allison (b. 1925) sought advice from John Burdon Sanderson Haldane* (1892–1964), who suggested that these thalassemias occurred in regions with heavy endemicity of *falciparum* malaria and that a balanced polymorphism might be involved. Allison (1954a,b,c, 1955, 1956, 1964) concluded that (a) the homozygous sickle condition is a lethal childhood condition in Africa, (b) balanced polymorphism exists because the heterozygote is at an advantage principally owing to the strength of its protection against *falciparum* malaria, (c) malaria exerts its selective effect mainly through differential viability of subjects with and without the sickle-cell gene between birth and reproductive age, and (d) high frequencies of the gene for hemoglobin S, carried by the heterozygotes, are found only in areas where *falciparum* malaria endemicity is, or has recently been, high. Two distinct forms of the disease occur, α-thalassemia, and β-thalassemia. Considering only α-thalassemia, we find that one allele for the S-hemoglobin of the disease gives rise to the sickle-cell trait, which protects the bearer against *falciparum* malaria, whereas the presence of two alleles gives rise to the full disease of sickle-cell anemia.

The gene $Fy^-$ confers complete immunity on blacks against vivax (tertian) malaria; it is thus a favorable gene in regions where there are many competing causes of death in children, for example, bacterial infections such as bowel diseases, and parasitic infections.

Many of the balanced genetic polymorphisms are as yet unexplained, but more will probably be studied, in a manner similar to that above, as a response to a major environmental hazard.

Vogel and Motulsky (1982, p. 394) suggest that, owing to the very low mutation rates, possibly between $10^{-8}$ and $10^{-9}$, for single base substitutions in the DNA and the low overall human population in earlier centuries, it is quite possible that only one or two mutations were responsible for the great numbers of persons now possessing the sickle-cell gene!

[For examples of enzyme and protein polymorphisms in man, see Harris (1969) and Roberts (1975). For a world distribution of polymorphic genes, see Roychoudhury and Nei (1988). See especially Dacie (1988) for an obituary and personal bibliography of Hermann Lehmann (1910–1985). On p. 428 of this obituary, there are listed 81 "new" hemoglobins described by Lehmann and his associates; the number listed worldwide now appears to be over 300; see Smith, Hill, et al. (1983, pp. 100–140). It may be that polymorphisms other than sickle-cell anemias have often occurred as mutations. Note that insulin-dependent diabetes mellitus is also a polymorphism.]

# 5.7 Human Blood Groups

Before 1900, it had been known that transfusion of the blood of one species into the veins of a member of another species led to lysis of the transfused blood cells. In particular, it was known that lamb's blood transfused into human veins led to such hemolysis. Thus nonhuman blood could not be used clinically to treat blood loss. The result of transfusing human blood into a human recipient was not known.

Karl Landsteiner* (1868–1943) reported in 1901 that he had tested the red cells of each of 22 individuals against the serums of each of the other 21 individuals and obtained the results shown in Table 5.7.1, excepting the line and column AB, a blood group that did not occur in his small sample, but that was discovered by members of his group, Decastello and Sturli (1902). Landsteiner's interpretation was that blood cells containing A, B, or AB agglutinogens were agglutinated by the serum of any person not possessing any such factor; so O blood, containing neither A nor B, agglutinates all other blood cells, A serum agglutinates B and AB blood cells, B serum agglutinates A and AB, and AB serum agglutinates the blood cells of no other group. Landsteiner (1901) recognized the importance of his work for blood transfusion but it was only after Ottenberg (1908, 1937) that typing of the recipient and cross-matching of donor and recipient became standard practice for transfusion; Ottenberg also suggested that the blood groups were inherited.

The genetics of the inheritance of the ABO system of blood groups have been worked out by Bernstein (1930a,b), Thomsen, Friedenreich, and Worsaae (1930a,b), and Wiener (1939), of which this last gives the most detailed analysis. Von Dungern and Hirszfeld (1910) hypothesized that there were two independent loci with alleles, A and a, at one locus and B and b at a second locus, A and B being dominant to a and b, respectively. In either parent, the possibilities of contributing to the gamete are given by the generating function $(A + a)(B + b)$, so by the use of the device of Mendel the

TABLE 5.7.1. Interaction of red blood cells and serum.

| Recipient blood group | Donor blood group | | | |
|---|---|---|---|---|
| | A | B | AB | O |
| A | − | + | + | − |
| B | + | − | + | − |
| AB | − | − | − | − |
| O | + | + | + | − |

The line AB and column AB were not available to Landsteiner (1901).

genotypic possibilities are given by the terms of

$$[(A + a)(B + b)]^2 = (A + a)^2(B + b)^2$$

$$= (a^2b^2) + (A^2b^2 + 2Aab^2) + (a^2B^2 + 2a^2Bb) \quad (5.7.1)$$

$$+ (A^2B^2 + 2AaB^2 + 2A^2Bb + 4AaBb)$$

where the phenotypes are given by O, A, B, and AB within the four bracketed expressions respectively. Equation (5.7.1) is equivalent to column 1 of Table 15 of Wiener (1939). Neither A nor B agglutinogen can occur in the phenotype unless it appears in at least one parent. This is due to the hypothesized dominance. For frequencies, we have to modify the Mendel-type equation writing $(pA + qa)$ and $(\bar{p}B + \bar{q}b)$, $p + \bar{p} = 1$, $q + \bar{q} = 1$, into equation (5.7.1) where $p$ and $\bar{p}$ are the relative frequencies in the population for A and a, $q$ and $\bar{q}$ for B and b. With such a modification (5.7.1) would show that

Relative Frequency of AB = Product of the Relative Frequencies
of A and of B                  (5.7.2)

$$= p\bar{p}.$$

But, as surveys have shown, this holds for no actual population given in Table 20 of Wiener (1939). [It is assumed that the populations are in a state of equilibrium according to the Hardy-Weinberg law.] The hypothesis of von Dungern and Hirszfeld must be set aside; that hypothesis also implied that an O × AB mating could produce an offspring, AB; a properly authenticated case has never been given, however. Felix Bernstein* (1878–1956), therefore, hypothesized that there is only one locus and that the alleles are A, B, and R (Bernstein, 1925). Phenotypes O, A, B, AB then correspond to genotypes RR,AA and AR,BB and BR and AB, respectively. Suppose that the gene frequencies for A, B, and R are $p$, $q$, and $r$, respectively, in the population. Then the relative frequencies of offspring genotypes in the population are given by

$$(pA + qB + rR)^2 = p^2A^2 + 2pqAB + 2prAR + q^2B^2 + 2qrBR + r^2R^2$$
$$(5.7.3)$$

with frequencies for the genes still $p$, $q$, and $r$. Wiener's Table 23 can be obtained by the following: the frequency of O-mothers is $r^2$, so that the frequencies of the genotypes with mothers mated at random is given by

$$r^2R(pA + qB + rR) = pr^2AR + qr^2BR + r^3R^2, \quad (5.7.4)$$

or in terms of phenotypes $pr^2$A, $qr^2$B, and $r^3$O. With mothers of genotype AB and hence phenotype AB the contribution is

$$pq(A + B)(pA + qB + rR) = pqA^2 + pq^2AB + pqrAR$$
$$+ p^2qAB + pq^2B^2 + pqrBR \quad (5.7.5)$$

and so the phenotype frequencies for O, A, B, AB are 0, $pq(p + r)$, $pq(q + r)$, $pq(p + q)$, respectively.

Under the Bernstein hypothesis, estimates of $p$, $q$, and $r$ should add up to unity approximately or subject to slight differences due to sampling errors. Wiener's Table 21 shows that this is so. The hypothesis of von Dungern and Hirszfeld fails this test badly. Table 19 of Wiener warrants discussion. Here, with mothers of Group O, there are five exceptions to the rule that the offspring cannot be AB; there are in all 5,010 offspring but the five exceptions occur in only three surveys of sizes 108, 70, and 70—in total 248; under the null hypothesis that the birth of an AB is equally likely in all children, the expectation would be $248 \times 5/5010 = 0.25$. There is a lack of homogeneity between the results of the different observers. This test of hypothesis would have to be modified to get rid of the multiple comparison difficulty. Von Dungern and Hirszfeld (1911) showed that A was not homogeneous and following them it has become standard practice to call the alleles $A_1$, $A_2$, B, R. None of the reasoning above is vitiated by this discovery.

Other blood group systems have been defined—the MN, P, Rh complex, Lutheran, Kell, and others [the reader is referred to Race and Sanger (1975), Nora and Fraser (1974), or a genetics textbook]. From our point of view the most interesting is the Rh complex. Levine and Stetson (1939) discovered a special antibody in the serum of a woman who had just delivered a macerated stillborn child. Several related antibodies were found. There was initially much confusion but R.A. Fisher (1943, 1947) was able to interpret the findings by suggesting the presence of three loci and the order with which they would appear on a single chromosome (see also Vogel and Motulsky, 1982, p. 129). These examples show the power of the combinatorial methods applied to the genetics of an important clinical problem, namely the dangers of a woman becoming immune to a blood agglutinogen of her husband and hence to that of her fetuses or children, leading to repeated miscarriages.

Fisher had always been interested in the field of human biology and so when he was appointed to the Galton Chair in human genetics at University College, London in 1933 (Fisher, 1935a), he established a statistical methodology and a center for the detection and estimation of human (genetic) linkage. He realized the value of the human blood group genes as markers for use in such studies as the description of the human genome. In these and related serological and hematological problems, George Lees Taylor (1897–1945) and later Ruth Ann Sanger (b. 1918) and Robert Russell Race (1907–1984) became collaborators [see Fisher (1943, 1945) and Clarke (1985)].

(See §5.6 for blood groups as polymorphisms.)

## 5.8 Extinction of Families or Genes

This problem has usually been considered as the extinction of family names, passed down by the male members, rather than the extinction of families, because the female lines are more difficult to trace [see Heyde and Seneta

(1977) for references and a more detailed account]. It was well known that the names of many Roman clans no longer appear in the later histories of the Roman Empire, and the extinction of noble or other names in modern times was the subject of papers by the demographer and statistician Louis-François Benoiston de Châteauneuf (1776–1856). There was a general opinion that the names of noble houses persisted for no more than three centuries. A solution was given by Irénée-Jules Bienaymé* (1796–1878) that may be summarized as follows. Suppose that a subpopulation with male net reproduction rate equal to $\rho$ is being considered in a total population that has a net reproduction rate not less than unity. Then if $\rho \leqslant 1$, the subpopulation will become extinct with probability one. If $\rho > 1$, the probability of extinction is the unique nonnegative solution of $s = f(s)$, where $f(s) = \sum_{j=0}^{\infty} p_j s^j$, $0 \leqslant s \leqslant 1$ and $p_j$ is the probability of a male having $j$ sons. $f'(1) = \rho$, the male net reproduction rate (Bienaymé, 1845).

Bienaymé's solution appears to have been forgotten. Thus, according to Pearson (1914–1930, II, pp. 141 and 143), Francis Galton in correspondence with Alphonse Louis Pierre Pyramus de Candolle* (1806–1893) noted how the families of great men die out and that many had drawn the conclusion that genius is associated with sterility. Galton propounded the general problem in the *Educational Times* and received in answer one erroneous reply. Later, the problem was solved by Galton and Watson (1874), but, of course, Bienaymé (1845) has the priority. The problem has importance in genetics as a solution to the extinction of unfavorable genes.

## 5.9  Human Pedigrees

Human pedigrees have been kept in many societies, sometimes by oral tradition, and usually written for the nobility. Owing to various experiences, such as war, fires, floods, and so on, the records have often been incomplete and so are marred by gaps; these gaps have often been filled erroneously. The most successful analysis from a demographic point of view has been that of Sigismund Peller, dealing with records of the German nobility (see Peller, 1965). William Augustus Guy (1810–1885) wrote on the duration of life of sovereigns but their special risk of assassination reduces the usefulness of these data (see Guy, 1847).

We have already mentioned pedigrees of scientific importance—Maupertuis (1752), de Réaumur (1751), Dean (1971), and others in §§4.1 and 5.1. The success of animal and plant experimentation, particularly the *Drosophila* work, took the emphasis away from pedigrees, but the work of Garrod (1902), as in §5.2, showed the power of pedigree research in humans, in whom experimental work cannot be carried out. Pearson, Nettleship, and Usher (1911–1913) wrote a monograph on albinism in man and Pearson decided that a permanent repository for such monographs was required. He, therefore, founded *The Treasury of Human Inheritance* [see Pearson (1914–1930, IIIA, p. 345) and Pearson (1908–1909)].

Stern (1965) recalls that in 1933 Bernstein gave a seminar at the Californian Institute of Technology on human genetics with emphasis on blood groups. T.H. Morgan (see §4.9) suggested that his work could have been carried out by pedigree analysis, which came as a surprise to the speaker. Weinberg (1909, 1910) identified and solved problems of ascertainment caused by the circumstances of such collections of data and L. Hogben, J.B.S. Haldane, and R.A. Fisher refined the methods for their treatment.

The reason for the consanguinity of recessive defects became clearer.

## Rebuilding of a Pedigree

The Lambert pedigree of the inheritance of a skin disorder, *ichthyosis hystrix gravior*, had begun with a report of a boy aged 14 years in 1732 with an uncommon skin distemper. The investigation began as an exercise for Penrose's student Curt Stern* (1902–1981). This pedigree purported to show that the condition was determined by a gene on the Y-chromosome. A pair in the pedigree had gone on tour as the "astonishing porcupine man and his son." Penrose and his wife Margaret with great ingenuity and industry were able to construct a new pedigree, including discovery of the birth date—1716—of the first Edward Lambert. This new pedigree showed that the condition was not sex-linked as supposed, and should be regarded as due to a dominant autosomal gene (Penrose and Stern, 1958).

[For inheritance of abnormalities of the skin, see Cockayne (1933).]

Chapter 20 of Sturtevant (1965b) is devoted to the genetics of man and gives a condensed overview of aspects of heredity in man, including political aspects. The book contains many useful summarizations, for example, a chronology of genetic studies.

# 6
# Death Rates and Life Tables

## 6.1 Halley's Life Table

Edmond Halley* (1656–1743) made only one contribution to medical statistics [Halley (1693) reprinted in Halley (1942), Halley, Haygarth, et al. (1973), and partially in Smith and Keyfitz (1977)].

Halley (1693) notes that the number of persons in the population has been lacking in previous surveys; nor has there been a knowledge of the ages of those dying. The studies of Graunt and of Petty have been unable to establish standards because there have been more deaths than births in their cities, London and Dublin, respectively. A city must be chosen free of disturbance by adventitious increase from abroad or by decay from emigration. These defects are practically absent from the bills of mortality of Breslau, wherein age and sex of all those dying are available and also the number of births for each year from 1687 to 1691. For these five years 6,193 persons had been born and 5,869 died, or per annum 1,239 and 1,174 respectively, with a net increase of 65 per annum.

According to Greenwood (1948), Graetzer (1883) had analyzed the correspondence between Halley at the Royal Society of London and Kaspar Neumann at Breslau. Graetzer did not find a copy of the correspondence at Breslau but he extracted from the Breslau archives all the data that could have been sent to Halley. In the archives of the Royal Society, he obtained letters from Neumann to Halley. Georg Friedrich Richard Boeckh (1824–1907) gave a clarification of Halley's procedures (Boeckh, 1893). It is of interest that Graetzer (1883) found that mortality rates in Breslau in 1687 to 1691 were not greatly different from those in 1876 to 1880.

In passing it may be noted that, although Halley did not use life-table notation, he did use most of the common life-table functions: (a) the median age at death for persons aged $x$ years; (b) the age specific death rate, for example, "one per cent of those that are of those Ages"; (c) the probability of surviving one year or other time interval from a given age, the modern $p_x$; (d) some actuarial measures, for example, years' purchase of an annuity, that is, present value of an annuity, perhaps until death.

In the absence of a census, Halley could not compute age-specific rates, except in childhood, from deaths by age and populations at risk. At the youngest ages, Halley could proceed by a method now variously attributed to him, Pell, Farr, or Lexis; it could be referred to as the method of local generation death rates (see §6.2).

Halley (1693) was able to give an estimate of fertility in the form of what we would call a crude birth rate, that is, births per total population per annum, or more refined versions as births per population of women 15 to 44 years of age. Graunt had not had the appropriate data and his estimate had been wide of the mark.

## 6.2 Age-Specific Death Rates

Graunt (1662) and the other early demographers, lacking a knowledge of population numbers and of age at death, were unable to compute death rates directly. In §6.1, it has been noted that Halley (1693) computed $p_x$, and by implication, $q_x$. The first computation of the age-specific death rates, $m_x$, from the deaths in a year and the years at risk is attributed to Pehr Wilhelm Wargentin* (1717–1783) (Wargentin, 1766/1930) on the basis of deaths and known populations by age and sex for Sweden (see Lancaster, 1990, pp. 402–403).

To discuss the problem of the interpretation of death rates over age and over time, it can be supposed, for ease of exposition, that the death rates are available for either sex for single years of age and for single calendar years over a considerable length of time in some geographical region. The country or geographical unit can be supposed to be large, so that sampling errors are negligible. For definiteness, it might be supposed that the observations commenced in 1800 and this might be taken as the origin. Then for either sex, males for definiteness, a table could be constructed with calendar years as rows, $t = 0, 1, 2, \ldots$ and ages, $x = 0, 1, 2, \ldots$ as columns, and the entries into the body of the table are the death rates corresponding to the row year and column age. Then such a table can be read (a) by rows—this is the *calendar method*, for the rates for a calendar year can be read off by choosing the appropriate line; (b) by columns—this is the *narrative* or *historical method*; for example, mortality in the first year of life can be examined in different years in the first column of the table; (c) by diagonals—this is the *global cohort method*, meaning that the same cohort is being followed through life. That the experience of the cohort is being examined follows from the fact that, in passing along the diagonal, the age and the year each increase by unity and so the rates refer to the same birth cohort, although not exactly. Note that the term *cohort* is often used because many of the other words with the notion of collective, for example, *set, group, generation*, and *class*, are already used for other or more specific notions.

It comes as a surprise to many observers that the graphs of the calendar rates and of the cohort rates may differ considerably, although of course they would be identical in an era of unchanging death rates. Two problems arise: (a) The computation of the populations or the rates along the diagonal when the rates are required for a single year of age (or alternatively for a single year of experience), which may be termed the *local cohort method*. According to H. von Schelling (1949), Laplace wished to follow the dying out from year to year of individual annual sets of births (that is, cohorts as we should say), although we have never located the actual passage. (b) The computation and interpretation when the shape of the curve throughout life is considered, the *global cohort method* mentioned above.

Georg Friedrich Knapp (1842–1926), writing on the mortality in Saxony, enters practically upon the question of interpreting the rates, based on experience from the Duchy of Anhalt in 1860–1866. This calculation was therefore frequently called the Anhalt method, although at about the same time K. Becker had applied it in Oldenburg. In Holland, the method had been already proposed independently by van Pesch (1866). Knapp (1868, 1869) specified various groups ("*Hauptgesammtheiten*") of deaths: first, persons belonging to the same cohort, who died at a given age; second, persons of the same cohort, dying within a certain space of time; and third, persons who died at a given age in a certain space of time.

In each case, the problems (a) and (b) above can be considered graphically after the manner of Gustav Anton Zeuner* (1828–1907) (Zeuner, 1869; see also Pressat, 1972).

It has to be explained why the theory of Zeuner, Knapp, and others was allowed to fall into disuse. Possibly it happened because the death rates were almost stationary in the 1870s and so all reasonable interpolation methods would give approximately the same results, whether considered in a calendar or cohort method, and there would be no need for refinements in technique. Further, there would have been no diseases for which the methodology would have introduced a novel interpretation, as was to occur later with the analysis of the tuberculosis death rates.

The article of J. Brownlee (1916a) on tuberculosis in England and Wales did not lead to any quickening of interest; indeed, it was not mentioned in Derrick (1927) nor by Greenwood (1927a), an admirer of Brownlee, who in his discussion on Derrick's paper is reported to have said:

For example, he [Greenwood] had read a paper by Sir Robert Philip, who attached very great importance to the shifting of the age incidence, because he argued that, first of all, the effect of modern tuberculosis measures was to help the young, and that the change in the death-rate at the later ages would begin many years after the death-rate at the earlier ages had been affected. It seemed to him (the speaker) that almost any hypothesis could be made to cover those facts.... For the present, however, he ventured to feel that the method of extrapolation proposed was quite as romantic as any of the extrapolations which medical people like himself had attempted. (pp. 154–155)

Nor had Greenwood (1927a) noted the now-famous article of McKendrick (1926) in which the theory is explicitly stated.

However, the presidential address of Greenwood (1936a) spells out the theory in detail. The clinicians in the meantime had noticed the declines in the death rates from tuberculosis, greater in the younger than in the older adult age groups, as Philip (1924) had forecast. Andvord (1921), very unfavorably reviewed by Elderton (1922–1923), conceived of tuberculosis as a disease (infection) acquired in childhood and having its effects in later life and he considered that its mortality should therefore be studied by cohorts. Posthumously published, Frost (1939) cites Andvord (1930) and gives a good description of the method. It is a curious fact that Frost had read Sydenstricker (1927), who was basing some of his article on Brownlee (1916a), yet neither Sydenstricker nor Frost's literary executors noted that Brownlee had introduced the method; further, it seems also that Brownlee (1916b) had not realized the importance of his own work.

Kermack, McKendrick, and McKinlay (1934a,b) reintroduced the topic into the medical literature in a discussion of total mortality. They found that the mortality experience could be expressed as a product of two factors, one depending on the year of birth (and naturally the country) of the cohort and the other on the age of the members of the cohort. In general, it may be stated that for total mortality, a cohort having a favorable experience at the early ages will also have a favorable experience throughout life. This is an important conclusion, for it contradicts the thesis so often brought forward that hygienic measures actually weaken the population in the generation (cohort) to which they are applied, and that if many infantile deaths are avoided, the generation will suffer greater mortality in the future. We note that this argument is applied with even greater force to the case of mortality from tuberculosis. It is evident that the method of generations or cohorts must be applied to the study of any of the chronic diseases of long standing if fallacious conclusions are not to be drawn from official statistics and other forms of data. The generation method was not received into medicine as easily as it deserved, yet it brings an order into the study of the tuberculosis death rates over the decennia late in the nineteenth century and early in the twentieth, when the curves of death rates show great changes when examined by the calendar methods. However, when they are examined by the generation method they are seen to be similar for each of the generations, although pitched at a lower level for the later generations; Springett (1950, 1952) has graphs from various countries, justifying the hypothesis of Kermack, McKendrick, and McKinlay.

Cancer of the lung, now widely believed to be due to the abuse of tobacco particularly in the form of cigarettes smoked at high temperature, is an example of a disease of long standing that is on the increase (see §13.7). When the increases in mortality from cancer of the lung were considered in the 1940s and 1950s, an anomaly was observed; the age-specific mortality rates from the other cancers in any epoch could be seen to be increasing throughout life, but the graph of cancer of the lung, say in England and Wales, 1950, was seen to have a maximum at age 70 years with some decline in the higher ages. This phenomenon was noted by Kennaway and Waller (1953/1954). They showed that the anomaly was a generation effect and was not present in the rates of the cohorts. The biological explanation is that successive cohorts had been increasingly heavy smokers and that in a cohort there was a lag period between the initiation of smoking and the deaths from cancer, with increasing risk of lung cancer appearing in each cohort.

In 1912, George Udny Yule* (1871–1951) was appointed lecturer in statistics at Cambridge. He was also statistician to the School of Agriculture there.

Yule (1934) was concerned with the computation of various indices, especially the standardization of the death rates; he showed that, unless the target and standard populations were very different in age distributions, the direct and indirect methods of standardization gave closely concordant results. He also saw that it was unnecessary to think of the standardized rates as happening in an ideal population; so he introduced the equivalent average death rate by using as a standard a population uniformly distributed over ages 0 to 65 years, convenient for standardization for comparisons between occupations. He recommended that death rates be recorded without decimals and as rates per million per annum whenever possible—a suggestion that would have saved the World Health Organization the printing of, perhaps, many millions of rates beginning with zero in their publications. The Registrar General of England and Wales had good reason to receive advice in the proceedings of the Royal (after 1886) Statistical Society on topics related to his own publications. Possibly, Yule was the greatest of all these "strolling players."

A.J.B.R. Auget de Montyon (1733–1820), often cited as the first great French demographer, made tabulations of mortality, calculated expectations of life, and so on, drew conclusions from them, and suggested suitable large-scale administrative action to counter undesirable factors, possibly causative. Thus he selected six parishes and computed expectations of life at various ages. He was able to show advantages of persons living in three parishes in high land or open country, as opposed to others living in three parishes in low and marshy areas.

## 6.3 Life Tables

For easy reference in the text, we give here the notation of the life tables, discussed in my *Expectations of Life*, Lancaster (1990, pp. 38–46) and in more detail in Smith and Keyfitz (1977). We may begin with a death rate, $m_x$, for each age $x$, $x = 0, 1, 2, 3, \ldots$; $m_x$ is usually defined as deaths per thousand per annum. An elementary formula yields $q_x$, the probability of death of a person of precise age $x$ dying before the precise age $x + 1$. Then $p_x = 1 - q_x$ is the complementary rate of survival to precise age $x + 1$; $l_x$, the survivors to precise age $x$, $l_{x+1} = p_x l_x$. The actuaries usually take $l_0 = 100,000$ but for theoretical work it is more convenient to write $l_0 = 1$. The expectation of life at age $x$ is given by equation (6.3.1). In some experimental or observational work we may be able to calculate the set $\{l_x\}$ directly. Suppose $n = n_0$ animals come under inspection and that in the unit epochs of time there are $n_x$ deaths, $x = 0, 1, 2, \ldots$. The survivors will be $n_0, n_0 - n_1, n_0 - n_1 - n_2, \ldots$. A division through by $n_0$ will give an observed table of $l_x$, commencing with unity.

Although Halley was the first to construct a life table from population data, Johan de Witt* (1625–1672) gave a law of mortality and constructed a table to be used for the computation of annuities; his law may have been founded on observations, freely used. De Witt (1671) assumed that of 128 persons reaching the age of 4 years, one death took place every half-year so that 28 would survive to age 54 years. Similarly, in the next ten years one death took place every nine months, so that $14\frac{2}{3}$ survive to the 64th birthday. In the following decade, one-half die in each six months, so that $4\frac{2}{3}$ will survive, and finally one-third die in every half-year until after 7 years all have died out. De Witt used what is now termed $l_x$ and gave a correct method for the computation of annuities; such actuarial and financial side issues can be passed over here as not of medical interest. The implied $q_x$ rates do not form an increasing series, for the rates increase to 0.18 at age 53 years and then decline to 0.14 at age 54.

Antoine (the elder) Deparcieux* (1703–1768) gave a formula for the expectation of life corresponding to

$$\overset{0}{e}_x = \sum_{i=x}^{\omega} l_i/l_x - 0.5 \tag{6.3.1}$$

and showed that adult expectations of life had been increasing over the previous half century (Deparcieux, 1746). He also calculated a life table for monks entering a monastery at ages 17 to 25 years and observed until death; this corresponds to a cohort life table. Deparcieux also calculated life-table values by the observations of deaths only, but this led him to an approximation to the life-table values, which is rather inaccurate if the population observed is not stationary (see Westergaard, 1932).

On technique, Joshua Milne (1815) gave a notation for the life table and discussed under what conditions a life table could be constructed; his work is summarized in Chapter 4 of Smith and Keyfitz (1977).

Daniel Bernoulli (1766) introduced the notion of age as a continuous variable in his paper on the elimination of smallpox by inoculation (variolation); this device has advantages as many results can be stated simply by integrals and the theory of differentiation can be applied as in Bernoulli's problem to estimate the effect of abolishing smallpox.

Johannes Daniel Gohl (1665–1731) gave tables of deaths by cause. [See Fenaroli et al. (1981) for early life tables and Imhof et al. (1990) for life tables constructed on German parish data. See Lancaster (1990) for other parish life tables and Barinaga (1991) for some modern comparisons.]

## 6.4 Laws of Mortality

The actuaries had desired to find a closed formula for $l_x$ or $\mu_x$ of the life table, to ease the burden of computation. Some of these formulas have been discussed in Westergaard (1932), Smith and Keyfitz (1977), and more briefly by Lancaster (1990). With Greenwood (1928) we can agree that some excellent fits to data have been obtained but no special meaning can be assigned to the parameters estimated.

According to Eyler (1980), William Farr was especially indebted to Thomas Rowe Edmonds (1803–1889), an actuary who published a work on life tables (Edmonds, 1832); in it, he announced as an important discovery that the age-specific mortality rate was related geometrically to the age. Indeed, the logarithm of the rates could be fitted by three straight lines; alternatively, the rate could be said to rise or fall by a constant ratio in each of three stretches of age, namely falling by 32.4% each year from 6 weeks of age to 6 to 9 years, where it reached a minimum, then increasing by 2.99% per year from 15 to 45 years of age, and increasing from 45 years onward by an annual rate of 7.9%. Edmonds believed that the two points of intersection between the three lines would be subject to change according to the population studied. Such a rule of thumb can hardly be called a "law of nature."

It seems probable that applied mathematicians brought up in the Newtonian and Laplacian solutions for the motions in the solar system were, and are, unduly biased in their treatment of biological data. Thus Brownlee (1919, p. 43) found that Makeham's formula "implies that the substances or capacities on which life depends decay according to the law of the uni-molecular reaction, that is that the amounts present at the end of equal intervals of time can be represented by the terms of a geometrical progression."

We notice in passing that there was a popular aspect of the belief that life events would follow determinable laws. Thus, number mysticism applied to health created in 1633 the belief that the human body had a seven-year cycle and every seventh year was critical to health, with particular danger at 49 and 63 (see Westergaard, 1932, pp. 32–33); Kaspar Neumann (1648–1715) proved that the phases of the moon had no influence on health.

## 6.5 Fertility

In the demographic literature, the term *fertility* is used in the sense of offspring borne, whereas *fecundity* is used for the potential for bearing offspring. It is thus possible for fecund women not to be fertile. There have been various methods of defining and estimating indices of fertility.

According to Farr (1885/1975, p. 87), no trustworthy birth statistics existed in England and Wales prior to the Civil Registration Act of 1837; conditions may well have been more favorable in other European states, where there were fewer difficulties about varying religious faiths. At first it was believed that the crude birth rate, namely the proportion of registered births in a year to the population at all ages, was sufficient for most purposes. In the absence of rapid population-size movements, it indeed may have sufficed to compare the birth rates between successive years, although comparisons between different populations may have been invalid if the age and sex structure differed. The Census Report of England and Wales for 1881 used the females between the ages of 15 and 45 years as a divisor.

The age specific fertility rates,

$$f_x = \frac{\text{(Number of Births to a Specific Age Group)}}{\text{(Years of Life Spent by Females of the Age Group)}} \qquad (6.5.1)$$

appear to have first been computed by Henric Nicander (1744–1815) for the 16 years ending with 1795 for Sweden and Finland. Joshua Milne (1776–1851) in 1815 reproduced his table, and that article is reprinted in Smith and Keyfitz (1977). For the eight age groups, 15–19, 20–24, 25–29, ..., 45–49, 50 +, the proportion of women being confined within any year is the reciprocal of 40.8, 7.8, 4.6, 4.3, 5.4, 10.6, 46.5, 1776.0, respectively, and the age-specific fertility rates are per thousand per annum 25, 127, 216, 230, 185, 94, 21 and 0.6, respectively.

Boeckh (1886) defined the gross reproduction rate (GRR) as

$$\text{GRR} = \sum f_x, \qquad (6.5.2)$$

where summation is over all the years of life and $f_x$ is the rate of female births per annum to a female passing through life in some given population. If age groups span 5 years, the corresponding contribution to the GRR would be $5f_x$.

He defined the net reproduction rate (NRR) as

$$\text{NRR} = \sum l_x f_x, \qquad l_0 = 1 \qquad (6.5.3)$$

where $l_x$ is the proportion of survivors to age $x$ years.

The two results can be given in the integral form by assuming that $f_x$ is a continuous function of the age, $x$. Then equations (6.5.2) and (6.5.3) become

$$\text{GRR} = \int_0^\infty f_x \, dx \qquad (6.5.4)$$

and

$$\text{NRR} = \int_0^\infty f_x l_x \, dx. \qquad (6.5.5)$$

There is an approximate relation between the two rates

$$\text{NRR} \cong l_m \text{GRR}, \qquad (6.5.6)$$

where $m$ is the mean age at confinement of the mothers.

Boeckh (1886) could then give the rule that schedules of birth and death rates imply that a model population is increasing, stationary, or decreasing, if and only if NRR is greater than, equal to, or less than unity, respectively.

## 6.6 Fibonacci Speculations

Leonardo da Pisa (or more usually Fibonacci)* (*ca* 1170–*ca* 1250) introduced a rabbit problem that illustrates well the general points to be covered in the next section. It is assumed that a pair of rabbits requires a month to mature and then reproduces itself every month; there is assumed to be no mortality. Let us call the total population $a_i$, $i = 0, 1, 2, \ldots$. At time 0, there is one immature pair, $a_0 = 1$; at time 1, there is a mature pair, $a_1 = 1$; at time 2, there is a mature pair plus an immature pair, $a_2 = 2$; at time 3, there are two mature pairs and one immature, $a_3 = 3$; at time 4, there are three mature pairs and two immature, $a_4 = 5$; in general

$$a_k = a_{k-1} + a_{k-2}, \qquad k = 2, 3, \ldots. \tag{6.6.1}$$

An easy computation yields the series of Fibonacci numbers, 1, 2, 3, 5, 8, 13, 21, 34, 55, 89, ....

The new pairs are the offspring of pairs that are present at time $(k - 2)$. This relation can be represented by the matrix relations,

$$\mathbf{MA} = \begin{bmatrix} 1 & 1 \\ 1 & 0 \end{bmatrix} \begin{bmatrix} a_{k-1} \\ a_{k-2} \end{bmatrix} = \begin{bmatrix} a_k \\ a_{k-1} \end{bmatrix}, \qquad k = 1, 2, \ldots \tag{6.6.2}$$

and in general

$$\mathbf{A}_n = \mathbf{M}^n \begin{bmatrix} 1 \\ 0 \end{bmatrix} = \begin{bmatrix} a_n \\ a_{n-1} \end{bmatrix}. \tag{6.6.3}$$

There is a general theorem for such iterations that states that $a_k$ tends to grow like $A\lambda_1^k$ where $A$ is a constant and $\lambda_1$ is the larger root, $\frac{1}{2}(1 + \sqrt{5})$, of the equation

$$|\mathbf{M} - \lambda \mathbf{1}| = \begin{vmatrix} 1 - \lambda & 1 \\ 1 & -\lambda \end{vmatrix} = 0, \qquad \text{which can be written as } \lambda^2 - \lambda - 1 = 0.$$
$$\tag{6.6.4}$$

Indeed, the limiting ratio between successive terms is $\frac{1}{2}(1 + \sqrt{5}) = 1.6180\ldots$, and already the ratio of the 10th Fibonacci number to the 9th is $89/55 = 1.6181\ldots$

The rabbit population thus comes to have a stable age distribution, immature pairs in a proportion $a_{k-2}/a_k$ and mature pairs in a proportion $a_{k-1}/a_k$, and the total population increases in the ratio $1.6180\ldots$ per generation.

For later applications it is convenient to generalize the matrix $\mathbf{M}$. Let $\mathbf{n}_t$ be the vector, whose elements are the numbers at age 0, 1, 2, ... years in the population at time $t$ years precisely. Let $f_0, f_1, \ldots$ be the age specific fertility (females only counted) at ages 0, 1, 2, ..., and let the $p$'s be from an appropriate life table. The relation between the population vectors in successive years can be represented by the matrix equation

$$\mathbf{n}_{(t+1)} = \begin{bmatrix} f_0 & f_1 & \cdot & \cdot & & f_k \\ p_0 & 0 & & & & \\ \cdot & p_1 & \cdot & & & \\ & & \cdot & \cdot & & \\ & & & \cdot & 0 & \\ & & & & \cdot & 0 \\ \cdot & & \cdot & \cdot & \cdot & p_{k-1} & 0 \end{bmatrix} \qquad \mathbf{n}_t = \mathbf{M}\mathbf{n}_t \qquad (6.6.5)$$

From repeated operations with $\mathbf{M}$ we obtain

$$\mathbf{n}_t = \mathbf{M}^t\mathbf{n}_0, \qquad (6.6.6)$$

where $\mathbf{n}_0$ is the initial population vector. The theory of canonical forms of matrices or a direct expansion of the determinant of the matrix, $|\mathbf{M} - \lambda\mathbf{1}|$, can now be used to determine the ratios between the subpopulations in successive years and the asymptotic age distribution.

## 6.7 Euler's Stable Populations

Leonhard Euler* (1707–1783) made two substantial contributions to population dynamics. The first of these appears in *Die göttliche Ordnung* of Johann Peter Suessmilch (1707–1767) and is reprinted in translation in Smith and Keyfitz (1977). The mathematical model specifies that a population is initiated by one couple of age 20 who give birth to additional couples at ages 22, 24, 26, and who die at age 40. The implied fertility is high—six births to each woman during her life and each couple is represented by three couples in the next generation. The net and gross reproduction rates are equal to 3.

A modern demographer would consider females only, without worrying about legitimacy. Since mortality is being neglected until the end of the child-bearing age, assumed to be age 41 precisely, it is assumed that $l_{41} = 0$ and consequently $l_x = 1$ for $x = 0, 1, 2, \ldots, 40$. We would now have the same recurrence relation between females as we had for pairs, $(B_t \equiv B(t))$,

$$B(t) = B(t - 22) + B(t - 24) + B(t - 26). \qquad (6.7.1)$$

By 1760, Euler and Joseph Louis de Lagrange* (1736–1813) knew how to solve linear recurrence relations such as (6.7.1) or the Fibonacci series, which has been solved in §6.6. Euler's recurrence relation can be written in the form of equation (6.6.5) by setting all $f$'s in the first row as follows: $f_{22} = f_{24} = f_{26} = 1$ and all other $f$'s equal to zero and every $p_s = 1$ for $s = 0, 1, 2, \ldots, 25$. All other elements in the matrix are set to zero. Thus a matrix of size 27, say, is obtained. The equation

$$|\mathbf{M} - \lambda\mathbf{1}| = 0 \qquad (6.7.2)$$

can be solved; then the limiting ratio between two consecutive population sizes is the largest root of this equation.

Using this method, Euler determined that the recurrence relation (6.7.1) would imply that ultimately the population would increase approximately geometrically so that

$$P(t + 1) = \lambda P(t) \equiv 1.04696 \ P(t) \tag{6.7.3}$$

where $\lambda$ is the greatest positive root of

$$\lambda^{26} - \lambda^4 - \lambda^2 - 1 = 0. \tag{6.7.4}$$

The numerical results were given in Suessmilch (1761–1762). From that table, the ratios, $P(282)/P(280)$, $P(284)/P(282)$, ..., $P(300)/P(298)$, were 1.044, 1.087, 1.133, 1.117, 1.177, 1.164, 1.134, 1.100, 1.067, and 1.041, respectively. Suessmilch used the table for some speculations on the antediluvian and postdiluvian populations of the world. The ratio seems to converge to its ultimate value rather slowly but a comparison of Euler's series with the Fibonacci series in §6.6 shows that in the former there are 12 numbers at 2-unit distances, whereas in the latter there is a single unit between generations.

In his second contribution, Euler (1767) sets up a mathematical model with assumptions: (a) the schedule of $\{p_x\}$, or equivalently of $\{l_x\}$, of the life table is fixed; $l_0 = 1$. Further the $l_x$, $x > 100$, are so small that their effect is negligible and so they can be set equal to zero; (b) the births are increasing geometrically, that is,

$$B_{t+1} = \lambda B_t, \qquad B_0 = B \quad \text{and so } B_t = \lambda^t B.$$

Under these hypotheses, the numbers of survivors to the 100th year after time 0 are given by

$$\lambda^{100} B \left( 1 + \frac{l_1}{\lambda} + \frac{l_2}{\lambda^2} + \cdots + \frac{l_{100}}{\lambda^{100}} \right). \tag{6.7.5}$$

Further, the numbers of population by age are given by the terms of the expansion

$$P = B \left( 1 + \frac{l_1}{\lambda} + \frac{l_2}{\lambda^2} + \cdots + \frac{l_{100}}{\lambda^{100}} \right). \tag{6.7.6}$$

Euler (1767) points out that if the ratio $B/P$ is known and $\{l_x\}$ is given by the mortality hypothesis, $\lambda$ can then be calculated by interpolation from trial values, say $\lambda_1, \lambda_2, \ldots, \lambda_k$.

Suppose now that there has been a constant schedule of $l$'s over a hundred years and that the births in a year are $B$ (our $B_t$), the deaths $D$, the population $P$, and by hypothesis the population $\lambda P$ in the following year; then the increase in population can be evaluated in two ways as $\lambda P - P$ and also as $\lambda B - D$, so that

$$\lambda P - P = \lambda B - D, \tag{6.7.7}$$

and so

$$P = \frac{\lambda B - D}{\lambda - 1}. \tag{6.7.8}$$

By equations (6.7.6) and (6.7.8) there follows

$$\frac{B - D}{\lambda - 1} = B\left(\frac{l_1}{\lambda} + \frac{l_2}{\lambda^2} + \cdots + \frac{l_{100}}{\lambda^{100}}\right), \tag{6.7.9}$$

from which $\lambda$ can be calculated. [Note that the series of (6.7.9) lacks the unit of the series in (6.7.6).] Euler notes further that from $\lambda$, $B$ and $D$, the total population $P$ can be calculated. Euler (1767) ends with the conclusion that the numerical problems of population depend on a knowledge of fertility and mortality, if these have reached a steady state for the area; armed with estimates of these parameters, most questions likely to be proposed could readily be answered.

It seems remarkable to us now that Euler did not combine the hypotheses of his two contributions. To do so, he needed only to postulate some schedule of fertility rates, $\{f_x\}$, where only female births are counted, to obtain a recurrence relation,

$$B(t) = \sum_{t=15}^{50} B(t - x)l_x f_x. \tag{6.7.10}$$

where it is assumed that births are negligible except at ages 15 to 50. Since $\{f_x\}$ and $\{l_x\}$ are specified, equation (6.7.10) can be solved and the ratio $B(t + 1)/B(t) = \lambda$ determined. He would have then obtained the result that, asymptotically, births and so the population in every age group increase in the ratio $\lambda$ and so the age distribution takes a stable form.

Comparing this theory with that of Leslie (1945), we see that the Euler theory is identical but he is using artificial data; Leslie's work can be seen to be an avoidance of the more difficult modern mathematical theory—a discretization of the general procedure discussed in the next section.

It is worthwhile to mention the sequence of authors, Euler, Suessmilch, Thomas Robert Malthus* (1766–1834), and Darwin (and Wallace), interested in the growth of population. D.M. Simpkins in her biography of Malthus (*DSB*, IX, pp. 67–71) mentions that Quetelet and Verhulst showed that the European censuses of the early nineteenth century indicated how early exponential increases in population in European states later tended toward a logistic curve. However, this is a general property of the purely logistic curve.

## 6.8 Logistic Theory of Population Growth

Some history of the dynamics of population has been given in §§6.6 and 6.7 without much reference to the improvements in estimating the schedules $\{l_x\}$ and $\{f_x\}$. Lorimer (1959, p. 152) points out that the methods of Euler (1761,

1767) were neglected for many years. Ludwig Ferdinand Moser (1805–1880) pointed out the impropriety of using observed deaths in an actual population as equivalent to deaths in a stationary life table population ("Halley's method," as he contemptuously called it) (Moser, 1839); indeed, his criticism of current estimations of the age-specific mortality rates strengthened the theory of the construction of life tables, influencing both Quetelet and Georg Friedrich Knapp. Moser noted that an exponential rate of births could be found in the United States but not in France, so that it was not a universal law. Quetelet (1835/1842, pp. 169–172) set up the hypothesis "resistance ... to the unlimited growth of population in proportion to the square of the velocity with which the population tends to increase." This hypothesis led Verhulst (1838, 1845–1847) to develop the logistic law. He states that if the population, $p(t) \equiv p$, is growing in a geometrical progression, the differential equation is

$$\frac{dp}{dt} = mp. \tag{6.8.1}$$

If the growth of the population is retarded by the increase of population, the differential equation is

$$\frac{dp}{dt} = mp - \phi(p), \tag{6.8.2}$$

where $\phi$ is some positive and increasing function of $p$. The simplest hypothesis is to set $\phi(p) = np^2$, $n > 0$. The solution of this equation is

$$t = \frac{1}{m}[\log p - \log(m - np)] + \text{constant}, \tag{6.8.3}$$

or equivalently

$$p(t) = \frac{be^{at}}{1 + ce^{at}} = \frac{b}{e^{-at} + c}. \tag{6.8.4}$$

Relations can be obtained $p(-\infty) = 0$, $p(+\infty) = b/c$, which define the asymptotes; the curve has a point of inflection when $p(t) = \frac{1}{2}b/c$, that is, at a point for which the population is equal to half the ultimate population; the slope at the point of inflection is $ab/(4c)$.

[For applications of the logistic curve to demography, see Verhulst (1838), Pearl and Reed (1920), Lotka (1934–1939). The logistic curve has also been applied to the theory of growth of bacterial colonies.]

## 6.9 Integral Equation of Population Growth

Sharpe and Lotka (1911) begin their paper by making the following statements: The age distributions of populations are variable but not unlimited. Even if the initial population were of irregular form, the "irregularities" followed through for some years would no doubt be smoothed over. It seems that (subject to constant schedules of fertility and mortality and to zero migration) there must be a limiting "stable" type about which the actual distribution varies, and toward which it tends to return if disturbed therefrom by any agency. Previously, Lotka (1907) had determined a "fixed" distribution form which the population would tend to approach; it remained to determine whether after some disturbance the distribution would return to the fixed form, so that the fixed form was also "stable."

The analysis of Sharpe and Lotka (1911) was not completely satisfying to the mathematicians. The problem may be regarded as fully solved by Feller (1941). [See Samuelson (1976) for a detailed discussion of the problem of the integral equation of population growth.]

There have been many applications of the theory to demography. Perhaps the best known are those of Kuczynski (1928, 1930). Robert René Kuczynski (1876–1947) set out to show that there was an impending decline in western European populations, although these same populations were actually increasing at the time of writing.

It was soon appreciated that the theory could be applied to a great number of problems, in particular biological population theory including evolution, for example, Provine (1971), and even more generally to the theory of aggregates. Here we may especially mention Leonhard Euler, Alfred James Lotka* (1880–1949), Haldane, Fisher, Sewall Wright (1889–1988), and William Feller* (1906–1970).

## 6.10 Human Population Statistics

There has recently been great interest in living populations, especially in relation to evolutionary problems. Yet only in the human species is there a mass of data, in which theory can be readily matched with observations. Most essays on population growth have dealt with the human species. Notable series have been amassed by the statistical bureaus of England and Wales and of Sweden, among others. Here we point to some of the principal scientific facts established by Graunt. From the point of view of population, there has been also an interest in genealogies (see §5.9).

John Graunt* (1620–1674) did important work in London before the establishment of the official series. Let us recall some of his conclusions:

1. *Statistical regularity.* Graunt (1662) found that the number of deaths per year in London in the absence of epidemics tended to be regular. Later authors, including Quetelet, were also aware of this feature.
2. *Sex ratio at birth.* Graunt found that one hundred times the ratio of male to female births, the masculinity rate, was 108, which is not far from the modern Australian figure of approximately 105. Graunt realized that differential mortality of the sexes would give approximately equal population numbers of the two sexes in the childbearing years. Suessmilch (1761–1762) was to make much of this later.

3. *Urban and rural death rates.* The urban death rate was higher than the rural one. In fact, the population of London was only maintained by migration from the country. Graunt also recognized the influence of season and years on the mortality. He noted that the death rates were more variable in the country. This can probably be explained by the fact that some diseases that were endemic in the city were not always present in the country, and so acted as epidemic diseases there.

4. *Infant mortality.* Graunt's greatest achievement was possibly to obtain an estimate of the infant mortality without a knowledge of the age at death. Thirty-six percent of all infants born alive died before their sixth birthday.

Graunt's work is important because he looked at unpromising material critically and judged what conclusions could be drawn from it, and so is an example to all later workers in the field of medical or vital statistics.

If we accept Graunt as the author of the *Observations*, the importance of William Petty* (1623–1687) as a demographer lies in his views of what statistics should be collected and what special investigations should be carried out by the State.

The human population has rarely been subjected to catastrophic accidents, famine, plague, floods, and so on, as have the nonhuman populations. Often its problems have been treated as actuarial or demographic. Some features of interest are founder effects, important in humans and nonhumans. In a few cases genetic effects, such as the establishments of the gene for sickle-cell anemia, can be discussed as in §5.6 and the gene for diabetes as in Lancaster (1990, p. 223). Some observations, practical and ethical, can be made on man in the human population but his long life is a hindrance.

[For Graunt, see Ogle (1892) and Hald (1990). For Petty, see Strauss (1954). For parish records, see Lancaster (1990, especially Chapter 38) and for plague in England, see Mullett (1956).]

# 7
# Evolution

## 7.1 Introduction

This subject has become an important field of research, to which we cannot hope to do full justice in the space of a chapter; it is the study of the distribution of the genes in natural populations and their changes over time [see Mayr (1982, p. 553) and §4.3].

After tables of mortality and fertility rates had been developed in Sweden and in England and Wales, mathematics could be applied, and the theory reached a definitive stage in this century. There is not such a body of factual material available for any other species, but the wish remained among biologists to use such a theory to aid speculation on the course of evolution.

There were some difficulties in this general scheme, the most notable being the lack of a time scale. For the importance of a time scale, see Johanson and Shreeve (1989), a popularizing account of the development of the forerunners of modern man; in their team there was a geologist to advise about conditions in the past and to identify the date or era of the fossils. Special chemical tests for dating have also been devised.

## 7.2 Establishment of a Biological Time Scale

Fisher (1953) gave several examples of the use of statistics in biology. Gérard Paul Deshayes* (1797–1875), a French paleontologist and malacologist, had described 1,074 species of mollusks in Tertiary strata in the neighbourhood of Paris, of which 660 had not previously been described. After comparisons of the numbers of species present in any formation, he found that he could divide the Tertiary Age into three epochs, according to the percentage of species held that survived to the present or were identical with present-day species. In the oldest (or earliest) formations, there were about 3% of species identical with those of the present; in the formations of the middle epoch, 19% of species were identical with present-day mollusks; and in the latest formations, 52% were identical with those of present formations.

Charles Lyell* (1797–1875) took up this stratigraphical reasoning with Deshayes's cooperation. Further analysis showed that rocks of each period yielded certain characteristic species, not found in the rocks of the other periods. Thus there had been major changes in Tertiary molluscan faunas. In the third volume of his *Principles of Geology* (1833), and with the help of William Whewell* (1794–1866), Lyell named the Tertiary ages Pleistocene (90–95%), Pliocene (35–50%), Miocene (17%), and Eocene (3%), from the Greek words for fullest, full, middle, and dawn, combined in each case with the Greek for recent.

Fisher (1953) gave the naming of the epochs by Lyell, on the basis of a statistical study of their fossil content, as an example of how scientists forget the origins of important developments in their field of interest. Indeed the appendix to the 1833 edition of Lyell's *Principles*, with details of the classifications and of the calculations based on the numbers counted, was deleted from later editions, although the relevance of the shells of Deshayes is recognized in the third edition in the text. Other methods have made the dating more precise, and comparisons between regions have become possible.

The data collected by Deshayes and similar studies by others provided powerful arguments for the reality of evolution against the concept of the "unchanging harmony of the designed world" (see Mayr, 1982, p. 483).

Although we can all be agreed on Lyell's uniformitarianism as the general rule for natural events, some extinction can come from catastrophes from outside the world, for example, the mass extinction at the end of the Mesozoic Era, especially of the dinosaurs (see Stanley, 1987, p. 133).

## 7.3  Static Nonhuman Populations: Weldon

Walter Frank Raphael Weldon* (1860–1906) was the son of Walter Weldon (1832–1885), a journalist and an industrial chemist, who discovered and developed an important process for the regeneration of manganese peroxide used in the manufacture of chlorine. Raphael took a variety of courses at University College, London and worked under Edwin Ray Lankester* (1847–1929); he later transferred to Cambridge University where he studied zoological topics under the influence of Francis Maitland Balfour* (1851–1882). In 1888, his thoughts turned from morphology and embryology to problems of variation and correlation, and in the next year he read the *Natural Inheritance* of F. Galton (1889c) and realized that organic correlation (in the sense of Cuvier the relation between organs) could be treated mathematically. In 1890, he was appointed Jodrell Professor of Zoology at University College, London and Fellow of the Royal Society, which Pearson (1906) says was on the strength of his first two biometric papers. (See §19.2 for Weldon's remarks, as a founder of *Biometrika*, on the necessity for measurement in biology.)

Weldon saw that he must obtain a deeper knowledge of probability and read the original articles and books of the great French authors. He studied mathematical distributions relevant to biology and even carried out random sampling experiments, for example, on the throws of $n$ dice of two series of $n$ dice with precisely $m$ of them held in common. Some of this work survives anonymously in an unlikely place, §18 of R.A. Fisher (1925 and later editions).

Weldon (1906) wrote an essay as part of Strong's (1906) *Lectures on the Method of Science*, in which he obtains by sampling a joint distribution of two binomial variables with parameters $(\frac{1}{2}, 12)$ with six random variables held in common. He points out the analogies with Weismann's chromosome theory but mentions neither Mendel, nor Yule (1902), nor Sutton (1903).

Weldon (1890) was the first to apply Galton's method to animals other than man. Single metrical variables had normal distributions and he computed correlation coefficients between pairs of variables. He believed that

A large series of such specific constants would give an altogether new kind of knowledge of the physiological connexion between the various organs of animals; while a study of those relations which remain constant through large groups of species would give an idea, attainable in no other way, of the functional correlations between various organs which have led to the establishment of the great sub-divisions of the animal kingdom. (Weldon, 1892, p. 11, cited by Pearson, 1906)

Following Galton's discovery of the value of the bivariate normal distributions and K. Pearson's generalization of them to higher dimensions, Weldon hoped to obtain a characterization of populations; in particular, he worked on the shrimp, *Crangon vulgaris*, by computing means, variances, and covariances (equivalently, standard deviations, and correlations). Here he was entering dangerous ground, for he began to seek the effect of an increase in $X_1$, say, on the joint distribution of $X_2, X_3, \ldots$ (see Pearson, 1903). Pearson's treatment yielded an interpretation of the change, that the form of joint distribution was maintained, that there would be continuous change, and that the extent of the change could be unlimited. These computations gave results contrary to the opinion of Jenkin (1867) and his closed sphere of variation (see §4.4).

Alas, a reading of Pearson's praise of Weldon shows clearly why the ideas of Pearson and Weldon were in error. The applied mathematician looks for simplicity; it is a well known feature of much applied mathematics that questions correctly posed often have a simple solution. Pearson's assumption that the theory of heredity must be simple is false. Pearson and Weldon gave no biological model for their theory of heredity and so, if time is used as a variable in their mathematical models, it leads to the view that homoscedasticity holds for the infinite plane; but no test was ever carried out to verify that view. Under Mendelian theory, assuming no mutation, there can be no development past that associated with an optimal set of genes.

[See Provine (1971) for controversies, and §6.5 for Cuvier.]

## 7.4  Numerical Results of Selection

R.C. Punnett, a poultry geneticist and collaborator of W. Bateson and later J.B.S. Haldane, invited Henry Tertius James Norton (1886–1937) to give a discussion of the replacement of one gene by another allele at a locus under various hypotheses, especially free mixing (panmixia). Norton's table was given in Punnett (1915, p. 155) and is reproduced here complete with Punnett's explanation as Table 7.4.1. Norton unhappily suffered very bad health, dying early, but is the author of a general paper on this topic (Norton, 1928).

[Norton's table] affords an easy means of estimating the change brought about through selection with regard to a given hereditary factor in a population of mixed nature mating at random. It must be supposed that the character depending upon the given factor shews complete dominance, so that there is no visible distinction between the homozygous and the heterozygous forms. The three sets of figures in the left-hand column indicate different positions of equilibrium in a population consisting of homozygous dominants, heterozygous dominants, and recessives. The remaining columns indicate the number of generations in which a population will pass from one position of equilibrium to another, under a given intensity of selection. The intensity of selection is indicated by the fractions 100/50, 100/75, etc. Thus 100/75 means that where the chances of the favoured new variety of surviving to produce offspring are 100, those of the older variety against which selection is operating are as 75; there is a 25% selection rate in favour of the new form.

The working of the table may perhaps be best explained by a couple of simple examples.

In a population in equilibrium consisting of homozygous dominants, heterozygous dominants and recessives the last named class comprises 2.8% of the total: assuming that a 10% selection rate now operates in its favour as opposed to the two classes of dominants—in how many generations will the recessive come to constitute one-quarter of the population? The answer is to be looked for in column B (since the favoured variety is recessive) under the fraction 100/90. The recessive passes from 2.8% to 11.1% of the population in 36 generations, and from 11.1% to 25% in a further 16 generations—i.e. under a 10% selection rate in its favour the proportion of the recessive rises from 2.8% to 25% in 52 generations.[1]

It is assumed that the species has sufficient size so that at each step from one generation to the next, the resulting numbers with a particular allele will be negligibly different from the expectation. The general conclusion to be drawn is that an allele with any positive advantage will come to predominate in numbers.

In a series, subsuming Norton's work, Haldane (1924) stated that he had verified Norton's calculations and theory. Now Norton is little remembered but his table (our Table 7.4.1) had an important influence on Četverikov, whom we will now discuss.

---

[1] From Punnett, R.C. *Mimicry in butterflies*, pp. 154, 156. Copyright © Cambridge University Press, Cambridge, 1915. Reprinted by permission of Cambridge University Press, New York.

TABLE 7.4.1. The change brought about through selection of a given hereditary factor in a population of mixed mating (panmixia).

| Percentage of total population formed by old variety | Percentage of total population formed by the hybrids | Percentage of total population formed by the new variety | A. Where the new variety is dominant | | | | B. Where the new variety is recessive | | | |
|---|---|---|---|---|---|---|---|---|---|---|
| | | | 100/50 | 100/75 | 100/90 | 100/99 | 100/50 | 100/75 | 100/90 | 100/99 |
| 99.9 | 0.09 | 0.000 | | | | | 1,920 | 5,740 | 17,200 | 189,092 |
| 98.0 | 1.96 | 0.008 | | | 28 | 300 | 85 | 250 | 744 | 8,160 |
| 90.7 | 9.0 | 0.03 | 4 | 10 | 15 | 165 | 18 | 51 | 149 | 1,615 |
| 69.0 | 27.7 | 2.8 | 2 | 5 | 14 | 153 | 5 | 13 | 36 | 389 |
| 44.4 | 44.4 | 11.1 | 2 | 4 | 12 | 121 | 2 | 6 | 16 | 169 |
| 25.0 | 50.0 | 25.0 | 2 | 4 | 12 | 119 | 2 | 4 | 11 | 118 |
| 11.1 | 44.4 | 44.4 | 4 | 8 | 18 | 171 | 2 | 4 | 11 | 120 |
| 2.8 | 27.7 | 69.0 | 10 | 17 | 40 | 393 | 2 | 6 | 14 | 152 |
| 0.03 | 9.0 | 90.7 | 36 | 68 | 166 | 1,632 | 4 | 6 | 16 | 165 |
| 0.008 | 1.96 | 98.0 | 170 | 333 | 827 | 8,243 | | 10 | 28 | 299 |
| 0.000 | 0.09 | 99.9 | 3,840 | 7,653 | 19,111 | 191,002 | | | | |

From Punnett, R.C. Mimicry in Butterflies, p. 155. Copyright © Cambridge University Press, Cambridge, 1915. Reprinted by permission of Cambridge University Press, New York.

## 7.5  Russian Naturalists: Četverikov

The Russian geneticists worked more in natural populations than did the American or English. Because of controversy in which some geneticists argued that genetic findings in the laboratory might be misleading because conditions were not natural, Četverikov (1926/1961), who can well be taken as the representative of the Russian naturalists, wished to see how populations in the wild developed under processes of mutation and selection. To the geneticists of those days, natural populations appeared to be remarkably uniform.

Sergeĭ Sergeevič Četverikov [Chetverikov]* (1880–1959) came from a well-to-do and intellectual family of merchants in prerevolutionary Russia. His younger brother Nikolaĭ (1889–1973) studied mathematics under Aleksandr Aleksandrovič Čuprov* (1874–1926) and became a leader in Soviet mathematical statistics. During Sergeĭ's preparation for a career in engineering, he learned about Darwinism and natural history; about 1895, he began to collect butterflies and planned to become a zoologist despite the opposition of his father, which was later withdrawn.

In 1907–1909, as a graduate student in the Department of Comparative Anatomy in the University of Moscow, he completed a thesis on the carapace of *Asellus aquaticus*, a widely distributed isopod crustacean, and analyzed their adaptive functional significance in the struggle for existence. His later essay, *Waves of Life*, drew attention to the implications of fluctuating population size in nature. In 1915, he published a thesis on the fundamental factor of insect evolution, which answered the question of why vertebrates evolved toward larger sizes whereas early fossil insects evolved toward smaller body sizes. On the basis of arguments from the theory of similitude, he concluded that the "fundamental cause" of the evolution of insects was their outer chitinous skeleton,

thanks to which they were in a position, by continuously diminishing the size of their body, to conquer for themselves an entirely independent place among the other terrestrial animals, and not only to conquer it, but to proliferate in an endless variety of forms and thereby acquire a tremendous importance in the general economy of nature. *Thus, their smallness became their strength.* (Četverikov, 1915, p. 24)

In 1921, Kol'tsov invited Četverikov to join the staff of his Institute of Experimental Biology. In 1919, word had reached Russia of the works of W. Bateson on poultry, W.E. Castle on guinea pigs, and T.H. Morgan on *Drosophila*. In the Institute a strong group was formed studying systematics, evolution, biometrics, and genetics. A visit from Hermann Joseph Muller* (1890–1967) in August 1922 brought them cultures of *Drosophila*. Četverikov's group then decided to work on the theory and practice of Morganist genetics. The group was encouraged by Četverikov to read Western science and a seminar was arranged to exchange ideas. A sign of progress was the publication of Četverikov (1926). Much criticism was received from Russian biologists, the Lamarckians in particular; here Četverikov (1926) pointed out that we are not only completely unable to produce desirable mutations artificially, but are even unable to influence the frequency of their occurrence. (p. 170)

To what conclusion does our examination of Norton's table lead? Many of these conclusions have a very great importance for a correct understanding of the evolutionary process.

First of all, we see that because of the effects of free crossing and selection, under the conditions of Mendelian heredity, every, even the slightest, improvement of the organism has a definite chance of spreading throughout the whole mass of individuals comprising the freely crossing population (species). Here Darwinism, insofar as natural selection and the struggle for existence are its characteristic features, received a completely unexpected and powerful ally in Mendelism.

One of the most substantial difficulties of Darwinism has always been the difficulty in imagining the process by which minute improvements of the organism [were effected], the importance of which for survival appeared, generally speaking, completely negligible. The living organism, in the course of its individual life, from the egg until its death, is exposed to such infinitely diverse influences of the surrounding environment, is faced with danger from the most varied causes so many times that it seems that the small advantage which a slight improvement of the organism can offer would be completely submerged in the thousands of deaths threatening it from all sides. Let this advance help it to escape a concrete danger A; it will all the same be destroyed by danger B, or C, or D, etc., and, thus, it would take an extremely advantageous concurrence of circumstances for the organism to survive and to pass on by inheritance its small advantage. And in later generations, the same struggle and accidental basis of survival will threaten its descendants.

Now, because of Mendelism, our understanding of this process has changed. By virtue of the properties of free crossing *nothing is lost of that which is acquired by the species*. No matter how small the newly arisen improvement be, hundreds of thousands of generations, perhaps, will pass, but finally, sooner or later, it will emerge and gradually *become a property of all the individuals of the species*.

Another important conclusion from the inspection of Norton's table is that transformational *change of a freely crossing population—species, the replacement of the less adapted form by the more adapted one, in a word, the process of adaptive evolution of the species, always proceeds to the end*. It is a matter of indifference whether the better adapted form is dominant or recessive, whether the intensity of selection is 50 per cent or one per cent. Once the transformation has started, once the species has moved from a dead stop, the process automatically proceeds farther, until *the whole species changes in toto*, or until selection ceases.

This conclusion is very important for an accurate understanding of the role of various features in the evolutionary process. *Under conditions of free crossing, that is, until there is isolation* (in one of the forms indicated above), the struggle for existence and natural selection can continuously alter the physiognomy of the species, can disseminate more and more new adaptive characters through the *whole mass* of individuals of the species, can perfect any features of its organization, *but never under these conditions does the species give rise to a new species, never will there be a subdivision of the species into two, never will speciation occur*. (Četverikov, 1926, pp. 183–184 of transl.)[2]

---

[2] Copyright © *Proceedings of the American Philosophical Society* 1961; 105:167–195. Reprinted with permission.

[Note: In the 1961 version of Četverikov's paper there are errors in the displayed formulae on p. 178.]

After the rediscovery of Mendel's work, there were many experimental geneticists and population naturalists, whose past work had been in distinct fields of a very varied nature, which led them to accept conclusions not at all mutually consistent. For example, there were doubts about whether Mendelian theory was consistent with evolutionary theory. Perhaps, Četverikov (1926) was the first to write about the possibility of a consensus, later to be termed the *evolutionary synthesis*. The principal scientists involved were Četverikov, known to the West through Dobzhansky and Timofeeff-Res(s)ovsky, R.A. Fisher, J.B.S. Haldane, and Sewall Wright, these last three using approaches more mathematical than those of Četverikov. The following is a fuller list, suggested by Simpson (1949, p. 278 of 1966 edition) with some additions: in Great Britain, R.A. Fisher, J.B.S. Haldane, J.S. Huxley, Cyril Dean Darlington* (1903–1981), Conrad Hal Waddington (1905–1975), Edmund Brisco Ford (1901–1988); in the United States, S. Wright, H.J. Muller, Theodosius Grigorievich Dobzhansky* (1900–1975), E. Mayr, Lee Raymond Dice (1887–1977), George Ledyard Stebbins (b. 1906); in Germany, Nikolaĭ Vladimirovič Timofeeff-Res(s)ovsky* (1900–1981), Bernhard Carl Emmanuel Rensch (b. 1900); in the Soviet Union, S.S. Četverikov, Nikolaĭ Petrovič Dubinin (b. 1907); in France, P.G. Teissier; in Italy, Adriano Antonio Buzzati-Traverso (b. 1913). The full details of some of the above are given elsewhere in this book.

As an example of the work of the Russian naturalists and the influence of Četverikov's paper of 1926, it was suggested that, despite the apparent phenotypic uniformity of wild populations, they may contain hidden recessive mutants (see Mark Adams, *DSB*, XVII, pp. 155–165). A colleague of Četverikov, N. Timofeeff-Ressovsky, in 1925 collected 78 pregnant female *Drosophila melanogaster* and from the progeny made brother-sister matings for the $F_1$ generation. Clearly there were in the $F_2$ a number of homozygous recessives, many lethal.

## 7.6 Speciation

Although Charles Darwin is famous for, among other things, his studies on the biology of the Galapagos Islands and the biogeography of the British Isles, we will concentrate here on the work of Alfred Russel Wallace* (1823–1913), to whom some authors are prepared to assign the title of founder of the studies of biogeography. Wallace was brought up in straitened circumstances and early worked for his brothers, principally William, a surveyor in the days when new roads and railway lines were being constructed. George (1964) mentions that Wallace took a keen interest in wild flowers and animals. For his early education he was taught by his father but from the age of 5 to 14 he attended Hertford Grammar School. His autobiography tells us that more important was his supply of books from a library managed by his father. He began a herbarium and learned the names of many plants. After the death of his father in 1843, he became unemployed but obtained a teaching post in Leicester. Here he read *An Essay on the Principle of Population* by T.R. Malthus. Here, too, he met Henry Walter Bates* (1825–1892), later to become a distinguished entomologist, evolutionist, author, and editor. Wallace suggested that they travel to earn a living by collecting objects of natural history in South America; they separated in 1850 but Wallace continued to explore the Amazon basin from 1848 to 1852. After a shipwreck on his voyage home to England, Wallace lost much of his exploratory material; this was to delay the pronouncement of his views on evolution.

Wallace had been anxious to collect information on the variation and evolution of species for he had been converted to the view that species arise through natural laws. He hoped that he and Bates would be able to supply scientific details and perhaps discover some evolutionary mechanism, in particular to apply the uniformitarian geological principles of Charles Lyell to the organic world, as against the catastrophic theory.

He found marvellous adaptation of the animals in South America to the environment. An important finding was the number of different species of the same genus to be found in the Amazon basin (George, 1964, p. 23), separated by a river or tributary. Previous zoologists (taxonomists) had not worried about the locales of the animals classified. [See the map facing pp. 15 and 16 of McKinney (1972) for Wallace's travels in South America.]

He was able to make a number of generalizations about speciation as a result of this study; for example, if a new species was found in two locales, he would be able to find representatives of the species in the locales between.

In March 1854, having decided to gather more material in Asia and the East Indies, Wallace left England for Singapore, spending eight years in the modern Malaysia and Indonesia [for a map of his travels see McKinney (1972, facing pp. 26 and 27)].

Darwin had already established two principles of the mutability of species from his work on the Galapagos Islands; he believed that the island fauna provided evidence for evolution. Further, he believed that there were two types of island, namely, oceanic and continental. Oceanic islands are of volcanic or coral origin and have never had close connection with any neighbouring land. The "continental" islands were separated from the nearest land by either shallow seas or narrow straits. Wallace in his *Island Life* discussed in detail the different modes of an island receiving colonizing fauna. He was able to divide the area of the East Indies into biogeographic regions according to their animal populations, his best-known work being on the Wallace Line, separating Oriental from Australian faunal regions. Modifications to the site of the Wallace Line have been necessary because of further surveys since Wallace's day.

His great losses in the shipwreck made Wallace reluctant to write on evolution, although, provoked by an article denying "organic progression,"

Wallace (1855) claimed that he had explained "the natural system of arrangement of organic beings, their geographical distribution, their geological sequence" and also the reason for their peculiar anatomical structures.

As an example of biogeography we may recall that in the same (1953) presidential address which we have cited in §7.2, Fisher mentioned the work on eels of Ernst Johannes Schmidt* (1877–1933), naturalist, oceanographer, physiologist, and, after 1910, director of the Carlsberg Physiological Laboratory in Copenhagen. He was not only an ichthyologist but also a biometrician, interested in the number of vertebrae and fin-rays of various species of fish. In some species there were significant variations even in the reaches of the single fjord. With the eels it was different; it seemed that they had uniform mean and standard deviation throughout the world, for example in the Azores, in the Nile, and in Europe generally. His researches led to the conclusion that the breeding grounds were partly in the Sargasso Sea. Adult eels spawn there and then die and the young migrate back to western Europe and the Mediterranean Sea (see Schmidt, 1922). Measurements of size of the elvers showed that some lost ground on the homeward run. Because of the lack of isolation of the eels breeding in the tropical seas, there was no tendency to a formation of a new species.

## 7.7 Evolutionary Synthesis

Let us review some of the particular points accepted in the evolutionary synthesis and expressed as statements in a mathematical model with the aid of Mayr (1982).

1. In a population not affected by mutations or differential mortality, the gene ratios under panmixia (free crossing) remain stable as formalized by the Hardy-Weinberg law where changes over time are due to sampling errors only (see §4.3).
2. Inheritance is Mendelian, that is particulate, and not blending as in any of the hypotheses of the Darwinian or Galtonian statements (see §4.4).
3. Although the genotype is particulate, the phenotype characters are distributed approximately continuously, since each can be represented as if it were the sum of one or more genotype variables and reacts with other variables in the genotype and a variable representing the environment and not heritable. So Mendelian predictions are based upon the identity of the germ cells rather than on the somatic characters of the parents (see §4.6).
4. The processes of genetics are Markovian for, given the genotypes of the parents, the distribution is independent of the genotypes of earlier ancestors (see §4.6).
5. The behavior of the Mendelian factors parallels that of the chromosomes in sexual reproduction (see §4.9).

6. Linkage may be evident (see §4.10).
7. There is only one kind of variation, namely mutation, although its effects are variable (for example, see §§5.5–5.7).
8. There is no soft inheritance, that is, modification of the genotype by effects from the environment.
9. A single gene may affect several characters of the phenotype (pleiotropy).
10. Natural selection is effective.
11. Saltationary theories are not accepted. Changes come about by continuous processes; for example, from the reptiles there developed both the birds and the fish.
12. Rates of evolution vary between different organisms and populations; evolutionary changes occur rapidly in small populations.

[We leave a critical discussion to Ernst Mayr (1982), especially pp. 535–570.]

We may consider the evolutionary synthesis and its importance for medical and veterinary therapy. In my *Expectations of Life* (1990), I have briefly mentioned the problem of drug resistance to antimalarial drugs. At the time of preparation of that book, there was not a proper appreciation of the possibility of the development of new races or species; for example, besides resistance of patients to antimalarial drugs, the "golden staphylococcus" has been observed and many other similar incidents have been recorded.

Neanderthal man was discovered in 1856 and is now known to have been widely distributed throughout the world with the exception of the Americas and Australia. At first, he was not regarded as being in the direct line of the human race. Also some remains had been found in Belgium in 1829 and on Gibraltar in 1848 and later recognized to be Neanderthal. After heated debates, the Neanderthals are believed to be in the line of descent of *Homo sapiens sapiens* by many leading anthropologists. Choice has had to be made, which would have been more acceptable if proper measurements had been made and the sense of personal opinion had been avoided. One authority suggested that Neanderthals and *Homo sapiens sapiens* had produced fertile offspring without identifying the two classes. [See Constable (1973, including a foreword by R.S. Solecki) and Birdsell (1981).]

[See Mayr (1991) for an account of Charles Darwin, R.W. Clark (1984) for a popularization of Darwin's theories, and Johanson and Edey (1981) for a modern version of human evolution.]

See also the work of Leslie Clarence Dunn* (1893–1974) (*DSB*, xvii, pp. 248–250).

# 8
# Infectious Diseases and Microbiology

## 8.1 Introduction

This chapter gives a necessary background to Chapters 9 through 12. As to Chapter 16, it can be remarked that no scientific classification of disease was possible until the etiology of the microbial and parasitological diseases had been worked out. Indeed, it was easier to classify the causal organisms than the human disease syndromes. Further, once the microbe had been determined, the complete and incomplete syndromes could be established; thus, for example, all the syndromes caused by the tubercle bacillus could be unified.

The reality of contagion had to be established. Hippocrates and Galen, whose works dominated medical thought on epidemics until modern times, were clearly anticontagionists. Although Fracastoro introduced the notion of a *contagium animatum* in 1521 and Leeuwenhoek had revealed a microscopic world in the next century, academic medical opinion by the beginning of the nineteenth century regarded the contagionist doctrines as discredited (Ackerknecht 1948); however, inoculation and vaccination experiments and observations gave support to a contagionist theory. Liebig's chemical theory of infection is mentioned because its consequences could not be distinguished from those of a later bacteriological theory by statistical methods; it is largely forgotten now but it was favored by Farr, considered seriously by Snow in an essay, and many of Semmelweis's statements have their origin in Liebig's theory, features often overlooked by medical historians.

A comprehensive account of the development of bacteriology would be out of place but some themes are mentioned. Bassi is recognized as the "founder of the doctrine of parasitic microbes." It was necessary for Pasteur, Cohn, and Tyndall, among others, to demolish the notion of spontaneous generation of bacteria or other life forms, before such microbes could be established as the cause of disease. Further developments in theory were also required. Here Koch, modifying some conditions of Henle, gave criteria to be satisfied by organisms before they could be considered as the cause of disease (see §8.8).

## 8.2 Ancient Doctrines on Contagion and Epidemics

The most ancient traces of the hypothesis that diseases can be transmitted by touch (contagion) are to be found in the writings of the Egyptians and the Jews, according to Karl Friedrich Heinrich Marx (1796–1877) in his book, *Origines Contagii*, cited in Bulloch (1938/1960); the views are most clearly set out in *Leviticus*, chapters 13–15; the Jews thus had beliefs that disease could be spread by contact, although they also introduced notions of supernatural intervention. Such views of contagion were widely spread among the ancients [see also Adams (1844–1847) and Singer and Singer (1917–1918)].

Later there arose a rival notion that pestilence is due to natural, especially cosmico-telluric, phenomena, such as eclipses, comets, earthquakes, tidal waves, and other, perhaps, meteorological events. Hippocrates* (460–*ca* 370 B.C.) favored such a notion, for he believed that atmospheric conditions affected the workings of the body and so led to the prevalence of certain diseases; in his doctrines the state of the air is regarded as a cause of the disease and the air is affected by miasmas inimical to mankind, but, as Greenwood (1921) points out, the Hippocratic doctrine differs from the modern idea of *contagium animatum* in that it does not assign any active role to the contagium in the successive recipients. It regards air as the main cause of disease; indeed when the air is contaminated by miasms, persons become ill. So the school of Cos, or alternatively of Hippocrates, contributed remarkably little to epidemiology, even though its members strove to promote a strictly scientific medicine (that is, to exclude magic and religion, to observe, by the senses, sight, hearing, touch, smell, taste, and to reason). Further, they believed that chance does not exist, that the principle of causality holds without exception; they refused to tie medicine to philosophical principles (see R. Joly, *DSB* VI, pp. 418–431), but, of course, Hippocrates has been influential, indeed inspirational, in clinical medicine. Unfortunately, the doctrine once known as the "epidemic" theory, that epidemic (in its original sense of an effect falling on the people) disease is due to specified or unspecified atmospheric, worldly or cosmic effects, persisted into times when it was no longer useful. The epidemiological doctrine, coming down to modern times, was due to Galen* (129/130–*ca* 200). Galen's doctrine on the "epidemic constitution," influential even into the twentieth century, can be summarized as follows:

The generation of a herd sickness depends on the interaction of three sets of factors: (a) *katastasis*, an atmospheric factor, which we could term the specific factor; (b) *crasis*, an internal factor corresponding to the natural susceptibility of the herd and its members; and (c) a *procatarctic* factor, for example, life-style. Of Galen's factors, only the first is objectionable to modern scholars, for it leads to the miasmatic doctrine of herd sickness or "epidemic theory," *Galenic* as we have termed it, which is almost self-fulfilling.

It is necessary to speak of Galen's first factor, for variations on it have been important in later developments. Thus for Thomas Sydenham* (1624–1689), the unanalyzed and mysterious terrestrial or cosmic determinants of an "epidemic constitution," the *katastasis* of Galen's doctrine, were central to the understanding of epidemic disease. Further, he thought that acute illnesses in a particular epoch "bore a common stamp" according to Greenwood (1932, pp. 8–9), who cites Sydenham as saying,

the aforesaid species of disease, in particular the continued fevers, may vary so enormously that you may kill your patient at the end of the year by the method which cured sufferers at the beginning of it. This is the state of the case. There are different constitutions of years due to some hidden inexplicable change in the bowels of the earth when the air is contaminated by such effluvia as predispose and determine the bodies of men towards some disease or other; heat or cold, dryness or moisture, are not the causes; this state of affairs endures so long as the particular constitution is dominant and then yields its place to another. Each of these constitutions is characterised by a particular kind of fever, not seen under other circumstances, and fevers of this class I term *Stationaries*. (*Medical Observations* I:ii:5; translation of Sydenham by R.G. Latham)

Sydenham's younger contemporary John Freind* (1675–1728) in Freind (1717) pointed out that Sydenham himself treated continuous fevers in the same way regardless of the epidemic constitution, although advocating the need for wholly different treatment (see Greenwood, 1932).

Galen's doctrine of *katastasis* is self-fulfilling in that the only evidence we have of the "epidemic constitution" is the epidemic that it is believed to cause. Nevertheless it has played a part in controversies even since the bacteriological revolution of the 1880s. The doctrine is anticontagionist, since the insensible or unstated effects of the "epidemic constitution" are credited with the causation and no other factor is seen to be active.

## 8.3  New Approaches to Contagionism

With the revival of learning, a new tradition was established; people began to observe and to record their observations.

Giovanni Boccaccio (1313–1375) in the *Decameron* mentions that to touch the clothes or objects used by those ill of the plague causes the communication of the disease. Bulloch (1938/1960) cites Jacopo da Forli (1350–*ca* 1414) and Alessandro Benedetti* (*ca* 1450–1512) as believing that plague could be passed on by touch; they believed also that infection could be taken up by the clothes or other objects and be retained so that the clothes or objects remained infective. Such contagionist doctrines were revived by many authors, of whom Girolamo Fracastoro* (*ca* 1478–1553) is usually cited.

Some diseases can be spread by contagion, in the following ways:

1. by simple contact, as in scabies and leprosy
2. by fomites, such as clothes and sheets
3. without direct contact or the intervention of fomites.

Greenwood (1932) states that Fracastoro was the first to state the principle of *contagium animatum*, a living agent reproducing itself, which he says is a "conception wholly different from that of the contagion of the Greek and Arabist writers." Bulloch (1938/1960) and Cecil Clifford Dobell* (1886–1949) do not believe that Athanasius Kircher* (1601/2–1680) should be given credit here. Certainly they reject his microscopic observations [see Dobell (1932, pp. 141 and passim)].

It should be noted that many of the words used in the theory and practice of epidemiology have changed their meaning since the bacteriological (or organismal) revolution. Thus, infection stems from the Latin *inficere*, which has the sense of to dip in, to stain, to spoil, and so on; these meanings, to spoil or corrupt by noxious influence and then to fill (the air with noxious influences) and so after the rise of bacteriology, to infect with disease, almost always of microbiological type, do not imply in the earlier centuries any knowledge of the etiology.

To see the words *epidemic* and *contagious* used in their prebacteriological sense, we can do no better than to mention a passage from an anonymous review, possibly by the author, of a paper by W. Macmichael (1825) entitled "The Progress of Opinion on the Subject of Contagion."

A mode of transmission, or perhaps merely of origin, was suggested by Giovanni Maria Lancisi* (1654–1720) in his book, *De Noxiis Paludum Effluviis. Libri Duo* (1717), in which he remarked on the association of malaria in Rome with the surrounding swamps, and attributed the spread of the disease to the mosquitoes.

## 8.4 Inoculation, Variolation, and Vaccination

*Inoculation*, a word borrowed from horticulture where the "eye" is transferred from plant to plant, is a term used for any injection of foreign matter into the human body. It can be accidental. Thus Claude Pouteau (1725–1775) in about 1744, while a medical student at the Hôtel-Dieu of Lyons, pricked his finger during the conduct of an autopsy. His finger became inflamed and gangrenous. According to Wangensteen and Wangensteen (1978), he was convinced that wound infection was contagious and was perhaps "the first to write intelligibly" on it; he also identified the "erysipelatous nature of puerperal fever." His work was later reviewed by Jacques-Mathieu Delpech (1777–1832), who treated the wounded of Napoleon's army at Montpellier and observed the outbreak of hospital gangrene in the crowded wards. He and his colleagues agreed that suppurating wounds harbored contagious matter of an animal nature.

Alexandre-François Ollivier (1790–ca 1855), a 21-year-old surgeon's aide in Napoleon's army in Spain, tested the hypothesis of contagion of hospital gangrene by having himself inoculated by lancet with the pus of a case; he sustained a severe spreading cellulitis with some scarring. His superiors ignored or depreciated his reports. Ollivier (1822) published a treatise documenting his own researches and recalling the work of his predecessors, Pouteau, Delpech, and John Pringle* (1707–1782).

Before 1800, some surgeons were aware of the dangers of wounds at operation or autopsy, they recognized the relation between erysipelas, wound infection, and puerperal sepsis, and they knew the value of cleanliness, sterilization of instruments, and the use of clean dressing materials. Further, they understood the value of isolation of the septic from other cases. Yet because there was no biological model, that is, a bacteriological theory, to account for these findings, surgeons in general and medical academics rejected the conclusions on contagion to be drawn from the older studies; such persons prefer incidents with a given cause, for then it is easy and intellectually satisfying in an exposition to proceed from cause to effect, whereas in the absence of a cause their exposition may falter for they must plead ignorance or perhaps as in modern times mention a "cryptogenic" cause. Thus we find the academics in the Imperial and Royal Hospital in Vienna opposing the institution of a commission to study the disgracefully high maternal mortality there.

A more systematic approach to the problem of inoculation was initiated in the early nineteenth century. Bulloch (1938/1960) points out that much of the basic work was carried out by Marie Humbert Bernard Gaspard (1788–1871) in 1808 and published by him in 1822 and 1824. Many other authors injected a variety of substances into a number of animals and measured the effects on the recipients. This was of importance for the testing of the Liebigian hypothesis and for determining the mode of action of bacteria.

Variolation, or the deliberate infection with smallpox, was often carried out in the times before vaccination, sometimes by inoculation, at other times by inhalation of dried crusts from the vesicles of an active case. Similarly vaccination, the inoculation of cowpox vaccine, was often carried out after its introduction by Jenner in 1796. Clearly, these unplanned experiments proved the possibility of contagion and the specificity of each of the two infections. They also led to suggestive experiments. Thus Jean-Baptiste Auguste Chauveau* (1827–1917) in 1868 caused active vaccinia lymph to stand, so allowing the leukocytes present to settle out; they were shown not to be carriers of the virus. The soluble constitutents in the active lymph were then removed by the addition of water and so the lymph was separated into soluble material and solid particles, which contained infective material. John Scott Burdon Sanderson* (1828–1905) confirmed and extended the work of Chauveau (1868) and so gave weight to the theory that the cause of infection (contagium) was corpuscular rather than fluid. Jean-Antoine Villemin (1827–1892) in 1865 injected tuberculous material from a human case, producing characteristic tubercles in rabbits and showing that fresh animals could be infected from them. This technique of demonstrating the presence of tubercle bacilli is still useful in cases where the number of bacilli present in the lesions is small.

Inoculation experiments, planned or unplanned, were important first in bringing out the specificity of infectious diseases and second in introducing the notion that the disease-causing agent or organism multiplied, for if animals were inoculated in series, there would remain after a number of inoculations only a minuscule amount of the original material injected into the first animal.

## 8.5  Liebig's Chemical Theory of Infection

We now consider scientific explanations of the allied processes of decomposition, eremacausis (defined as a slow combustion taking place in the presence of air and water), putrefaction, and fermentation given by Johann Justus von Liebig* (1803–1873). These explanations continued to be taken seriously even into the twentieth century and delayed the acceptance of the bacteriological doctrines; Farr, as is noted in §8.1, was greatly influenced by the Liebig theories.

For reasons not yet made clear, Liebig remained opposed to the idea that the yeast bringing about fermentations was alive (see Fruton, 1972); he clung to this idea until the end of his life (see Vallery-Radot, 1919, p. 175). Liebig (1846) wrote:

The greatest errors and mistakes arise from this, that in appreciating morbid conditions, things which generally occur together are regarded as determining one another, the one as the cause of the other. Thus, for example, for the right understanding of morbid conditions and for the proper choice of remedies, there is no opinion so destitute of a scientific foundation as that which admits, that miasms and contagions are living beings, parasites, fungi, or infusoria, which are developed in the healthy body, are there propagated and multiplied, and thus increase the diseased action, and ultimately cause death.... By the recognition of the cause of the origin and propagation of putrefaction in complex organic atoms, the question of the nature of many contagions and miasms is rendered capable of a simple solution, and is reduced to the following.

Do facts exist, which prove that the state of transformation or putrefaction of a substance is propagated likewise to any parts or constituents of the living body; that by contact with the putrefying body, a state is induced in these parts, like that in which the particles of the putrefying body themselves are? This question must be answered decidedly in the affirmative.

It is a fact, that subjects in anatomical theatres frequently pass into a state of decomposition, which is communicated to the blood in the living body. The slightest wound with instruments used in dissection excites a state which is often dangerous or even fatal.

The fact, observed by Magendie, that putrefying blood, brain, bile, eggs, &c., laid on recent wounds, cause vomiting, lassitude, and death after a longer or shorter interval, has never, as yet, been contradicted.

It is a fact, that the use of several kinds of food, as flesh, ham, sausages, in certain states of decomposition, is followed in healthy persons by the most dangerous symptoms, and even by death.

These facts prove, that an animal substance in a state of decomposition, can excite a diseased action in the bodies of healthy persons; that their state is communicable to parts or constituents of the living body. Now since, by products of disease we can understand nothing else than parts or constituents of the living body, which are in a state of change in form or composition, it is clear, that by means of such matters, as long as this state continues, as long as the decomposition has not completed itself, the disease will be capable of being transferred to a second or third individual. (pp. 188, 203–204)

Liebig's idea that an unorganized fluid of chemicals could act in a way that others would regard as a vital process had an analogue in pathology, where a humoral theory of disease was favored.

It should be noted that epidemiology according to the Liebig theory would be little different from that under the bacteriological theory, so that the "cold water" surgeons and the hygienists, including Farr, could justify their methods of disease control equally readily under either theory; for example, filth could be thought to contain Liebig's ferments. The importance of Liebig's theory to us here is that the chemists, particularly Liebig, were responsible for a point of view that denied the relevance of bacteria to infective disease. It is clear that William Farr (1885/1975, p. 317) was far more impressed by Liebig than by Pasteur. With the ferments acquiring a supposed importance, the ideas of the antibacterial surgeons, of F. Nightingale, and of the sanitarians become more readily understood. Semmelweis working in Vienna expressed in many places Liebigian ideas such as "decomposed-organic-animal matter" (see also Snow's remarks in §11.1).

It should be noted that, although Liebig opposed the germ theory of infection, the modern theory of infection by viruses is Liebigian, in that viruses are not now supposed to be "alive"; indeed, the tobacco disease virus in 1935 was shown by Wendell Meredith Stanley* (1904–1971) to take crystalline form. [See Brown (1984) for a discussion of the distinction between viruses and the ambiguous "other living things."]

Specificity is lost in the Liebigian theory.

# 8.6  Specificity of Infectious Diseases

We must recognize at once that we cannot expect to find any comprehensive theory of infectious disease before the scientific period. There was first of all the difficulty of knowing which diseases were infectious in the modern sense. Human disease is so complicated and diverse in its manifestations that it took centuries to sort out the phenomena into more or less distinct entities—and in fact this process of sorting out is still going on. Even if in the Middle Ages it had been possible to recognize clearly the diseases we now call malaria, scurvy, tuberculosis, rickets, typhoid fever, and that common mediaeval occurrence when the rye harvest was poor—ergot poisoning—it would still, I think, have been impossible to say which, if any, of these diseases were infectious.

It is interesting that during the classical period the idea of contagious disease was much more clearly expressed by lay writers than by medical writers and philosophers, and right up to the nineteenth century there was a general tendency to regard contagiousness as a rather unimportant secondary phenomenon and to theorize on the underlying primary causes which might be responsible for the initiation of epidemics.[1] (Burnet, 1941, p. 607)

Thomas Fuller (1654–1734), a physician of Sevenoaks, Kent, England, was a pioneer exponent of the specificity of the individual acute exanthemata. Fuller (1730) wrote:

The Particles which constitute the material and efficient Cause of the Small-Pox, Measles and other venomous Fevers are of specific and peculiar Kinds; and as essentially different from one another, as Vegetables, Animals and Minerals of different Kinds are from one another. And since it is most certain that no Effect can be produced but by its own proper Cause, I am hard to believe that the Small-Pox or Measles can be produc'd by such Things as have no matter of Affinity with them; such are Fevers of any other Sort; cadaverous Steams from them that dy'd of other Diseases; from putrefy'd Carrion; Exhalations from fermenting Minerals: Vapours out of deep Vaults that had long been shut up, from Tempest, Thunder, Earthquakes, nor from foul Ways of Living, Nastiness, corrupt Meats and Drinks.

Nobody ever yet saw a Miliary Fever or Measles or any of the Under Species beget a true Small-Pox, or any of its Sorts, nor on the contrary: and nobody was ever defended from the Infection of anyone Sort, by having had another Sort.... The Pestilence can never breed the Small-Pox, nor the Small-Pox the Measles nor they the Crystals or Chicken Pox, any more than a Hen can breed a Duck, or a Wolf a Sheep, or a Thistle Figs and consequently one Sort cannot be a Preservative against any other Sort.

What could be done to investigate a contagious disease before Pasteur and Koch? Some authors recognized the possible etiological significance of the animalcules observed under the microscope by A. van Leeuwenhoek. To account for his clinical observations on the importance of contagion, Marc Anton Plenčič* (1705–1786) set out in his *Opera Medico-physica* the theory that disease organisms are both constant and specific, so that a given animalcule always causes the same disease in a specific host; these organisms would "breed true" and that consequently this would be a feature of the diseases, too. In his time the infections tended to be lumped together as fevers, often qualified by some adjective(s) such as low, remittent, and so on; Plenčič's ideas would lead naturally to a classification of disease by cause. Plenčič considered that only a minute amount of material would be needed to transmit such a disease because of the numbers of animalcules in a small volume and the possibilities of rapid multiplication. Plenčič saw that such a theory could apply to men, to animals, and to plants alike. This has seemed very plausible theorizing to some critics but lacking in observational support to

---

[1] Copyright © *The Medical Journal of Australia* 1941; 2:607–612. Reprinted with permission.

others. V. Kruta (*DSB*, XI, pp. 37–38) points out that the theory of Plenčič attracted little attention; nor were the first six editions of Topley and Wilson (1925–1975) at all favorable. Nevertheless such theorizing has often helped to direct research into important new channels.

Many diseases have only been recognized as specific entities in recent times. Gaspard Laurent Bayle (1774–1816) introduced the concept that a development in time of a lesion specifies rather than alters a disease. In 1801, he defended the specificity of smallpox, although it may have appeared in either its discrete or its confluent form, by showing that there were cases of contagion from one type to another. This idea was important for tuberculosis, and Théophile-René-Hyacinthe Laennec* (1781–1826) showed that the various lesions such as phthisis, spinal caries, and so on were indeed the signs of a single specific disease.

Pathological studies also gave credence to the specificity of disease; thus Pierre Bretonneau* (1778–1862) believed that it was "the *nature* of morbid causes rather than their *intensity* which explains the differences in the clinical and pathological pictures presented by diseases" according to Bulloch (1938/ 1960). Thus diphtheria and enteric fevers were specific phlegmasias, that is, inflammations, accompanied by fever and associated with particular signs and symptoms; on the other hand, other causes could be responsible for some of the clinical features. Bretonneau (1826) believed that there was only one true diphtheria and moreover that the germs (*germes reproducteurs*) might well differ in their degree of infectivity. He believed that the "*principe générateur*" of enteric fever could remain in existence, preserving its pathogenic properties and could later cause the same disease in other individuals. Bretonneau's work on diphtheria was read at the Académie Royale de Médecine in 1821 and published in 1826. In 1829 he read a memoir to the Académie on enteric fever, but could not be persuaded to publish any further work (see Rolleston, 1924). His researches on enteric fever and specificity were made known through his students. He demonstrated the special lesions in the Peyer's patches in the ileum (small intestine) and defended the concept of a specific transmissible agent, or at least poison. Bretonneau's *Traités de la dothinentérie et de la spécificité* were finally published posthumously in 1922. The French observers also distinguished enteric fever from simple gastroenteritis on pathological grounds.

Many workers gave evidence for the distinction between enteric fever and typhus, but the clearest demonstrations were given by William Wood Gerhard (1809–1872) in 1837 in the United States and by Johann Lucas Schoenlein* (1793–1864) in Germany (see Schoenlein, 1839a). Schoenlein referred to the two diseases as *typhus abdominalis* and *typhus exanthematicus*, respectively. These findings were confirmed by William Jenner (1815–1898) in Britain in the years 1849–1851.

A new approach to specificity was suggested by the work of David Gruby* (1810–1898), who described a number of fungi causing skin diseases in man and showed that the diseases could be produced experimentally; further there was a specificity, whereby distinct diseases could be produced by distinct organisms. [See his biography by V. Kruta (*DSB*, V, pp. 565–566), Ainsworth (1976, pp. 163–172), and Kisch (1954).]

Another hint on the specificity of disease came from the introduction of quinine (cinchona bark), the first effective remedy against an infectious disease, which had been introduced into Europe in the first half of the seventeenth century; its effective use countered the old doctrine that acute sicknesses must be treated with remedies against dyscrasias and so by blood letting, vomitings, and purgings.

## 8.7 Microbial Causes of Disease

Agostino Maria Bassi* (1773–1856) may well be regarded as the founder of the doctrine of parasitic microbes and a precursor of Cohn, Pasteur, and Koch. In 1833, he was investigating the silkworm disease, known as muscardine in France, in which the affected worms become covered, after death, with a white efflorescence of a hard consistency. This characteristic efflorescence appeared after death but Bassi showed that, by taking the worm in the early stage, removing a portion of the epidermis, passing the worm through a flame, and touching the underlying tissues with a heated pin, he could transmit the disease to another, clean silkworm. Further, Bassi found that the *Botrytis* could survive three years outside the living body of any silkworm and then infect fresh silkworms (Bassi, 1835–1836/1958). A Milanese botanist, G. Balsamo-Crivelli, identified the infective organism as a cryptogam, later to be known as *Botrytis bassiana*. Jean-Victor Audouin* (1797–1841) confirmed Bassi's views. Vittadini (1852) was able to grow the *Botrytis* on artificial media and recognized its spore. Schoenlein identified another fungus, now known as *Achorion schönleinii*, as the cause of the human skin disease, favus (see Schoenlein, 1839b). This organism was named and shown to cause a contagious disease by Robert Remak* (1815–1865), then in Schoenlein's laboratory. [See Bulloch (1938/1960, p. 166) for a number of discoveries of parasitisms by fungi before 1853.] But to some extent this was a dead end, for few serious human infectious diseases are due to fungi. Some suggestive observations were made by the French microscopist Alfred Donné (1801–1878). Donné (1837), making microscopic examinations of mucus and of pathological discharges, especially of the human genitalia, found pus cells in gonorrhea, and in chancres he found spirochetes thought by Bulloch (1938/1960) probably to be *S. refringens* (now *Treponema pallidum*). He also found the flagellate protozoon now known as *Trichomonas vaginale* in mucus.

Friedrich Gustav Jakob Henle* (1809–1885), at a time when contagionist doctrines were at their lowest ebb according to Ackerknecht (1948), considered the possibility that living organisms were responsible for some infectious diseases. As George Rosen (1910–1977), the translator of Henle's review remarks, Henle (1840) was catholic in his choice of data, ranging from veterinary medicine (liver flukes) to fetal pathology (intrauterine smallpox) (see Rosen, 1938). Henle (1840) states that in infectious disease, the morbid matter increases during the incubation period and that if it is a *contagium vivum* it probably belongs to the plant kingdom. Epidemics and endemics (that is, endemic diseases) attack the most diverse individuals and do not remain confined to the one species, so that epizootics and epidemics may occur simultaneously; there is, further, no need to assume any pathological predisposition.

Individual infectious diseases seem to be similar in very different climates. The epidemic may migrate with persons. Air seems to have little effect, for example, fumigation and vaporization have not been effective; miasmata have not yet been identified as living; miasmatic diseases, for example, ague, are never contagious [in the sense of Fracastoro (1) and (2) (see §8.3)]. Henle pointed to "the ability to multiply by assimilating foreign materials, known to us only in living organic beings"; he says that no chemical multiplication would resemble yeast fermentation. The infecting dose is small, for example, a drop of diluted chickenpox is infective by inoculation. A precise typical course of the disease points to an independent temporal development of the infective agent. Infective diseases attack both the strong and the weak. Sneezing in early cases of influenza indicates entrance to the body via the nose. Henle was unfortunate in that there were no differential stains to indicate parasites in the tissues. Moreover, on p. 947 of the translation, he remarks:

Finally to these observations I have to add my negative experiences which I was able to make, occasionally, to be sure, on typhoid cadavers, on smallpox and vaccinia contagia, on desquamated skin of scarlatina and other skin diseases. I can surely affirm, that neither any of the known infusoria nor a plant of the yeast type nor *Botrytis bassiana* are to be found in the contagia mentioned.[2]

There had been a question as to whether the disease was a living entity reproducing itself. Henle remarks that this is not so; it is the causal agent which is reproduced.

After Henle's review, new data began to be collected. Thus in 1855, Ferdinand Julius Cohn* (1828–1898) gave a careful study of a disease in animals (the housefly) caused by a fungus. Gruby, already mentioned in §8.6, discovered an animal parasite in the blood of a frog and named it *Trypanosoma*. He also, with Henri Mamert Onésime Delafond (1805–1861), discovered a parasitic hematozoon in dog microfilaria, and showed that it could be transmitted by inoculation. It is of interest that Gruby had worked at the University of Vienna with Joseph von Berres (1796–1844) and Karl von Rokitansky in the early 1840s.

---

[2] Copyright © *Bulletin of the History of Medicine* 1938; 6:907–983. Reprinted by permission of Johns Hopkins University Press, Baltimore.

The history of the identification of the anthrax bacillus is rather less straightforward than that of muscardine. It appears that Casimir Joseph Davaine* (1812–1882), in a consultation with Pierre-François-Olive Rayer (1793–1867), saw in 1849 the large rod-shaped bacilli in the blood of a sheep suffering from anthrax; in 1865, Davaine gave a remarkably logical defence of the thesis that these bacilli were indeed the cause of the disease; further, he injected material from a malignant pustule of man into a guinea pig, causing the characteristic signs of the disease and observing the bacillus in the blood (Théodoridès, *DSB*, III, pp. 587–589).

Before bacteria could be accepted as causes of disease, the doctrine of spontaneous generation had to be discredited, for a classification of spontaneously developing organisms could hardly be established. F. Cohn, already mentioned, was a distinguished botanist and microscopist who had worked on algae and fungi; from 1870 he turned his attention to the biology of the bacteria and founded a new journal, *Beiträge zur Biologie der Pflanzen*, which would accept contributions in this new field. He established the fixity of bacterial species and devised a classification of bacteria. Even as late as 1870 there had been debate as to whether there was one type of bacterium or many; here backwardness was due to the technical difficulties of obtaining pure strains in culture. Cohn experimented with the effect of heat on bacteria and showed that while many bacteria could not survive heating to 80°C, *Bacillus subtilis* could survive boiling because of the existence of spores—an important insight for the spontaneous generation debate. We note that this problem has its origins in ancient times (see Farley, 1977). Here attention is centered on the bacteriological implications.

The necessary techniques and experiments to discredit the doctrine of spontaneous generation were devised and carried out largely by Louis Pasteur, F. Cohn, and John Tyndall* (1820–1893). Pasteur (see Geison, *DSB* X, pp. 383–384) had early recognized the importance for medicine of his work on fermentation and spontaneous generation; moreover, he had already in 1865 supported Davaine's claim that the anthrax bacillus was the cause of anthrax in a discussion at the Académie des Sciences. He had hoped in 1867 to begin a study on anthrax but was delayed until 1877 for he wished first to destroy the doctrine of spontaneous generation and second to obtain a collaborator, as he felt that he was "insufficient" for work in medical or veterinary fields, a problem overcome by the enlistment of Jules Joubert (1834–1910); later assistants were Pierre-Émile Duclaux* (1840–1904), Charles Édouard Chamberland* (1851–1908), Pierre Paul Émile Roux* (1853–1933), Louis Ferdinand Thuillier (1856–1883), and Adrien Loir (?1862–1941). However, Pasteur (1878) had explained his principles:

Impressed as I am with the dangers to which the patient is exposed by the germs of microbes scattered over the surface of all objects, particularly in hospitals, not only would I use none but perfectly clean instruments, but after having cleansed my hands with the greatest care and subjected them to a rapid flaming ... I would use only lint, bandages and sponges previously exposed to air of a temperature of 130 to 150°C.; I would never use any water which had not been subjected to a temperature of 110 to 120°C.[3]

In 1879, he demonstrated the streptococcus as a cause of puerperal sepsis to the Académie Royale de Médecine. Pasteur recommended boric acid as a nonirritant antiseptic agent. Heinrich Hermann Robert Koch* (1843–1910) in 1878 had made his study on wound infections and medical researchers were becoming more confident and more capable of finding a bacterial cause.

In the meantime, Koch (1876) had demonstrated his work on anthrax at Cohn's laboratory and published it in Cohn's *Beiträge*. Koch (1877) described in the same *Beiträge* his techniques for dry-fixing thin films of bacterial culture on glass slides and staining them with aniline dyes following the advice of Carl Weigert* (1845–1904), who had in 1871 demonstrated the presence of bacteria in the tissues by his staining methods. Koch indeed was a great innovator, introducing many new techniques for culturing bacteria and obtaining high-grade microphotographs of them in culture and in the tissues. The followers of Pasteur and Koch were to determine the cause of all the common bacteriological diseases in the years before 1920.

As an example of the difficulties in accepting the contagionist doctrines, Jean André Rochoux (1787–1852) stated that experience of the typhus and yellow fever of the Napoleonic wars had more than anything else undermined the belief in contagion and declared in Rochoux (1832) that the experience with cholera discredited sanitary measures such as quarantine. C.E.A. Winslow (1943, p. 182) wrote, "Until the theory of inanimate contagion was replaced by a theory of living germs and until to that theory were added the concepts of long-distance transmission by water and food supplies and, above all, of human and animal carriers–the hypothesis of contagion simply would not work." Welch (1925) wrote that in 1848 official opinion was hostile to the germ theory of disease. Yet the erroneous "filth theory of the generation of epidemic disease" was a useful guide to preventive action. Furthermore, as Ackerknecht (1948) has remarked, few anticontagionists died as a result of their self-experiments.

## 8.8 Koch's Postulates

Whether an organism can be considered as the cause of a disease is determined by its satisfying the reasoning of Henle (1840), as modified in the famous postulates of Koch (1882), who first stated them informally. We may write them as follows:

---

[3] G.L. Geison in *Dictionary of Scientific Biography*, vol. X, p. 389b. Charles Scribner's Sons, New York. Copyright © American Council of Learned Societies, 1974. Reprinted with permission.

1. The organism should be found in all cases of the disease in question, and its distribution should be in accordance with the lesions observed.
2. It occurs in no other disease as a fortuitous and nonpathogenic parasite.
3. The organism should be cultivated outside the body of the host repeatedly in pure culture and the organism so isolated should reproduce the disease in other susceptible animals (see also Koch, 1890.)

Evans (1976) points out that these rules were not to be used dogmatically; in any case difficulties arose when the virus diseases were investigated. At first, and even after Koch's time, it was assumed that the presence of an organism was both necessary and sufficient for the disease. It was later found that healthy carriers exist for some pathogens; their presence is not sufficient for the disease. With the redefinition of many diseases caused by organisms their presence is necessary, although technical methods may fail to demonstrate them in individual cases. In diphtheria, patients died of myocarditis although the bacilli could not be found in the heart; it was proved that a soluble toxin could explain the difficulty. Streptococci were found in the lesions of cases of puerperal sepsis; that they were also found in the normal vaginal flora was held to be evidence against the bacterial causal hypothesis. Later, it was shown that only one group of pathogenic hemolytic streptococci was responsible for such infections and that the normal flora did not contain hemolytic streptococci, which points to the necessity of an accurate diagnosis of the organisms seen or cultivated. Technical difficulties may prevent the fulfilment of Postulate 3; thus the bacillus of leprosy has never been cultured although its presence has been demonstrated in the human lesions and in armadillos. There have also been difficulties with the cultivation in vitro of protozoan parasites and viruses. The clause, insisting that the bacteria must be cultivated over several generations, was to ensure that there was not a carriage of parts of the original host to the test animals.

When it became known that some diseases were caused by filter-passing viruses, later abbreviated to viruses, it became clear that Koch's postulates were unduly restrictive. Rivers (1937) suggested modifying them: Postulate 1—a specific virus must be found associated with a disease with a degree of regularity; Postulate 2—the virus must be shown to occur in the sick individual not as an incidental or accidental finding but as the cause of the disease under question; Postulate 3—could not be applied because viruses need living culture media. Further, at least one natural plant disease needs the action of two viruses and there are viruses that lie latent in both humans and in animals, which may cause further difficulties. There may be no susceptible laboratory animal to satisfy Koch's Postulate 3.

Evans (1976) suggests that with the development of various laboratory techniques other observations can be used; for example, if a virus is the cause of a disease, then antibodies against it will be developed in the body, so that a rise of antibody concentration (titer) is confirmatory evidence. He believes that, although Koch's postulates have been useful in the past, the general scientific community is now able to bring many different considerations to the solution of the problem of causation of disease by microorganisms. [As a modern instance, see Aurelian, Manak, et al. (1981).]

Koch's postulates were successful in persuading observers of the truth of the hypothesis that bacteria were indeed the cause of disease and in rejecting facile claims that a given organism was the cause of a given disease. An illuminating story of the final acceptance by William Osler (1849–1919) in 1886 of the malaria plasmodium as the cause of malaria is given in Cushing (1940) and Russell (1955); Osler was at first sceptical because of the previous reporting of micrococci that could not be cultured and successfully injected to cause new cases of disease; he was also sceptical of the claims of Charles Louis Alphonse Laveran* (1845–1922), who had reported ameboid forms in the blood. Nevertheless, he examined many cases of malaria, finding not only the ameboid forms but also crescentic bodies not visible in non-malarious patients, and became convinced of the validity of Laveran's claims.

As a result of the discovery of bacterial and other parasitic causes of disease, it became possible to link together various disease syndromes according to their cause; thus, the local lesion and secondary rash of syphilis could be linked to such later manifestations as aortic aneurysm, tabes, and general paralysis of the insane; tuberculosis infection could be shown to be the cause of many syndromes such as acute miliary tuberculosis, lymph gland infections, pulmonary tuberculosis, spinal caries, and so on. Another result of the bacteriological revolution was the establishment of the doctrine that certain diseases "bred true"; indeed, it could have been of little value to classify diseases into separate entities if their properties were not constant. Without this doctrine, even the description of the development of an epidemic becomes confused as in the writings of Charles Creighton* (1847–1927) (see Creighton, 1965). Nevertheless, we must allow for different syndromes to be caused by the same organisms, for example, *Streptococcus pyogenes* causes erysipelas, infection of wounds, and puerperal infections. Some of the difficulties with *Myco. tuberculosis* disappear when microscopic observations are available, when it is seen that the lesions in different organs exhibit the same sequence of microscopic changes.

Scabies is not a disease causing mortality but it is of interest in showing how medical theorists ignored evidence linking organisms with disease. It is due to a mite, *Sarcoptes scabiei*, the largest form of which is about a quarter of a millimetre in length and just visible to the naked eye. The female form lays eggs in the burrows in the horny layer of the skin; the eggs develop into larvae, and later nymphs and adults wander over the body and may enter hair follicles causing intense irritation. Arabian authors in the ninth and tenth centuries knew that the disease was associated with an animalcule that could be picked out with the point of a needle. Busvine (1976) points out that this was known in the medicine of Western Europe, for example, S.F. Renucci in 1834 was able to find the female adults consistently in their burrows. Yet even as late as 1863, leading dermatologists were denying a role in the disease to the mite. Henle (1840) gives a proper explanation of why the mite is to be considered as the causal organism of the disease and not a mere concomitant. Later F. Hebra, in Hebra and Kaposi (1866–1880), also gave a correct account.

The establishment of organisms as a cause of infective disease was also important for therapy. In 1891, Dimitriĭ Leonidovič Romanovskiĭ (1861–1921), using an eosin-methylene blue stain, was able to detect morphological damage in the malaria parasites of patients being treated with quinine. He stated that quinine cured malaria by damaging the parasite more than the host, a novel concept not appreciated at the time but later to become a leading motive of P. Ehrlich.

After the triumph of parasitology and of microbiology in the years around 1880, advances could be made in many fields, practical and theoretical; first, organisms could be accepted as the cause of many diseases and note of this made in their definition; thus, fever could be seen not as a disease but as a symptom (Wilson, 1978). Second, the disease so defined could be seen as an entity so that an epidemic could not be imagined as starting as scarlatina and ending as measles, for example. Third, direct action against the organism might be possible. Fourth, the chain of infection might be broken by a variety of methods. Fifth, microbiology gave the true reason for the virtue of cleanliness.

[For microbiology, see Bulloch (1938/1960), Clark (1961), Ford (1939), Foster (1970); for parasitology, see Faust (1955), Foster (1965), Garnham (1971), Lechevalier and Solotorovsky (1965), and Scott (1939).]

## 8.9  Contagion and the Bowel Infections

In this section, there are a number of examples in which geographical site has been an aid to finding some properties of infective diseases.

John Snow* (1813–1858) was the son of a farmer (see Richardson, 1887/1936). He was sent to a private school at York, where he showed great proficiency in mathematics. In 1838, he became Member of the Royal College of Surgeons and "passed the Apothecaries Hall," and so became duly qualified in medicine. He had invented a pump for the resuscitation of newborn children and was an ingenious designer of such instruments. He noted in 1846 that the difficulties with ether anesthesia were due to faulty apparatus. Soon he had made an appropriate apparatus and gained a great practice as an anesthetist, administering anesthesia at two of Queen Victoria's confinements. In 1848, he turned his thoughts to the cause and method of spread of cholera on which he wrote a small pamphlet, but he included the substance of this and of his later works in a single paper (Snow, 1853/1855/1936). He shows that cholera can be spread by fomites; for example, the first patient with cholera died in a room on 22 September 1854 and the next lodger in the same room suffered a typical attack on 30 September; he cites other cases where cholera was spread by fomites from person to person, it being unreasonable not to believe that there were cause and effect. He also notes that being present in the same room with a patient might not be sufficient for spread and gives evidence that persons washing their hands after close contact with the patient or fomites are not usually attacked. He gives evidence that the principal cause of symptoms and of death is dehydration and that the "morbid material or cholera poison" must multiply in the alimentary system. Further, it must have "some sort of structure, most likely that of a cell" (p. 15). "It is no objection ... that the structure of the cholera poison cannot be recognised by the microscope" for neither could the matter of smallpox or syphilis.

Snow believes that his article proves the modes of spread, independently of such visual aid. He points to lack of personal cleanliness being a great factor in aiding the rapid spread of cholera, with lack of water and light being subsidiary aids to its spread. To explain the rapid and widespread nature of a cholera epidemic, Snow calls on the mixture of cholera evacuations with the "water used for drinking and culinary purposes, either by permeating the ground, and getting into wells, or by running along channels and sewers into the rivers from which entire towns are sometimes supplied with water."

Snow (1853/1855/1936, pp. 28–57) details local outbreaks of cholera following fouling of wells, including the Broad Street pump epidemic. He gives a map, 30 inches to the mile, centered on the pump, showing the location of cholera deaths in the epidemic; in general, the density of cases falls sharply as distance from the pump increases but one block wholly within 100 metres of the pump has no deaths; indeed, the brewery hands drank no water. By a careful follow up, he was able to show that persons dying some kilometres from the pump had drunk water transported from it. Snow was able to abort the epidemic by having the handle of the pump removed. Snow says he knows of no instance in which cholera has been

generally spread through a town or neighbourhood, amongst all classes of the community, in which the drinking water has not been the medium of its diffusion. Each epidemic of cholera in London has borne a strict relation to the nature of the water-supply of its different districts, being modified only by poverty, and the crowding and want of cleanliness which always attend it.

We may summarize the results of Snow's enquiries as follows. Briefly, the customers of two companies, A = Southwark and Vauxhall Water Company and B = Lambeth Water Company, supplying water to the southern suburbs of London suffered high death rates from cholera in 1849. By 1853, B had begun to receive its water from a reach of the River Thames not contaminated by the sewage of the city of London. In one area supplied by B alone, there were no deaths. Snow followed up this lead in an epidemic in 1854, and ascertained that 286 deaths had occurred in 40,000 houses supplied by A and 14 deaths in 26,000 houses supplied by B. Here Snow was able to obtain a satisfactory control series because nearby houses in some suburbs were supplied by different companies. The argument is that given an equal probability of a consumer being infected from the water of A or B, a concentration of cases in the consumers of A has occurred; this would be a very rare event, under the null hypothesis of equal probabilities, and therefore, the null hypothesis is rejected.

We have seen in §8.6 that Bretonneau had the idea of specific diseases and in particular had shown that enteric fever (typhoid) was a specific disease. He had been one of the first to also give evidence that it was contagious; he believed that it was distinct from a second disease, exanthematic typhus. William Budd* (1811–1880) in a country practice at North Tawton, where he was the only medical practitioner, witnessed an epidemic commencing 11 July 1839. Budd (1873/1931) noted that although there were unhygienic privies and pigsties in close proximity to the houses, typhoid had not been epidemic for some years in the village of "some eleven or twelve hundred souls." He witnessed fresh infections in almost every household where the infection had already occurred. His conclusion that the disease was contagious was strengthened by three persons leaving the place where they had become infected and proceeding to unaffected places; here they were responsible for infecting new cases. Budd saw other cases where infected persons had passed through neighbourhoods unaffected by the epidemic to give rise to fresh cases in distant areas. He concluded that typhoid fever "is an essentially contagious or self-propagating fever. If need were, it would be easy to show, by the doctrine of probabilities, that to attempt to explain [the facts] on any other principle would be absurd" (p. 25). He noted that the disease had a latent period, that immunity was conferred by one attack against a future attack, and there was an immunity of "large numbers of persons" against an attack although freely exposed to the "fever poison." Budd gave a striking example of spread by water in Richmond Terrace, Clifton; the inhabitants of 13 houses drew their drinking water from a pump, found to be "tainted by sewage" in late September. There were cases of typhoid in all of the 13 houses using the tainted water, whereas the remaining houses were free of the disease.

It is commonly believed that the theorists of the nineteenth century were slow in accepting such evidence. After the Century of Progress Exposition, held in Chicago in 1933, it was reported that two cases of amebic dysentery had occurred at a certain hotel $X$ in Chicago; later it was found that more cases of amebic dysentery had occurred at hotels $X$ and $Z$. Some skilfully designed questionnaires were then sent to guests at $X$ and $Z$ and four control hotels; later it was learned that 256 out of 9,919 guests at $X$ and $Z$, as against 3 out of 12,380 at the four control hotels, had suffered symptoms of amebic dysentery, clearly a statistically highly significant result. It was determined that in hotel $X$, a sewage leak had been occurring into the water later to be used at table (Bundesen, Connolly, et al., 1936).

A complement to the work of Snow and Budd mentioned earlier in this chapter is the following account, cited by Dudley (1931):

On Oct. 22nd, 1862, during operations in China 211 men out of the 500 which formed the ship's company of HMS *Euryalus*, proceeded to take a party of soldiers up river in flat-bottomed boats. The naval contingent drank the river water, but the army party took supplies of good water from Shanghai. Cholera broke out on Oct. 25th. The expedition returned to the *Euryalus* on the 29th, after four days' [sic] absence. During the following fortnight there were 39 cases of cholera and 54 of dysentery, and 29 men died. Not a case of these intestinal infections occurred among the soldiers. All these cases, with two exceptions among the nursing staff who got "choleraic diarrhoea," were among the sailors who joined the military expedition. (p. 510)

Some 44% of the naval landing party developed cholera or dysentery after the return to the ship, while less than 1% of those who remained on board suffered either disease. Note the existence of the two control groups, soldiers and naval personnel, remaining on the ship.

[See Lancaster (1990, p. 72) for the value of pure water for the cities and see also Goubert (1989).]

## 8.10 Contagion by Air, Direct Contact

Panum (1940) is a translation of Panum (1847), the observations of Peter Ludwig Panum (1820–1885) on the measles epidemic in the Faeroes in 1846. He found then that the catarrhal prodromal stage began usually 10 to 12 days after infection and that measles is most infectious at 14 days during the stage of eruption and efflorescence; further, the infectious period is limited to one or two days. Of the 98 old persons, alive during the previous epidemic of 1781, none was attacked in 1846. Measles could be avoided by the quarantine of houses or of whole villages, so that it seemed that infections were only possible after direct contact or passage within the air "most nearly surrounding the patient." Panum (1847/1940) thus concludes that measles is "not miasmatic but purely contagious in character" and further that he has found no evidence that measles can arise spontaneously. Schleisner (1851), investigating an epidemic in Iceland, which could have been contemporaneous with Panum's epidemic in the Faeroes, remarks that measles is contagious since the country is divided up by mountain ranges and the farms are widely dispersed, so that passage from farm to farm can be readily traced.

Contagion for many other diseases is by the air but it is usually not possible to demonstrate that this is the only method of spread and there are few examples. It seems, on looking at the standard texts, for example, Topley and Wilson (1983–1984), that aerial spread of infection was usually only established for a disease after bacteriological methods had been developed.

## 8.11 Vector-Borne Infections

Although intelligent guesses had been made as to the role of insects in the transmission of disease, Busvine (1980) believes that no standard textbook of medicine had given any specific reference to this role before 1871. Patrick

Manson* (1844–1922) in 1878 described the development of the nematode worm *Wuchereria bancrofti* in the body of a mosquito and later he showed that the mosquito is the vector of filariasis. Quite independent was the discovery of a vector-borne disease of cattle by Theobald Smith* (1859–1934). Smith graduated with a Ph.B. degree from the Albany Free Academy in 1881 and from the Albany Medical College in 1883 in first place. Smith (or Schmitt) had spoken German in the home and so was able to read the works of Koch and others of the German school and to publish in German periodicals. At graduation, he felt unready for private practice and became assistant to the chief of the veterinary division of the United States Department of Agriculture. Here Smith taught himself Koch's culture-plate methods. Among his other discoveries, he wrote on *Investigations into the Nature, Causation, and Prevention of Texas or Southern Cattle Fever* (1893), published over six years after he had first observed "small round bodies" in the red blood cells of the affected cattle.

Smith designed a series of critical experiments to prove that the agents of the disease were carried by ticks and that these agents could be passed on from an infective mother through an egg (transovarian passage). Tick-infected cattle from the southern states where the disease was prevalent were brought to graze on farms in Maryland and divided into two groups, Field A containing cattle with ticks present and Field B containing cattle with ticks removed. The southern cattle were fed separately but in each field ran with northern cattle. In Field A (with ticks) the northern animals developed "red water" (Texan) fever, but in Field B there were no infections of the northern cattle. Later cattle from Field B were taken to Field A where they became infected. Next, ticks collected from an enzootic area in North Carolina were introduced into a clean Field C, where they infected cattle. To prove the transovarian passage of the infective agent, larvae were hatched from the eggs of infected ticks and placed on the ears of clean northern cows, which later developed the "red water" fever. This was an important discovery and later strengthened the support for Ross and others introducing malaria as a vector-borne disease.

In 1895, David Bruce* (1855–1931) proved that a biting fly of the genus *Glossina* was the vector of nagana, a fatal African disease of horses and cattle, and later that *Glossina* species were also the vectors of the human sleeping sickness. Ronald Ross* (1857–1932) in 1897 announced that anopheline mosquitoes carried malaria, an important discovery to be confirmed and amplified by Italian workers.

A disease is said to be vector-borne (often written as transmissible) if the infecting agent is usually spread by vectors (insects, arthropods). If the disease is spread only by vectors, it is said to be strictly vector-borne, otherwise it is said to be facultatively vector-borne; for example, plague considered as a human disease is facultatively vector-borne since it can be spread by the human flea, *Pulex irritans*, and as a respiratory disease without the intervention of the flea as a vector. Yellow fever, typhus and malaria are strictly vector-borne. We are excluding from consideration here experimental transmission and accidental transmission by syringe in treatment or in drug injection.

Arthropods transmit pathogens mechanically or biologically. In mechanical transmission, the vector carries the infective agent without any development in the agent; usually this is a case of interrupted feeding as in transmission by the domestic fly. There are three varieties of biological vector transmission: (a) the organisms undergo cyclical changes and multiply as in malaria, (b) the organisms undergo cyclical changes but do not multiply as in bancroftian filariasis, and (c) the organisms undergo no cyclical changes but multiply as in bubonic plague. These three modes are known as cyclo-propagative, cyclo-developmental, and propagative transmission.

Many human diseases are now known to be vector-borne; a convenient table of these is available in Beaver, Jung, and Cupp (1984, pp. 558–563), which lists the taxonomic group of the species, the pathogenic agent, the role of the arthropod, and the method of human exposure. Malaria of the human species and bartonellosis may be the only vector-borne human diseases for which no animal host is known (see also Stanley, 1980).

Most of the vector-borne diseases of man being anthropozoonoses, it is necessary for us to consider briefly the general conditions for the spread of a vector-borne zoonosis, a problem in animal ecology.

## 8.12  Zoonoses and Anthropozoonoses

Diseases of animals are known as zoonoses; when they infect man they are known as anthropozoonoses but less formally also as zoonoses. Anthrax was the best known of the zoonoses and possibly few others had been recognized before the end of the nineteenth century. Bubonic plague was first clearly stated to be an anthropozoonosis by Roux (1897). The full implications have only been worked out in this century.

Epidemics of plague in man are unstable because the disease is too acute, causing death or an immunity of variable length; the same can be said of the domestic black rat, *Rattus rattus*, and of the wild brown rat, *Rattus norvegicus*, the brown rat being the more resistant. History shows that even the brown rat cannot perpetuate an epizootic, for Europe became plague-free in the late nineteenth century. Perpetuation of plague in the wild has been revealed by the work of Evgeniĭ Nikanorovič Pavlovskiĭ (1884–1965) and his school in the Soviet Union and by a group of workers sponsored by the World Health Organization. According to Baltazard, Bahmanyar, et al. (1960), in a nidus in Iranian Kurdistan, there was a complex interaction between the plague bacillus and four species of wild rodents of the genus *Meriones*, of which two were highly resistant and two highly susceptible. Perpetuation there depended on a balance between resistant and susceptible species; the continued presence of the bacillus in the same place (nidus) is linked to the sedentary nature of the resistant species, living in deep permanent burrows that act as reservoirs for infected fleas. As far as the epizootic is concerned, the epidemic or pandemic resulting is an epiphenomenon. Human plague occurs when the disease spreads from the nidus directly to man or to the brown and black rats and finally to man. Clearly this is a chance effect depending on some local contacts between wild rodents and the rats. The mathematical theory of epidemics may assist us to understand the various steps in the development of the pandemic but not to predict the future course.

Yellow fever is also an anthropozoonosis, transmitted in the urban epidemics from man to man by the mosquito, *Aedes aegypti*, but the real disease is the zoonosis of monkeys in Africa and America. There seems little doubt that the zoonosis was initially African with the infection being spread by the *Aedes* species.

Many zoonoses are vector-borne and this fact gives them a geographical distribution, for both host and vector will require special conditions for survival. Usually the distribution is limited—in the temperate regions especially, it may be spotty and even highly focal; in the hot, wet continental masses, there may be many vectors and host species, and the disease may be widespread. Man is not necessary for the perpetuation of a zoonosis in any nidus that is, natural habitat (see §§11.1 and 11.3), although man has often transported both vector and vertebrate host; examples are provided by the transport of rats and or fleas infected with the plague bacillus on sea routes. Moreover, man often transforms the ecology of an area by engineering works.

# 9
# Puerperal Sepsis

## 9.1 Early Incidents and Opinions on Puerperal Sepsis

Much information is available in the historical review of Fleetwood Churchill (1850). The first undoubted epidemic of puerperal fever occurred in Paris in the winter of 1746. Other epidemics appeared in Paris and Lyons. Twenty-four women died of it in 1760. Other epidemics appeared in 1830–1831, one midwife having 16 puerperal deaths among 30 patients. No such deaths occurred in the practices of other midwives at the same institution. Other incidents are reported. Churchill remarks on another series of cases, "*post hoc* is not always *propter hoc*" for puerperal fever was epidemic in Edinburgh in 1821–1822, a reference to the Galenic doctrine (see §8.2).

As an example of the difficulties in establishing contagion as important for the spread of disease before the rise of bacteriology, the history of investigations into puerperal sepsis is now examined. It is a favorable field because it affected only a small proportion of the population and its diagnosis was easy, there being high case fatalities. The incidence of fatal puerperal sepsis in the general population was low; indeed, the total maternal mortality rates for England and Wales were of the order of 5 per thousand confinements over the whole of the nineteenth century and possibly the contribution of puerperal sepsis was no more than 3 per thousand, so that a sequence of, say, three maternal deaths in a small country practice would have been a very rare event and, as we can see now, would overthrow the null hypothesis of no contagion. A study of Churchill (1850) or of the general literature of the disease shows that the practitioners varied in their beliefs on contagion.

The survey may begin with a reference to Charles White* (1728–1813). His works of 1773 and 1791 do not lend themselves to much epidemiological analysis; he was clearly on the contagionist side; his principles of hospital practice were enlightened and an inspiration to other hospital designers.

Alexander Gordon (1752–1799), in private practice near Aberdeen, concludes in 1795 from a study of the patients of midwives and himself that (a) childbed fever is spread by contagion; (b) the Galenic hypothesis (atmospheric or katastatic of §8.2) is irrelevant because some sequences of cases are specific to one medical attendant, and the cause could be carried to a new neighbourhood by the movement of such medical attendant; (c) some cases are due to the same cause as erysipelas (used at this time as equal in meaning to the modern "septic condition"). He mentions the simultaneous epidemics of puerperal sepsis and erysipelas.

Thomas Nunneley (1809–1870) like Holmes (1843, 1855, 1891) makes points that can be summarized by this statement that "puerperal fever is only one form of a diffused inflammatory action, which, when it is exhibited upon the surface of the body, is called erysipelas" (Nunneley, 1841, p. 89). Moreover, the two diseases possess a similarity of constitutional symptoms and postmortem findings; their epidemics are often simultaneous with many coincidences of childbed fever in the hospital and erysipelas in the nurses and servants; pus is deposited in various parts of the body; "erysipelas" follows autopsy wounds in childbed fever; the two diseases often coincide in one patient; childbed fever and erysipelas may produce each other in a second person. Perhaps because the title mentions only erysipelas, Nunneley's book does not appear in the bibliography of Holmes (1855).

Oliver Wendell Holmes (1809–1894) defines *epidemic* (really Galenic) influence for a widespread geographic incidence (Holmes, 1855). There may be a similar reasoning for connecting the cause of the disease with the *person* as with the *place*. Such results of observation could be *interpreted* in different methods and may be considered the effects of *chance*. Here Holmes considers a general case, not an observation:

A given practitioner, A., shall have sixteen fatal cases in a month, on the following data: A. to average attendance upon two hundred and fifty births in a year; three deaths in one thousand births to be assumed as the average from puerperal fever; no epidemic to be at the time prevailing. It follows, from the answer given to me, that if we suppose every one of the five hundred thousand annual births of England to have been recorded during the last half-century, there would not be one chance in a million million million millions that one such series should be noted. No possible fractional error in this calculation can render the chance a working probability. Applied to dozens of series of various lengths, it is obviously an absurdity. Chance, therefore, is out of the question as an explanation of the admitted coincidences; There is, therefore, *some* relation of cause and effect between the physician's presence and the patient's disease; Until it is proved to what *removable condition* attaching to the attendant the disease is owing, he is bound to stay away from his patients so soon as he finds himself singled out to be tracked by the disease. *Medical Classics* 1:255

Holmes (1855), or his unknown mathematical adviser, thus sets up a null hypothesis, finds that such sequences are associated with extraordinarily low probabilities and then suggests that it would be unreasonable to expect such sequences to appear under the hypothesis of noncontagion in the population of England and Wales over a span of fifty years (see §9.6).

It is difficult for a modern statistician not to praise the work of the mathematical consultant and not to believe that Holmes has successfully upset the null hypothesis (anticontagionist in this case), and that he has further confirmed contagionist hypotheses. Holmes thus believes that the disease spreads by contagion, that it is sometimes generated by contagion by erysipelas and conversely contagion to medical attendants and others may result in erysipelas. Autopsies on erysipelas and puerperal septicemia are especially dangerous sources. He does not specify the actual mode of transmission but includes manual transmission during diagnostic or surgical procedures as a possible mode.

The great virtues of Holmes's article are the literary style and the help provided by the unnamed "competent person," whose identity cannot now be determined according to a personal communication from Dr. R.J. Wolfe, Curator of Rare Books and Manuscripts, Boston Medical Library, who kindly examined the inventory of the O.W. Holmes papers at the Houghton Library, Cambridge, Massachusetts for me. We cannot determine whether this "competent person" knew of Maupertuis (1752) (see §4.1), Daniel Bernoulli (1735), or Michell (1767) (see §17.11). Often important information is yielded by simple mathematical methods.

Many obstetricians and others, for example Meigs (1856, p. 635), would not accept the contagionist views of Gordon, Nunneley, Holmes, Kneeland, and many others.

Since Lancaster (1994b) has closely analyzed the book *Die Aetiologie, der Begriff und die Prophylaxis des Kindbettfiebers* (1861) of Ignaz Philipp Semmelweis* (1818–1865) and shown that standard commentaries on it in the medical literature give erroneous opinions, it is unnecessary to mention his work further here.

## 9.2 Puerperal Sepsis in the Great Hospitals

It is necessary to stress the importance of the great hospitals for the production of high mortality rates from puerperal sepsis before, say, 1800. Many contemporary obstetricians were aware of this importance. Carl Eduard Marius Levy (1808–1865), Professor of Obstetrics at Copenhagen, had obtained statistics on the lying-in hospitals in London and Dublin (Levy, 1847); these tables were also published by Semmelweis (1861, Tables 25–30), who stated that Levy had published them in the *Bibliothek for Laeger* and that G.A. Michaelis had translated his article into German for *Neue Zeitschrift für Geburtskunde*, Bd. 27, Hft, 3, Seite 392. Further details are available in Churchill (1850). Christiaan Bernard Tilanus (1796–1883), a contagionist, wrote on p.310 of *Die Aetiologie* about his experience in a small institution with about 400 births annually. He used isolation and noted that the epidemic might cease if the number of admissions fell to zero on several days. He agreed that some infections came from the cadavers and so gave instructions limiting staff attendances at postmortem examinations.

In the special case of the Imperial and Royal Lying-in Hospital, part of the Allgemeines Krankenhaus in Vienna, it should be noted that Johann Lukas Boër (1751–1835), the first director of the obstetric wards, had returned from a tour of the hospitals of France and Great Britain with a resolution to follow the methods of Charles White. For the years 1784 to 1822, and so before the institution of K. von Rokitansky's pathology laboratory, there were 71,395 deliveries with 897 deaths, a maternal mortality rate of 13 per thousand, a creditable level when the more numerous possibilities of infection in any large institution are considered.

## 9.3 Maternal Mortality in England and Wales

It is desirable to discuss maternal mortality rates not dominated by the special influences and practices of the great hospitals; the solution is to study the rates in a particular country, England and Wales, chosen because clinical and official statistical opinions are readily available.

Boxall (1893), an obstetrician, gives the maternal mortality rates of the years 1847 to 1891 and notes that the mean was 4.85 per thousand live births and that the annual rates exceeded 6 per thousand in only 2 years, 1848 and 1874. Approximately one-fifth of the mortality was due to puerperal sepsis in most years but this proportion could rise in bad years. Before 1860, the mortality was slightly less in the provinces than in London, 5.00 as against 5.47, but after 1860 it was greater in the provinces, 4.89 against 3.74 per thousand.

The rates for England and Wales seem not to have been affected by the high rates in the lying-in hospitals, for they suffered few of those terrible epidemics of the great European metropolitan lying-in hospitals. Farr (1885/ 1975, p. 273) wrote:

Childbirth is of course a physiological process, and under favourable conditions, where the mother has been previously taken proper care of, is attended with little danger. Unfortunately English mothers do not escape scatheless; nor can this be expected under existing circumstances; 3875 mothers died during 1870 of the consequences of childbirth. But there is evidence of improvement. In the four years 1847–50 no less than 59 mothers died to every 10,000 children born alive; in the four years 1867–70 the deaths had sunk to 45. The error of collecting poor lying-in women into hospitals has been discovered, and to some extent discouraged; medical men have adopted wiser measures; they have taken greater precautions against infection, and midwives have been better taught. Still there is great room for improvement.

The maternal mortality from all causes was 5.9 per thousand in 1847–1850 and 4.5 per thousand in 1867–1870.

Bonney (1918–1919) complained about "the continued high maternal mortality" in the United Kingdom. He gave a long series of annual rates and pointed out that there had been little improvement over the years 1856–1917 in the maternal mortality rate; sepsis, toxemia and related diseases, hemorrhage, and embolism were the problems, in that order. Notwithstanding improvements in bacteriological knowledge there had been only a slight improvement in the death rates from puerperal sepsis. Full surgical skill and aseptic and antiseptic rigor were required.

Campbell (1924) stated that the contagious nature of the disease had been "fairly well recognised in this country" and that measures had been taken to prevent the spread of infection. The application of cleanliness and antiseptics had led to the practical disappearance of puerperal fever from lying-in hospitals, but there had been rather less success in the private practice of midwifery. Campbell (1927) noted that there were 694,563 births and 2,860 deaths of mothers in England and Wales in 1926. The causes of death and their contributions to the maternal mortality rate can be given as death rates per 1,000 live births, all causes 4.12, of which puerperal sepsis was responsible for 1.60, that is, about 39% of total maternal mortality. Campbell's table (p. 2) shows that slight improvements occurred over the years 1911–1926. Sulfonamide drugs began to reduce the rates from 1935 onward, and then after World War II, penicillin dramatically reduced the maternal mortality.

## 9.4  Puerperal Sepsis: Bacteriology

Karl Mayrhofer (1837–1882), an assistant in the wards of Division 1 of the Vienna Lying-in Hospital a few years after the departure of Semmelweis, is said by some to have demonstrated bacterial elements, then called vibriones, in the lochia in 1863, although this has not been accepted by Bulloch (1938/ 1960) as a definite proof. Stronger evidence was produced by L. Coze and V. Feltz of Strasbourg in 1866–1869; they inoculated the blood of parturients containing "bactéries aux chaînettes" into rabbits, causing infections; they failed to culture them and Bulloch states that, although they had the merit of directing attention to the possible bacterial nature of the disease, they cannot be said to have proved their point.

The priority for the discovery of the streptococcus as a cause of puerperal sepsis is usually given to Pasteur (1878). Other organisms have also been incriminated; but it is now recognized that members of group A of the species, *Streptococcus pyogenes*, are commonly the causal organism. Later work by the Colebrooks, Dora, Leonard, and Rachel, and by others has made the classification of group A into serological types more precise, greatly extending the ability to trace the source of the infection, which may be the parturient's throat, nose, or skin, and similarly for the medical attendants, nurses, and family contacts. Thus a clear picture of events leading up to infection can be formed.

## 9.5  Medical Statistics and Maternal Mortality

The failure of statistical analysis to convince the obstetricians and academic physicians that puerperal sepsis should be regarded as an infective disease must come as a shock to modern theorists. Let us consider various hypotheses that were available in 1860, say.

1. *Epidemic constitution and its atmospheric-cosmic-telluric theory.* This variant of the Galenic doctrine was widely held. It could have been dismissed as self-fulfilling and hence useless (see §8.2); it could be shown to be inconsistent with the data, as was demonstrated by Gordon, Nunneley, Holmes, and many others. Belief in the "epidemic" theory (in the sense of Galen) lingered on and even J.Y. Simpson gave credence to the idea that there was some common influence working in some years over extensive geographical regions (Simpson, 1851).

2. *Miasmatic theory.* Statistical reasoning did not succeed in eliminating the theory of miasmatic causes, possibly because miasms and bad odors were confounded with uncleanliness in any infection and so any general attack on one factor would affect the other.

3a. *Contagion and Liebig's fermentation theory.*

3b. *Contagion and organismal (microbial) theory.*
    It would have been impossible to distinguish between these two by statistical methods; indeed, it often makes little epidemiological difference to think of Farr's "varioline" and "syphiline" (as in Farr, 1885/1975, pp. 245, 249–250) as chemical compounds or as living organisms.

4. *Bacteriological theory* (the culmination of 3b). The bacteriological theory triumphed with the investigations of Pasteur and others, as has been noted in §9.4. Once accepted, this theory gives a satisfying explanation of the various epidemiological investigations of the observers, such as White, Gordon, Nunneley, and Holmes. The statistics then become data for operations research and quality control but are superseded by the more refined bacteriological methods as a source and test of hypotheses.

## Treatment and Prophylaxis

Although the folklore has it that Semmelweis was alone in his improvements, reference to Wangensteen and Wangensteen (1978) and to Graham (1950) will show that many others had instituted appropriate measures of isolation, reduction of hospital size and hospital crowding, and even mildly antiseptic measures against infections. The maternal mortality rates for England and Wales show that infection outside the great hospitals could not have been common, since total maternal mortality rates remained low; moreover, there were only moderate improvements in the maternal mortality rates between the bacteriological discoveries around 1880 and 1935.

## 9.6 Analysis of Contagion

The statistical analysis of contagion is a special case of the modern theory of clustering; it is concerned also with the problem of the reference set (see §17.11).

In §9.1, we mentioned the help given to O.W. Holmes by an unnamed "competent person" in the analysis of series of puerperal fever cases. The problem can be simplified by considering a series of coin tossing; it is required to test whether a coin is true or unduly weighted toward heads (say). If an experiment is carried out and 10 heads are obtained in 10 tosses, we should be inclined to reject the hypothesis of a true coin, for the probability is 1 in $2^{10}$ and so 1 in 1,024. On the other hand, if a sequence of 10 successive heads occurred during a long series of tosses, we could not say that the result was significant without further enquiry, since every time two consecutive tosses produced tails followed by heads, there would be an opportunity to obtain a sequence of 10 heads. In the simple toss of 10 coins there is but one beginning and so there is one experiment in the "reference set"; in the long series of $N$ tosses the required initial sequence, tails heads, will appear approximately $0.25N$ times and so the reference set will contain approximately $0.25N$ entries. The expectation of the number of sequences of 10 heads will be roughly $\frac{1}{4}N\,2^{-10}$.

Holmes's mathematician had taken the size of the reference set equal to the number of births recorded in England in the previous half-century, about 25 million; he had computed under the null hypothesis of independence of probability of obtaining a certain sequence of deaths; he had then calculated that with such a reference set of 25 million, the probability of obtaining such a sequence is less than "one chance in a million million million millions", 1 in $10^{24}$ indeed. Anyone computing such probabilities for Holmes's example will surely conclude that the null hypothesis is to be rejected; Holmes says this in other words, namely "no possible fractional error in this calculation can render the chance a working probability."

Sometimes the whole data cannot be obtained; Dr. Armstrong's *Essay on Puerperal Fever* is cited in Holmes (1891, p. 135): "in all, forty-three cases occurred from the 1st of January to the 1st of October, when the disease ceased; and of this number forty were witnessed by Mr. Gregson and his assistant, Mr. Gregory, the remainder having been separately seen by three accoucheurs." It is difficult to see how many confinements were attended by these gentlemen. Let us suppose that it was 200. Then taking 0.01 to be the probability of an attack of puerperal fever, the expected number of cases is $0.01 \times 200 = 2$. The number of cases under the null hypothesis of no contagion can be computed, with the use of the Poisson approximation to the binomial, as if the cases arose from a Poisson distribution with parameter 2. The probability to be assigned is $e^{-2}\,2^{43}/43!$ which is of the order of $10^{-39}$. Dr. Armstrong lists six other such sequences.

It can be noted that Holmes's "competent person" had done his significance test not on actual data but on data designed to show how "significant" in the modern sense actual data would be.

# 10
# Wounds and Hospital Infections

## 10.1 Wounds and Hospitals

A hospital in Europe was originally an almshouse, providing refuge for the sick and poor, under the direction of a religious order. A modern definition of a hospital, requiring the provision of bed care with physicians in regular attendance, would imply that the Hôtel-Dieu de Paris became a hospital in the early thirteenth century. It was followed by La Charité in 1519 and the Hôpital St. Louis in 1607. The Hôtel-Dieu served largely as a home for the aged and chronically ill in the years 1366 to 1368, for 7,500 died per annum or about 20 per day out of a mean hospital population of 900. It was called upon in emergency; thus in 1592, 63,000 died of bubonic plague there and during the Fronde insurrection the number of sick and wounded rose to 4,000.

Hospital statistics have appeared sporadically over several centuries. Thus Wangensteen and Wangensteen (1978) cite John Woodall (1556–1643), a great English military surgeon, who reported in his *Surgeon's Mate* in 1639 that there had been not more than 20 deaths in 100 amputations. Alexander Monro (primus)* (1697–1767) reported in 1737 that 99 amputations had been carried out in the Edinburgh Hospital with 8 deaths—a case fatality of 8%. Simpson (1871) was later to point out that, in the years 1842 and 1843 at the same hospital, there had been 21 deaths in 43 amputations, a case fatality rate of 48.8%.

Many surgeons did not like the idea of statistics for their operations; nor were the hospitals always able to collect or publish relevant statistics. Pressure came from reformers and sanitarians outside the hospital system. John Howard (1726–1790), known for his work on prison reform, turned his attention to lazarettoes and hospitals. He visited many hospitals, calling the Hôpital St. Louis and the Hôtel-Dieu "the two worst hospitals," which agreed with the remarks of d'Alembert in the *Encyclopédie* that the Hôtel-Dieu de Paris was "the richest and most frightful" of them all. Jacques-René Tenon (1724–1816), surgeon and pathologist, had long been studying conditions in the Paris hospitals; the work of Tenon (1788) formed the basis of the report of a committee of the Royal Academy of Science, which included Antoine-Laurent Lavoisier* (1743–1794) and Pierre Simon Laplace (see Ackerknecht, 1967, p. 16). As a result, the Academy sent Tenon and Charles-Augustin Coulomb* (1736–1806), the physicist, to England to study hospitals, prisons, and workhouses. Tenon found, as did Malgaigne (1842) later, that mortality rates at the Hôtel-Dieu were highest among all hospitals. Tenon recommended that surgical wards should not be near autopsy rooms, that operations and preparations for them should be carried out in separate rooms, that obstetrical beds should not be mixed with surgical beds, and that special wards should be provided for contagious diseases. We may note in passing that Lavoisier played an important part in the struggle to improve the public health of the cities (Greenbaum, 1972); for example, he was early active in seeking a good water supply for Paris. He worked as a scientist and as an active political economist, the last interest causing his death at the guillotine.

James Young Simpson (1811–1870) in Simpson (1871) cited John Simon (1864) as believing that disease was spread by the air. He noted also that certain vibriones, mycelia of fungi and "bacteria etc." had been found floating in the atmosphere of hospitals and crowded dwellings, especially by de Chaumont, Frank, Hewlett, and other medical officers of the English army. Simpson continued:

Pasteur, Gratiolet, Lemaire, and others, maintain that the atmosphere is full (especially in localities where, as in hospitals, the air is otherwise impure and tainted) of living spores and germs of various infusoria, etc., which, when they find a proper nidus, lead on by their development to fermentations, putrefaction, suppurations, etc. (p. 384)

Pus cells also had been detected in the air. At this time, it is evident that aerial contamination was held to be the chief mode of contagion. This led to an emphasis on cleanliness. Simpson (1871, p. 386) went on to consider how there was some danger of contagion when two persons were in a ward because one might carry an infection; but the danger was increased when the number of persons in a ward was increased to ten and still more when there were several or many wards closely linked together. He referred to Farr's law that mortality depends on population density (see §11.6). He noted that many authorities had pointed to such crowding as leading to the observed high mortality of troops in barracks. Simpson referred to an experiment in which some lunatics were billeted out; it was found that 5.6% died annually in private dwellings, whereas 8.6% died annually in the lunatic wards.

The hospitals are of special interest in scientific work, of which there are mentioned Pinel's work on the use of statistics and classification of disease and Louis's work on individual diseases in §16.2, nosocomial (that is, hospital) disease such as childbed fever in Chapter 9, and the infection of wounds in §10.4. Statistics from the hospitals were used with great skill by many reformers and sanitarians, for example, Edwin Chadwick (1800–1890), John Simon (1816–1904), Florence Nightingale, and Stephen Smith (1823–1922).

## 10.2 Cold Water and Cleanliness

It will simplify the discussion if we consider the work of surgeons and obstetricians, who, without necessarily believing that bacteriology or other science would justify their procedures, were pragmatic observers and contagionists. They saw that under cleaner and more orderly conditions in the hospitals, the case fatality rates from surgery were lower. We may select some individuals from among such surgeons.

Charles White of Manchester was among the first to insist on absolute cleanliness during delivery and may be thought of as a pioneer in aseptic obstetrics; indirectly, his work had a great influence on Boër, obstetrician at the Vienna Hospital. John Hennen (1779–1828), a surgeon in the Peninsular campaign, is quoted in Cantlie (1974b, pp. 374–375) as saying "with cold water, one is never at a loss for a remedy." There was, according to Cartwright (1967, p. 85) a "cold water" school of surgeons, who believed in cleanliness and the use of cold water previously boiled, and who aimed to obtain healing of wounds by first intention, that is, without suppuration.

Ovariotomy was an excellent test of the absence of septic complications for it required no opening of a body cavity other than the peritoneum, was a simple operation with little danger of hemorrhage and so almost all mortality might be expected to be due to sepsis. The lesion was a simple cyst of the ovary but it was dangerous to life because finally it would fill the whole abdominal cavity. Thomas Spencer Wells (1818–1897) was the first to have an extended series, in which he reported both deaths and survival from the operation. He had become aware in 1864 of Lister's interest in the possibility of microbes being responsible for wound infections, but it appears that it was only in 1878 during his third series that he adopted a truly Listerian approach. By 1880, he had performed 1,000 ovariotomies, a fact that was celebrated in the popular press and in an editorial (Anonymous, 1880) of the *British Medical Journal*, which gave his results in four successive periods commencing in the early 1860s as follows: (a) 500 operations, 373 survivals, 127 deaths, case fatality rate 25.4%; (b) 300, 223, 77, and 25.7%; (c) 100, 83, 17, and 17%; (d) 100, 89, 11, and 11%.

Robert Lawson Tait (1845–1899) is a more interesting case. Tait (1886) reported 139 consecutive ovariotomies performed throughout 1884 and 1885 without a single death. He stated that he had abandoned Listerism to use "scrupulous attention to cleanliness of every kind and in all directions." Tait (1890) again derided the bacteriologic hypothesis, even though it had been well established by Pasteur and Koch and their respective followers by that time. Tait appears to have been the "Robert Lawson" taking part in a discussion on wound sepsis mentioned in §10.4.

Florence Nightingale (1820–1910) believed that the germ theory of disease was a "fad" of the German pathologists. Rosenberg (1979) has given a detailed examination of her views, which were shared by many of her contemporary medical theorists. In her model of disease, the body is a dynamic system constantly interacting with the environment. She had no concept of specific diseases. She certainly did not justify her campaign against filth, disorder, and contaminated atmosphere by calling on the possible existence of microbes. Nor did Charles Creighton ever accept the importance of microbes.

On her arrival in Scutari during the Crimean War, Nightingale found the medical records in disorder. She soon introduced an orderly plan of recording. Her studies of the records showed that many deaths need not have occurred and resulted in reforms in the treatment, transport to base, and evacuation to Scutari. After the end of the Crimean War, she used statistics to show that the mortality rates of men in barracks were higher than those of civilians at the same ages. Nightingale was a great publicist and popularizer; according to Kopf (1918), she used graphical methods with skill. We conclude these brief notes by adding that she can be thought of as a pioneer in operations research.

## 10.3  Military Wounds and Illness

Of wounds in the days before asepsis and anesthesia, those most subject to statistical analysis were in the military. The great problems arose when gunshot wounds were complicated by fractures of the bones. First there was the debate about whether amputation should be immediate or delayed. Jean Faure, who favored delayed amputation for comminuted fracture (that is, with multiple fractures of the bone), stated that only 30 or 40 men survived out of 300 immediate amputations after the battle of Fontenoy in 1745, whereas he had saved all of 10 men with delayed amputation. Wangensteen and Wangensteen (1978) point out that a debate between Faure and Boucher at the French Academy of Surgery in 1756 brought out the very high case fatality of the operation, whether immediate or delayed. Authors, however, were inclined to omit their failures in discussions.

James McGrigor (1771–1858) had a long and sometimes exciting life. After he graduated with an M.A. degree at Edinburgh in 1788, he studied medicine at Aberdeen, Edinburgh, and London; he purchased a commission as a surgeon in a regiment later to become famous. He had field and organizational duties in the West Indies, India, Egypt, Netherlands, and on the Iberian Peninsula, with two shipwrecks before June 1815, when he was appointed Director-General of the Army Medical Department, a post he held until 1851. Wellington stated that "he is one of the most industrious, able and successful public servants I have ever met with." McGrigor had hoped that a school for training army medical officers would be founded but this was to come only long after his retirement. According to Cantlie (1974a), McGrigor, from his extensive experience in the Peninsular War, regarded accurate returns on the living conditions of the soldier to be of the highest importance. These returns from 1817 are still extant at the Royal Army Medical College.

In the early years of the nineteenth century, death rates of the troops at home and so in barracks were 15 to 17 per thousand per annum as compared with a crude death rate of persons at the same age in the civil population of about 10 per thousand; death rates from typhus, lung tuberculosis, and other infectious diseases were high in the soldiers. Much use of the records was made by Henry Marshall (1775–1851), often referred to as "the father of medical statistics," after 1835. He was assisted by the legally trained Alexander Murray Tulloch (1803–1864), who later was Commissioner with Sir John McNeill to investigate the commissariat of the British Forces in the Crimean War. An example of the value of medical statistics in determining policy can now be given.

According to Cantlie (1974a, p. 441), Marshall and Tulloch (1838) showed that for European troops in the Windward and Leeward Islands the death rate was 78.5 and the rate for admission to hospital was 1,903 per thousand per annum for the years 1817 to 1836; in Jamaica the corresponding figures were 121.3 and 1,812. Admissions to hospital and deaths were 86,541 and 5,966 in the West Indies generally for the same period. The years 1819, 1822, 1825, and 1827 were "unhealthy years," with the highest mortality rate 259 per thousand per annum and with yellow fever as the principal cause of mortality. Case fatality rates were high: yellow fever, 3 in 7; typhus, 3 in 13; remittent fever, 1 in 9; common continued fever, 1 in 23; intermittent fever, 1 in 165. The high invalidity rates had formerly been put down to exposure to the high ambient temperatures. But these authors showed that, while temperatures did not vary between the islands very much, there were great variations in invalidity rates. Further, the mortality rates might be in one year twenty times the value in another year. Similarly, the theory that the troops became used to the climate (or "seasoned") was shown to be false. Such statistics revealed the high rates of mortality in overseas stations and led to increased public and official appreciation of them. As a result, the policy of posting for extended periods in one station was altered and troops were moved to successive foreign stations in turn; so declines followed in the death rates from 80 to 62 per thousand per annum in the West Indies generally and from 120 to 60 per thousand per annum in Jamaica. [See Cantlie (1974a, p. 442) for a table of death rates per thousand per annum in the years before 1836, ranging from 14 in the United Kingdom to 121.3 in Jamaica, 483 in Sierra Leone and 668 in the Gold Coast.]

A second series of reports commenced in 1848 but the Crimean War forced a discontinuance. After 1860 Annual Reports on Army Health were instituted.

However, in the Armed Forces, especially on peacetime duties, almost every case of serious illness in the troops will be referred to a military hospital and so it has often been possible to give incidence rates for disease, especially serious infective diseases, for troops in, say, the British Army. [For the Navy see S.F. Dudley (1931).]

## 10.4  Statistics of Amputations

We summarize the conclusions of Simpson (1871):

1. *State of aggregation.* Timothy Holmes (1825–1907) denied that the state of aggregation was relevant; yet Simpson (p. 341) was able to give a table from Holmes (1863) that showed that 1 in 4 patients in hospitals with 100 to 600 beds and 1 in 6 patients in hospitals with less than 100 beds died after amputation operations.
2. *Lack of homogeneity between classes of hospitals.* The differences in case fatalities of amputations between classes of hospitals, 67% in the hospitals of Paris and 40% in those of London, were not to be explained by the constitutions of those operated upon or the nature of the injuries or diseases for which they were admitted. Here Simpson (p. 343) was citing Holmes (1863) again.
3. *Metropolitan versus country.* The case fatalities were over three times as high in the large metropolitan hospitals as in the rural areas (Simpson, p. 344).
4. *Operating experience.* "In country practice increased experience in amputations gives a still higher ratio of success to the results of the operations." Simpson noted that in the cumulated experience of surgeons, those who had fewer than 6 amputations lost 1 in 7, those who had from 6 to 11 amputations lost 1 in 9, and those who had 12 or more lost 1 in 12 patients after amputations (p. 347).
5. *Homogeneous comparisons.* The country case fatality rates were more favorable in toto or by classes of patient (p. 347).
6 and 7. *Country practice.* Rural rates were even more favorable than the urban when slighter amputations were compared. This applied especially for amputations of the forearm (p. 348).
8. *Amputations of the forearm.* Deaths usually resulted from specific morbific hospital influences.
9. *Double amputations.* These were "very fatal" in all hospital practice. Nevertheless, rural case fatalities with double amputations were of the same order as with single amputations in the city (p. 353).
10. *Age.* Rural death rates from amputations at ages over seventy years were near to urban rates for all ages combined (p. 353).
11. *Causes of urban–rural differences.* These were not due to differences in constitution of the patients or in the nature of their injuries or diseases (p. 357).

After making seven more comparisons, which we pass over, finally Simpson (1871, p. 381) made the point that it was pyemia that was responsible for the great differences in case fatalities between metropolitan hospitals and rural surroundings, as follows:

The relative prevalence of this formidable and fatal complication [pyaemia] within the walls of St. George's Hospital, with three or four hundred inmates, and the district without, from which the hospital chiefly derives its patients, containing a population of thousands, or tens of thousands, is presented by a statement made by Mr. Holmes in one of his papers in *The Lancet* [T. Holmes, 1869]. From 1865 to 1868 there originated 81 cases of pyaemia within the walls of St. George's Hospital; while during the same period 9 cases applied for admission from without. Few facts perhaps could show more clearly the usual hospital origin and character of this dreaded disease.

A summary of Simpson's findings is given in Table 10.4.1.

Burdett (1882) noted that Simpson's statistics had been "seriously impugned" by some authorities on the counts that the reliability of the sources could not be proved and no details of the cases had been given. Burdett had set out to test the relevance of hospital size to case fatality rates from amputation in 1876 and had spent much time on the survey. (We may note that his work antedated the work on bacteria in wounds by Koch, Pasteur, and others but that Listerism was becoming recognized.) Out of 160 hospitals, 91 responded to his request by questionnaire for information on the results of all amputations since the hospital was founded. His tables and comments can be summarized by saying that Simpson's findings have been verified and that, in pre-Listerian days, the smaller the hospital the more favorable were the results.

In recent times, however, the "aseptic or Listerian" treatment of wounds had brought about a remarkable improvement. These observations are supported by the statistics. The case fatalities for all amputations in the larger (London) metropolitan hospitals were 41.6% in 1868, 37.8% in 1872, (25.7% in University College Hospital over 38 years). Burdett cited, for comparable series with some important exclusions, a case fatality rate of 4.36% in Germany during 1880, obtained from Max Schede (1844–1902) in Pitha and Billroth's *Handbook of Surgery*. Burdett (1882) believed that applications of Listerian methods had nullified the advantage of small size in hospitals. In

TABLE 10.4.1. Case fatality from amputations by hospital size (1862–1868).

| No. of beds in hospital | Limb amputations | Deaths from amputations | Proportion dying* | Percentage dying |
|---|---|---|---|---|
| 300+ | 2,089 | 855 | 1 in 2.4 | 40.9 |
| 201–300 | 803 | 228 | 1 in 3.5 | 28.4 |
| 101–200 | 1,370 | 310 | 1 in 4.4 | 22.6 |
| 26–100 | 761 | 134 | 1 in 5.6 | 17.6 |
| Under 26 | 143 | 20 | 1 in 7.1 | 14.0 |
| In homes | 2,098 | 226 | 1 in 9.2 | 10.8 |

* Simpson's figures.
Data from Simpson (1871, Tables 22–25, and p. 399).

the discussion following Burdett's paper, Robert Lawson (? Tait) remarked that amputation results for the Crimean War were worse than those for the Peninsular War some 40 years prior to it. A professor of political economy, Leone Levi (1821–1888) found the superiority of results from the cottage hospitals to be against "all economic principles of working on a large scale," for the metropolitan hospitals had greater resources of capital, better methods of treatment, skilled surgeons and supervision. Another speaker, a surgeon Mr. T. Moore, recalled the falls in mortality in Glasgow where Lister had introduced his first reforms and believed "germs" were fewer in the small hospital; he also stressed the importance of ward size rather than hospital size, an importance that was disproved by the author, Burdett.

By 1882, there had also been extensive discussion in France about the role of cleanliness in hospitals, the importance of organic material on the walls, the importance of "micrococci" (see Wangensteen and Wangensteen, 1978). The works and conclusions like those of Burdett (1882) and Simpson (1871) became justified by the bacteriologic research of Pasteur, Koch, and their followers and pass from scientific to operational research. As commentators on obstetrics have pointed out, the hospital routines, although improved, had not abolished deaths from puerperal sepsis before chemotherapy and antibiotic therapy (see §9.3).

## 10.5  Hospitalism

The military surgeons seem to have been the first to consider the importance of the size of the hospital, namely whether wounded soldiers should be treated in small or large units. James McGrigor favored "dispersion of wounded in many regimental hospitals rather than the concentration into a few large units" (see Cantlie, 1974a, p. 347; 1974b, p. 87).

J.Y. Simpson was one of the first of the civilian medical men to show the dangers of the large hospitals. Simpson (1868–1869, 1869–1870) attempted to show how crowding leads to greater disease incidence. He believed that contagion, widely accepted as the mode of spread of puerperal sepsis, was also of importance in surgical fever and cited his earlier observations (Simpson, 1850), adding that the matter had attracted little attention from the hospital surgeons. He believed that surgical patients in surgical wards often suffered infection from other patients by accidental inoculation. In other cases there might be general spread by air of inorganic and organic materials from the wounded.

Simpson, writing in 1848 in the *Edinburgh Monthly Journal of Medical Science* (see Simpson 1871, p. 290), commented on hospitals as follows:

There are few or no circumstances which would contribute more to save surgical and obstetric patients from phlebitic and other analogous disorders, than a total change in the present system of hospital practice. I have often stated and taught, that if our present medical, surgical, and obstetric hospitals were changed from being crowded palaces,—with a layer of sick in each flat,—into villages or cottages, with one, or at most two, patients in each room, a great saving of human life would be effected; and if the village were constructed of iron (as is now sometimes done for other purposes) instead of brick or stone, it could be taken down and rebuilt every few years—a matter apparently of much moment in hospital hygiene. Besides, the value of the material would not greatly deteriorate from use; the principal outlay would be in the first cost of it. It could be erected in any vacant space or spaces of ground, within or around a city, that chanced to be unoccupied; and in cases of epidemics, the accommodation could always be at once and readily increased.

Simpson (1871, p. 291) mentions that in statistics collected by J. Malgaigne and Ulysse Trélat (1795–1879), 803 died out of 1,656 patients submitted to amputation in the hospitals of Paris; the case fatality rate was almost 1 in 2 or 48.4%. A colleague of Simpson, Dr. Fenwick, found that of 4,937 amputations there were 1,562 deaths, nearly 1 in "every 3 or 4," a case fatality of 32%; Fenwick noted that "the assertion that one person out of every three who suffers an amputation perishes, would have been repudiated a few years ago as a libel on the profession, and yet such is the rate of mortality observed in nearly 5,000 cases."

# 11
# Epidemiologic Observations

## 11.1 Perpetuation and Fadeout of Epidemics

The perpetuation of disease did not become a problem for epidemiologists until after the establishment of the communicability of the infectious diseases, so that it might well be that John Snow was among the first to consider the subject. He writes (Snow 1853/1855/1936, pp. 156, 158):

The material cause of every communicable disease resembles a species of living being in this, that both one and the other depend on, and in fact consist of, a series of continuous molecular changes, occurring in suitable materials. The organized matter, as we must presume it to be, which induces the symptoms of a communicated disease, except in the case of the entozoa, can hardly ever be separately distinguished, like the individuals of a species of plant or animal; but we know that this organized matter possesses one great characteristic of plants and animals—that of increasing and multiplying its own kind. In the instances of syphilis, small-pox, and vaccinia, we have physical proof of this increase, and in other diseases the evidence is not less conclusive.... The extent of population and of intercourse has great influence over the epidemic character of communicable diseases. The various irruptive fevers are constantly present in London, and are only liable to fluctuations in their prevalence. In less populous districts, however, there are not enough subjects to support their constant presence. One or other of them is often absent for a number of years, and, when re-introduced, spreads to a great extent. There is one disease which neither the metropolis, nor the country at large, nor even the whole of Europe, will supply with victims except for a time. The cholera has been twice spread over the world within the memory of the present generation, and it seems to be dying out a second time everywhere but in the south of Asia. Fatal as it is to the human species, it is itself so difficult of support that the world seems scarcely large enough for it, and, were it not for its pastures in India, it would be in danger of passing altogether out of existence, like the dodo of the Mauritius. So far as can be learnt from what remains of ancient medical literature, the communication of diseases was not generally recognised till a recent period. Even Sydenham did not recognise the communicability of any acute febrile disease except the plague. He did not even recognise the communicability of small-pox.

Snow (1853/1855/1936) thus notes that there are two extreme cases of the balance between a population and its epidemic causes, giving as examples the perpetuation (or "constant presence") of infections in London and the fade-out (or "dying out") of cholera over Europe.

Some estimates can be given, as by later epidemiologists, from simple arithmetical considerations, of the minimum size of a community required to perpetuate an infectious disease, although they are usually shown by observations to be too low. Measles is taken to be the disease and it is assumed that the incubation period is 14 days and that a case is only infective for a short time at the end of this period. A small, closed, human community is considered, an island say, with an annual birth rate of 25 per thousand, to determine the smallest population that could perpetuate the epidemic of measles indefinitely. If the epidemic were deliberately and rigidly controlled, two new cases would be required roughly every fortnight to avoid accidents to the cases or chance failure of the virus to pass from person to person. About 50 new cases would be required per annum and so the population would need to be of the order of two thousand; but under ordinary natural field conditions, this regularity would not be observed and there would be some outbreaks with many cases, and so the estimate of the critical size for perpetuation would have to be greatly raised.

To convince some workers that the minimum size of a perpetuating population must be so large, simulations have been carried out by Bartlett (1960) but it is easy to find actual communities where measles has faded out, for example, in Australia. A consequence of a fadeout is that a large proportion of the population may be susceptible to the disease when it is reintroduced; so there may be numerous cases not only of children but also of adults. Thus in the disastrous epidemic of measles in Fiji in 1875, there was a loss of 10% to 20% of the population. A party of Fijians had been entertained by the government of New South Wales in Sydney and were returned home by fast steamer; they were met at a great gala in Suva at which delegates from the districts and islands of Fiji were present; these delegates returned home unknowingly taking the infection with them so that there was a great epidemic occurring simultaneously throughout the group [see Squire (1875–1877) and Corney (1883–1884)]. As in other such intense island epidemics, there were multiple cases in families and the whole community was in disarray. It has been suggested that these high rates of mortality have been due to genetic causes, but Neel (1977) believes that the excess mortality is perhaps 80% nongenetic in origin. [For further notes on measles, see Wilson (1962) and the symposium articles for measles in the same issue of that journal, and Black (1966). For an account of many "virgin soil" epidemics in America, see Hopkins (1983).]

## 11.2 Island Epidemics

What might be called the classic investigations or observations of island epidemics took place with respect to Iceland, isolated before the modern era of fast sea and air transport but sufficiently close to be observed by European physicians. There are available for Iceland official records of incidence of all the common infective diseases for the years since 1811, but there are longer records. Thus Schleisner (1851) found that 134 out of the 541 years from 1306 to 1846 were epidemic years. Typhus was regarded as endemic and prevailed almost every year. There was also a typhus-like fever occurring in famine years, probably relapsing fever. In 1846, after an absence of measles for 60 years, there was a measles epidemic and the infant mortality was 654 per thousand, with a crude death rate of 55.8 per thousand per annum.

A similar measles epidemic in the Faeroes in 1846, described by Panum (1847/1940), was the cause of much less mortality. Such epidemics, particularly in Sweden, were discussed by Ransome (1881–1882) and others in the *Transactions of the Epidemiological Society, London*, whose members found them a strange contrast to experience in the great European cities. [See also Ransome (1880) for epidemic cycles and Rolleston (1937) for references to the older literature of such "virgin soil" epidemics.]

Varicella can be perpetuated in quite small communities of the order of 1,000 or 2,000 persons, for the virus can remain dormant in the tissues of the peripheral nerves after clinical infection, indeed for many years. Thus an attack of varicella in childhood may lead to an attack of herpes zoster in old age. The virus in these cases has become reactivated and migrated down the nerve fibres to the skin, leading to the symptoms of herpes zoster due to irritation of the nerve fibres. A pustular rash follows and the virus is released and may set off a new epidemic of varicella. Garland (1943) is regarded by Gordon (1962) as having given the most definite discussion of the reality of the identity of the two forms of disease.

Pickles (1939) noted early the identity of herpes zoster and chicken pox; in two out of three epidemics experienced, herpes appeared first. He pointed out that the association between the diseases had first been reported by Bókai in Hungary in 1892 and noted in the *Lancet* [but see the reference to it in Hebra and Kaposi (1866–1880, 1, p. 379)].

## 11.3  Epidemics in Australia

For examples of the theory and speculations on infective diseases, mentioned in §11.1, we may turn to Australia, which can be considered for epidemiological purposes in the nineteenth century as a group of islands. The epidemics there are relatively well documented in the official statistics and, moreover, the early experiences of the infective diseases in Australia are recorded in monographs by John Howard Lidgett Cumpston (1880–1954), a medical administrator with an interest in epidemiology. Some tables and graphs of the deaths from the infectious diseases in Australia are given in Lancaster (1952, 1960, 1967), and bibliographies in Lancaster (1964, 1973, 1982a).

## Introduction of Infections

Australia before the days of fast steamer transport had an experience resembling that of an island, for its principal contact with the rest of the world was by sailing ship around the Cape of Good Hope. The trip from England took more than ten weeks. Congestion on board ships greatly enhanced the ease of spread of infections and the chain of infections on a ship would be short, perhaps only one, two, or three generations of the epidemic occurring before the susceptibles were exhausted. In particular, a ship leaving England would be quite unlikely to carry the infections direct to, say, Australia or the west coast of America. However, the ships would call at some South American port for fresh food and then proceed on to, say, Capetown and Australia. At these ports, the population might be too small to perpetuate an epidemic but would provide a risk of infection for the passengers of the ships, so from time to time there might be infections reintroduced at these intermediate ports and then conveyed on by ship to the next port and finally Australia, for example. Indeed, according to Donovan (1970), measles was first introduced into Australia in this way in 1834, forty-six years after the first European settlement. It faded out within a year and was next observed to be epidemic in 1850 (Cumpston, 1927). Fever hospitals may have suffered their own private epidemics, where susceptible nursing trainees, coming from the country, passed the disease in a chain quite unrelated to the surrounding population.

## Scarlatina

The first case of scarlatina in Australia was observed in 1833 in Tasmania according to Cumpston (1927), with further cases in Victoria and New South Wales during 1841. These Australian epidemics came as a great surprise to the early commentators such as Morris Birkbeck Pell (1827–1879), who wrote that in long-settled countries, the mortality rates are as constant "as the planets in their orbits" (Pell, 1878), whereas in Australia there have been extraordinary changes in the rates from year to year. Indeed, in Victoria there were four departures, due to measles and/or scarlatina, of the crude death rates from their usual level between 1860 and 1900, larger than that later to be caused by the influenza epidemic of 1919.

## Rubella and Congenital Defects

Rubella has epidemiological properties like measles but since it does not cause deaths, its incidence cannot be traced as easily as can measles from the official statistics. Its fadeout from Australia at different times seems to be well accepted, although the evidence is not as firm as could be desired. It seems that it is, with difficulty, perpetuated in populations of moderate size. Its reintroduction into Australia in the late 1930s led to a dramatic discovery. Norman MacAlister Gregg (1892–1966) found that, in some 75 cases of congenital cataract occurring in infancy, a history of an attack of rubella in early pregnancy could be obtained from the mother (Gregg, 1941). Swan, Tostevin, et al. (1943) found that other congenital defects, especially deafness, not uncommonly followed such attacks of maternal rubella. These findings caused much surprise and Gregg's work was widely acclaimed. Some epidemiologists believed that a mutation must have occurred in the virus, for previously it had appeared in clinical medicine as a mild disease, important only in the differential diagnosis of acute exanthemata, and lacking in major complications. Doubts were cast on this interpretation by Lancaster (1951, 1954) and Lancaster and Pickering (1952), who demonstrated epidemics in 1898 and 1899 in all the mainland states of Australia and in New Zealand. The explanation of these epidemics was that rubella was not perpetuated in Australia with its rather small population and so women could reach childbearing age without having an attack of rubella. Susceptibles to the disease therefore increased and when the virus was reintroduced an epidemic could appear and some women at the critical stage of pregnancy could be infected together with the fetus. [For historical reviews, see Aycock and Ingalls (1946), Ingalls, Babbott, et al. (1960), Ingalls, Plotkin, et al. (1967), and a symposium in the *American Journal of Diseases of Children* (1965, 110, 345–476).]

## Smallpox

Although the smallpox and measles viruses are not closely related in the laboratory sense, they have some important epidemiological properties in common—incubation periods of approximately a fortnight and a strong immunity following the first attack, so they are not perpetuated in small populations. An epidemic of smallpox in the aborigines was observed near Sydney in 1789, the year after first European settlement; it was possibly initiated on the northern coastline and carried down by the aboriginal populations (see Fenner, Henderson, et al. 1988). Smallpox was later introduced by ship on several occasions (see Cumpston, 1914).

## Bubonic plague

Often an infectious disease can be studied more effectively in a country that is not the "home" of the disease. For example, in India there were great difficulties in tracing the source of the infection because of the many possible pathways, the existence of other diseases, the illiteracy of the population, and so on. These difficulties were less in South Africa and Australia. So it was that important observations on bubonic plague were made in Sydney by John Ashburton Thompson (1846–1915), where the first case of plague was reported on 19 January 1900 (Cumpston and McCallum, 1926). At this time, according to Hirst (1953), plague was generally thought to be an essentially human infection, transmitted directly or indirectly from case to case. Thompson and Tidswell (1903) reported that they could find no evidence of such a course of infection. Exact particulars were obtained that 289 cases lived in 276 dwellings and of these dwellings 266 suffered only a single case; similarly in a later epidemic 139 cases occurred in 124 dwellings and in 115 of them there was only one case apiece. The authors constructed a map of the plague cases and decided that the common locus for the infection was likely to be a workplace, hotel, or other meeting place. They made a study of the distribution of the human epidemic with respect to a rat epizootic, proved also to be due to the plague bacillus. The human epidemic was shown to occur in the vicinity of the rat epizootic. In 1904, a trained staff was able to find plague rats in close proximity to all 12 reported human cases. Thompson was led to suspect that transmission of plague was by the flea. The critical observations confirming this hypothesis were made in Queensland by Ham (1907).

## 11.4  Epidemics in Wensleydale

William Norman Pickles (1885–1969) was a general practitioner and the Medical Officer of Health of Aysgarth in Wensleydale, Yorkshire. This dale was largely cut off from the surrounding country so that Pickles, who had lived there almost all his life, was familiar with the whole area and with many of its inhabitants. Early in his practice, he had observed a gipsy woman washing her dirty linen at a faulty well; this incident, followed by a serious epidemic of enteric fever, left a "lasting impression of the unique opportunities of the country doctor for the investigation of infectious disease" (Pickles, 1939). Later he was to read in Budd (1873/1931):

It is obvious that the formation of just opinions on the question how diseases spread may depend less on personal ability than on the opportunities for its determination which may fall to the lot of the observer. It is equally obvious that where the question at issue is that of the propagation of disease by human intercourse, rural districts, where the population is thin, and the lines of intercourse are few and always easily traced, offer opportunities for its settlement which are not to be met with in the crowded haunts of large towns. This is one of the cases in which medical men practising in the country have for the acquirement of medical truths of the highest order, advantages which are denied to their metropolitan brethren, and which constitute, on the whole, no mean set-off against the greater privileges of other kinds which the latter enjoy.... Having been born and brought up in the village, I was personally acquainted with every inhabitant of it; and being, as a medical practitioner, in almost exclusive possession of the field, nearly every one who fell ill, not only in the village itself, but over a large area around it, came immediately under my care. For tracing the part of personal intercourse in the propagation of disease, better outlook could not possibly be had.

Pickles (1939) also traces the spread of *Sonne* shigellosis (dysentery) by fomites. In 1928, an epidemic of 250 cases of epidemic catarrhal jaundice occurred; a case was observed on April 16 and then five cases between May 14 and 19; in 40 cases it was possible to determine the length of incubation, which ranged from 26 to 35 days. Such data could hardly have been collected in a city or crowded rural region as there would have been too many competing epidemic chains. Another disease in which Pickles made accurate estimates of the incubation period was epidemic myalgia. In the investigations of all these diseases, Pickles was aided by his knowledge of the population and by its relative isolation.

[For obituaries of Pickles see H. and McC. (1969), H.-S. and H. (1969), and Pemberton (1970).]

## 11.5  Epidemics in Schools and Closed Communities

In a sense, the observations recorded in this chapter can be regarded as counterexamples to those differential equations of the epidemic theory in which the number of new infections is proportional to the product, susceptibles × infectives. In Boston, Wilson (1947) and Wilson, Bennett, et al. (1939) observed that the first case in a sibship was usually an older child, attending school or having other such opportunities to become infected.

Halliday (1928), studying the incidence of infectious diseases in Glasgow, found that in congested "often slum" areas children played in a common courtyard or "land" and this could determine an extended family or partially closed community into which the more mobile older children could introduce new infections, thus exposing the younger children to a much greater risk of infection than would be experienced by those living in better conditions in less congested housing estates; this difference in the number of contacts caused differences in the ages of acquisition of the common childhood infections. As the case fatality rates were higher at the younger ages, gradients in incidence and case fatality rates were evident. In the congested areas and housing estates, *measles* was essentially a disease of infants; in the better residential areas it was a disease of schoolchildren, and in the houses of the well-to-do public school class it was a disease of later childhood, 14 to 18 years, the disease being often contracted at boarding school.

Another type of small closed communities has been provided by the sailing ships in the days of large overseas migrations (see §11.3).

In his service in the Royal Navy, Sheldon Francis Dudley (1884–1956) was able to make many interesting observations on the progress of infectious diseases in the Navy generally but especially in the children of the Royal Hospital School at Greenwich. Here the children were $11\frac{1}{2}$ to $15\frac{1}{2}$ years old, joining the school in three separate groups annually. Detailed medical records were kept of each boy so that the incidence of illnesses could be followed throughout their stay of three years. With the aid of the Schick test, the state of immunity with respect to diphtheria could be determined for each child. In 1922, Dudley was able to relate the level of immunity to diphtheria among the boarders to the length of residence in the school, rising from 60% in new entrants. There were two epidemics with 61 cases in 1919 and 94 in 1921. In 1922, the level of immunity was 85% for those children who had been boarders at the time of the 1921 epidemic and 95% for those who had been boarders for both epidemics. Dudley (1923) concluded that these high levels of immunity were due to the occurrence among the boys of latent or subclinical infections, where the doses of bacteria were too small to become established and produce overt symptoms. Later, Dudley (1926, p. 57) made a general statement on the spread of droplet infection in semi-isolated communities, that the "chief practical lesson ... is that the individuals of a community should be isolated from each other to the greatest extent possible in their sleeping quarters. Especially is this the case in those communities to which many susceptibles are frequently added" (see also Dudley, 1936).

## 11.6  Urbanization and other Geographical and Social Factors

Graunt (1662) showed that urban death rates were higher than the rural; he believed that London only maintained its size by migration from the countryside. Forbes (1971) has confirmed this finding, for the deaths in Aldgate, London over the years 1560 to 1630 greatly exceeded the births even if the deaths in the epidemics of plague were excluded. This fact has a curious consequence noted by Russell (1948), that London acted as a kind of sink for the surplus population of the countryside. Many of the exceptional persons or misfits migrated to the city from the counties in the sixteenth and seventeenth centuries and as a result the country populations were able to maintain their own peculiarities. Many of the older demographers were aware of these great urban–rural differences in mortality. Suessmilch (1761–1762) comments that "the secret damage done to the state through towns must almost be considered equivalent to a pestilence" and Heckscher (1950) echoes that view by saying that comparisons between countries in the eighteenth and nineteenth centuries were really only comparisons between their degrees of urbanization.

William Farr (see §16.8), with his reformist ideals, was anxious to show the reality of such differences between rural and urban mortality; he assigned 63 deaths in London during 1838 to privation and starvation (Farr, 1839) but, after criticism by the Poor Law Commissioners and Chadwick, he gave less emphasis to these causes in later *Letters to the Registrar General*, stressing instead the importance of human ignorance or indifference.

At first, he was content to measure the differences between geographical units by means of the crude death rates. However, he saw that such comparisons might be misleading if the age distributions differed between such units. With the developments within the office of the Registrar General, he was able to give death rates within geographical units by sex, age, and cause of death.

Farr (1843) also speculated about the cause of such differences; thus he states "the existence in the atmosphere of organic matter is therefore incontestable."

Farr (1843) gave the formula

$$m' = m \cdot d \cdot o \cdot s \tag{11.6.1}$$

where $m'$ is the observed rate of mortality, $m$ is the "ideal" rate (that is, under perfect conditions), $d$ is the population density, $o$ is a measure of the number of organic particles entering the air in a given time, and $s = C/N$ where $C$ is the market price of the necessities of life and $N$ is the mean income. Of course, only when $C$ was greater than $N$ did low income influence the mortality. In the contrary case, hygiene was the important determinant but this was running ahead of supporting information.

Farr (1843) ranked the metropolitan districts by population density and female mortality rate ($m$) and found that the mortalities were proportional to the sixth root of the population density, writing

$$m' = m\sqrt[6]{(d'/d)}, \tag{11.6.2}$$

which is equivalent to

$$m = \text{constant} \times \sqrt[6]{d} = A\sqrt[6]{d}, \qquad \text{say}, \tag{11.6.3}$$

or

$$\log m = \text{constant} + \tfrac{1}{6}\log d. \tag{11.6.4}$$

Farr (1843) found that the relation held especially well for the ten healthiest and the ten unhealthiest districts. He believed that the population density was a measure of the "zymotic atmosphere" (here we are not to assume that he has bacteria in mind).

In his second *Letter to the Registrar General*, Farr noted that few writers appreciated the reality of the differences of death rates between geographical units; thus, "sufficient attention has perhaps not been paid to the great excess in the mortality of consumption, caused by the insalubrity of towns." Edward Headlam Greenhow (1814–1888) confirmed the importance of Farr's observation on tuberculosis; Greenhow (1857–1858) greatly extended the use of comparisons by cause and showed that the bowel diseases, that is dysentery, diarrhea, and cholera, were potent causes of the differences in death rates between geographical units.

The comparisons between geographical units in the Western world attained their greatest value in the late nineteenth century; they showed that there were real differences in mortality between units, that certain diseases (as we can now see, usually bacteriological or organismal) were particular causes of the differences, and that public health measures were effective in reducing the differences. The methods were strengthened by the use of the coefficient of correlation by G.U. Yule and others commencing in the 1890s [see the *Journal of the Royal Statistical Society* (*passim*), and §19.3). Such studies have continued up to the present but the problems are now more readily solved with the aid of bacteriological methods only available after 1880.

[For the importance of comparative statistics see Lancaster (1990), especially chapters 20, 25, 46, and 47.]

## 11.7 Topley's School

William Whiteman Carlton Topley (1886–1944) after graduation held various posts in pathology and bacteriology, interrupted by military service in World War I, during which he was a witness of the severe epidemic of typhus in Serbia. He was profoundly disturbed, like many of his contemporaries, by the pandemics of bubonic plague commencing in 1894 and of influenza commencing in 1918. He was not satisfied with the literature on the history of epidemics; he especially disliked the works of August Hirsch (1817–1894), writing in the pre-bacteriological age and those of Charles Creighton, who openly scoffed at the bacteriologists. It is clear that Topley had been working on some of the general problems of epidemics at Charing Cross Hospital in 1911 and again at Manchester in 1922–1927. The earliest of his joint publications with Major Greenwood was Greenwood and Topley (1925), a brief review of epidemiological concepts. Topley's appointment in 1927 as Professor of Bacteriology and Immunology at the London School of Hygiene, where Greenwood was already Professor of Medical Statistics, enabled them to cooperate closely and to develop a team which was to include Austin Bradford Hill (1897–1991), Graham Selby Wilson (1895–1987), Mary Joyce Wilson (née Ayrton) (d. 1976), William John Martin, and Herbert Edward Soper (1865–1930).

Greenwood, Hill, et al. (1936) summarized their conclusions on the artificial epidemics observed in the laboratories of the London School of Hygiene. To simplify the discussion, it may be helpful to treat only one of the infections, "mouse typhoid" due to *Bact. aertrycke*. This is a chronic disease and in Table 1 of their introduction it is shown that in about 90% of mice, autopsied between the 11th and 40th days inclusive, the bacterium can be isolated, and in over 80% in days after the 40th. A population of mice initially infected with *Bact. aertrycke* is maintained by immigration, which can be varied at will. There is no emigration, so that all removals are by death. The authors believe that (I.a1) in relation to secular time, the disease will never normally die out, although with small numbers of animals a few survivors at some time might happen to have no bacteria to pass on; (I.a2) the curve of deaths per day contains a succession of well-differentiated waves; with very low rates of immigration there will be well-separated waves of deaths and quiet intervals of few deaths; if the immigration rates are high, there will be minor fluctuations or an almost steady death rate; (I.a3) the average death rate is not highly correlated with the immigration rate and so, with a constant immigration, there is a tendency toward a constant population size and death rate;

(I.a4) this equilibrium is easily disturbed by intrinsic and extrinsic factors leading to violent fluctuations; (I.b1) if age is measured from the day of entry into the herd and time is measured by days, $l_{25}$ and $l_{50}$ are approximately 0.50 and 0.23 (see §6.3 for an explanation of the notation); at later cage ages the probability of dying within five days, $_5q_x$, decreases and between ages 40 and 60 days is approximately constant within sampling errors, so that at high cage ages, the mice can be said to be decreasing exponentially; (I.b2) the expectation of life limited to 60 cage days, $_{60}E_x$, is initially low but rises after about 30 days when it is greatly in excess of $_{60}E_0$, the expectation of life at entry but much below the control mice; most mice succumb to the prevailing disease; (II.a) selection by death of the most susceptible and natural immunization processes probably are the explanation of the changes in mortality in (I.b); (II.b) resistance probably varies in an individual mouse and low levels of resistance possibly account for the deaths; (II.c) infection occurs very early in cage life; (II.d) the state of equilibrium between migration and death is fundamentally unstable and subject to disturbances; (III) the levels of mortality, the proportion of immunizations to fatal infections, and the degree to which infection occurs are largely determined by the characters of the bacterial strain; virulence and infectivity vary independently between strains; highly potent "epidemic strains" possess both these characters; a strain of initial low epidemicity may change to a higher level of epidemicity during the epidemic; nevertheless, changes of this kind seem to have played no part in producing epidemic waves; (IV) artificial immunization can probably affect the course of the epidemics; no evidence was obtained that diet was effective in altering the epidemics; bacteriophage yielded no positive effect; (V) with mouse typhoid, dispersal must be carried out early to be effective.

## 11.8  Webster's School

The work of the American epidemiologists might begin with Webster (1946), an account of the work done in the laboratory of Leslie Tillotson Webster (1894–1943). He divides the history of epidemic sickness into three periods.

The first or preexperimental period, before, say 1876, we have already treated in §§8.1–8.7 and in Chapters 9 and 10, where the great problem was to convince clinicians of the reality of contagion without a biological model.

The second period began about 1876 when Pasteur and Koch made it evident that bacteria were the cause of many infectious diseases. This was the period of description of the bacteria and the diseases they caused by laboratory methods and in laboratory surroundings.

Webster says that the basic fact is that "certain sicknesses are caused directly when a specific microorganism comes into contact with a susceptible individual"; furthermore, there are two types of epidemic: (a) in susceptible populations exposed for the first time to a virulent microorganism, and (b) in populations in which the virulent microorganism is already established. An example of the first type is typhoid where the microorganism (microbe) has been introduced into a population, previously free of the infection, through the ingestion of contaminated food, milk, or water. Problems arise: (a) Why do some persons become ill, whereas others have a mild attack or escape infection altogether? (b) Why does the epidemic come to an end although many of the population remain healthy? (c) Did the virulence of the microbe change? (d) Did some individuals become immunized during the early stages of the epidemic? (e) Are other recognized factors involved? It had also been experimentally determined that animals could be immunized in the laboratory by repeated small doses of pathogenic organisms and therefore the epidemiologists believed that this phenomenon occurred in nature and so epidemics (both types) ended because part of the population became immune through chance exposure to small doses of the agent. "These explanations of epidemics drawn from experiments on individual rather than herd infection remained firmly entrenched until the advent of experimental epidemiology in 1920" (Webster, 1946).

Bacteriologists could reduce the "virulence" of disease-producing organisms by prolonged cultivation on certain laboratory media and then restore it by repeated animal passage; indeed, this was a preoccupation of Pasteur. From this experimental work, epidemiologists argued that virulence must fluctuate in nature. Hence they could attribute the rise and fall of epidemics in already infected populations, that is, of type b, to changes in the virulence of the specific agent.

Webster's third period began around 1920, when the two units were established around Topley in England and Webster in the United States. Webster (1946, pp. 80–82) cites Topley (1919) in his Goulstonian Lecture as follows:

We are thus left with a conception of the virus of a given disease being distributed fairly widely throughout the world as an apparently harmless parasite on the human host, but taking on during epidemic periods a new and sinister role, only to relapse again into comparative quiescence as the epidemic subsides.... While there seems little room for doubt that increased pathogenicity of the parasite must play an essential part in the rise of the wave of disease, it is much more difficult to decide on the relative importance of variations in the powers of parasite and host in bringing about its decline.[1]

Webster (1946) points out that these ideas have been greatly modified by later experimental data. Indeed, he submitted the animals studied to "challenge" at the beginning and end of laboratory epidemics. "The results ... showed a constancy and fixity of disease-producing power of a given strain of organisms under all conditions of natural infection" (p. 92).

---

[1] From Webster, L.T. Experimental epidemiology. *Medicine (Baltimore)* 25:77–109. Copyright © Williams & Wilkins, 1946, pp. 80–82. Reprinted with permission.

## 11.9 Rise and Fall of Epidemics

A few general statements can be made on the rise and fall of epidemics, but individual diseases have to be considered.

1. *Fadeout of epidemics.* We have discussed the fadeout of epidemics in §§11.1 and 11.2. The clearest cases arise in such diseases as measles, rubella, and smallpox, which are associated with a high degree of immunity after a single attack. Varicella was once supposed to act in a similar fashion but the virus can lie latent in the host tissues for many years and then erupt as an attack of shingles (see §11.2).
2. *Latency of infections within the host's body.*
   (a) Carrier states. There was also the problem of human carriers of the specific organism without obvious clinical evidence of the disease; Topley (1919) cited enteric fever as an example, with the disease lying latent for many years in the gallbladder, from which it could emerge to cause epidemics.
   (b) Reawakening of "microbial" activity, the return of an epidemic. In typhus, the infection can linger on after the patient is free of demonstrable clinical disease. The hygienic conditions of jails were formerly so bad that the disease could be perpetuated in them. It was recognized in 1934 that the *Rickettsia* can lie dormant in the human body for many years; cases of typhus, at first called Brill's disease, were found in migrants to the United States from Poland or Russia. It is easy then to explain epidemics during war, famines, and other civil catastrophes, at times when people are being herded together under poor hygienic conditions. It may be that some persons are just recovering from actual typhus infection. Transmission of *Rickettsia* is unduly easy and an epidemic results. The same can happen if there is a recrudescence of an old infection. The epidemic comes to an end by prevention of transmission and improved hygienic conditions. Louse-borne relapsing fever, due to *Borrelia recurrentis*, the first known human example of a living *contagium*, is a disease (possibly) exclusively of man; like typhus, it too appears in epidemic form in times of war, famine, and civil disorders. It may be that *B. recurrentis* and *B. duttoni*, causing a zoonosis, are not distinct species and that the epidemics can thus be explained by a perpetuation of *B. duttoni*.
   (c) Long life cycle. Malaria can lie latent within the bodies of the human hosts and cause the commencement of an epidemic if meteorological conditions are favorable, that is if there is a suitable abundance of the appropriate anopheline mosquito, often after a wet season; in Sri Lanka such an outbreak can occur if the season has been particularly dry, which favors the more dangerous anopheline vector.

3. *Anthropozoonoses.* Topley (1919) would have been interested in the solution of the origins of bubonic plague, yellow fever, and influenza, which can all be regarded as anthropozoonoses.

(a) Bubonic plague (see §8.12).

(b) Yellow fever. This is a zoonosis and epidemics only appear when large numbers of susceptible humans are present in the endemic areas. Up to 1910, say, the great human epidemics were usually in Europeans living, sometimes as military personnel, in tropical America or Africa. The epidemic only ceased when they were evacuated to *Aedes*-free territories. The local inhabitants were usually immune because of previous infections. In Gibraltar, epidemics would cease with the coming of cold weather and the disappearance of the vector mosquitos, *Aedes aegypti.*

(c) Influenza. A given strain of influenza-A virus cannot be perpetuated even by the whole world. The pandemic subsides. At some unspecified time there is a genetic mixing with another influenza-A strain in the body of some animal, and a new strain of influenza virus is formed. Owing to the abundance of life there, this new strain is usually developed in southeast Asia and spreads to the whole world. A new pandemic has arisen.

4. *Homogeneity of bacterial species.* Are bacterial species homogeneous with respect to virulence and other properties? Different types of the diphtheria bacillus can appear, that is, *gravis, mitis,* and *intermedius,* and various numbered types of *Str. pneumoniae* are known.

5. *Replacement of the special type of infecting organism.* This is relevant for diphtheria epidemics and possibly also for scarlatina. It has sometimes been relevant for smallpox, for variola minor immunizes the population against the more serious variola major, that is, smallpox of classical type.

6. *Periodicity.* Why do some diseases appear to possess the property of periodicity? The existence of periodicity, inherent in the infecting species, as Brownlee implied, has been rather overemphasized (see §12.3). Of course, weather conditions or social customs ensure that some diseases have an annual periodicity.

7. *Decline of the epidemic.* In purely human infections, after the epidemic has gone on for some time, the proportion of susceptibles falls until a new case itself generates less than one case on average; the rate of new cases falls and the epidemic ends. Occasionally a change in demographic, social, or ecological conditions has the same effect.

## 11.10 Myxomatosis in Rabbits

Topley and Wilson (1975) added a single reference to those of their Chapter 55 of the previous edition on epidemiology, namely the studies of myxomatosis in rabbits by Fenner and Ratcliffe (1965) in Australia (Frank John Fenner,

b. 1914). This work can be seen to be in the traditions of Topley and Webster, but on a grand scale (four continents); further, although it was an unplanned experiment, many efficient, capable, and interested persons were on hand to make precise biological observations and to test such features as the infectivity of the virus and the resistance of the host.

In 1896, a devastating epizootic occurred in rabbits, *Oryctolagus cuniculus*, being bred for the newly founded Hygiene Institute in Montevideo. Professor Giuseppe Sanarelli (1865–1940), its Director, found that the disease was infectious and highly lethal, producing symptoms unlike any experienced in Europe, including numerous mucinous tumors in the skin of the infected animals. Sanarelli named the zoonosis "infectious myxomatosis of rabbits," and suggested that the cause was a "filterable virus." Later research has shown that the virus is a poxvirus and that there are two varieties of the myxoma virus, the Californian and the South American. No further reports appear until 1909, when the zoonosis was observed in the European rabbits, bought in São Paulo, Brazil. After that time outbreaks appeared in European rabbits maintained in Brazil and in Argentina. For many years, myxomatosis was regarded as confined to South America. It was introduced into Australia in the early 1950s and has had its greatest impact there. It is best for us to study the introduction of the virus into Australia, which served as a "Topley" or "Webster" laboratory on a grand scale.

The myxoma virus had been enzootic in *Sylvilagus* species (that is, American species of rabbits) producing localized benign fibromas and being transmitted by arthropods. When transmitted to a European rabbit, *Oryctolagus cuniculus*, a very severe disease with widely generalized lesions, chiefly in the skin and testis, is produced; the case fatality rates are very high, especially in the young rabbit.

The myxoma virus was unknown in Australia until it was introduced, only for laboratory purposes, in 1926; it was again introduced in 1936 and studied intensively under Lionel Batley Bull (1889–1978) of the Australian Division of Animal Health [Council for Scientific and Industrial Research (CSIR)]. At this time, it was not known that the disease was transmitted by arthropods but Bull found that a native stickfast flea (*Echidnophaga myrmecobii*), which had adapted itself to the European rabbits, and several species of mosquitoes could transmit the disease. A trial release was made in a 36-hectare (approximately 89 acres) enclosure infested by stickfast fleas. The rabbit population was well-nigh exterminated within two months. However, trials in other areas were disappointing. The work was suspended in 1943. Francis Noble Ratcliffe (1904–1970) was appointed to lead the Wildlife Survey Section of the CSIR, Australia, in 1948. In 1950 trials were conducted in several sites along the Murray River; the results initially were again disappointing with the disease remaining localized and apparently dying out. In December 1950, myxomatosis flared up in one of the trial sites and then spread over southeastern Australia. It is now established as an enzoonosis of the Australian wild rabbits of the introduced European species, with epizootics flaring up from time to time in association with local and seasonal vector activity.

The survey of Fenner and Ratcliffe (1965) is of general interest because genetic changes appear to have taken place in both virus and host.

### Virulence of the Virus

In 1937, at the time of its discovery, the myxoma virus was believed to be of uniform and very high virulence for European rabbits. Soon after the introduction of such highly virulent strains into Australia and Europe, strains of lower virulence began to be recognized and confirmed by "challenge" tests, injections of standard doses into laboratory European rabbits.

### Resistance of Hosts

There were genetic changes in the hosts whereby resistant lines were selected naturally in areas having annual epizootics of the disease or selected artificially in the laboratory. Over a period of seven years case fatalities in the rabbits fell from 90% to 25%.

### Environmental Effects and Nutrition

High (low) ambient temperatures greatly diminish (accentuate) the severity of infections. Concurrent infections have variable effects in either direction. Malnourished rabbits may show negligible symptoms but die as rapidly as do controls. [For these three sets of conclusions, see the summaries of Fenner and Ratcliffe (1965, pp. 240–241, 250–251, and 143, respectively).]

## 11.11  Epidemiology after Topley and Webster

It may be said that the principal gain from the laboratory work of the English school under Topley and the American school under Webster was a sharpening of the concepts. Some workers continued the traditions of those schools, notably F.J. Fenner.

Great changes had occurred in the theory of epidemics. The epidemiologists had increased greatly in level of skill and numbers and there had been great advances in bacteriology and virology, so that the agent of a new disease could be more readily identified.

We may cite Greenwood (1927b), a principal colleague of Topley, in an abstract of Dudley (1926): "This is an important paper ... it is the first serious attempt to make a *continuous* study of epidemic disease in partly isolated human communities ... the experimental method applied not to *individuals* but to *herds* is the proper tool to use." The study of Halliday (1928) on the incidence of measles in Glasgow is in the same tradition. Many of the works can be regarded as answering special problems in public health or alternatively as showing that conclusions drawn from laboratory work had relevance to the outside world in the style of Strato (see §1.4). There have been many other studies, for example, Black (1966) and his colleagues examining primitive tribes, tribes passing into civilization, and epidemics in island populations. P. Haggett and colleagues have studied the spread of disease in Iceland and the northern islands. The greatest of all such studies have been on the pandemic of myxomatosis, of which Fenner and Ratcliffe (1965) have surveyed the Australian experience (see §11.10). There were many epidemiological studies in the work on the eradication of smallpox (see Fenner, Henderson, et al., 1988). An interesting unplanned experiment on the epidemiology of measles in monkeys, in whom measles is not perpetuated in the wild, is recorded by Meyer, Brooks, et al. (1962).

Many new diseases have been introduced into Europe in recent times (see Lancaster, 1990, p. 379). Some difficulties have been caused by unusual properties; thus *Legionella pneumophila* grows in culture media at temperatures above 45°C and is difficult to identify in tissue sections (Lancaster, 1990, p. 120).

Possibly the greatest present threat of disease to the world population comes from the human immunodeficiency virus (HIV), with its dreaded complication, acquired immune deficiency syndrome (AIDS). It, or closely related organisms, are now believed to have been endemic in chimpanzees, *Pan troglodytes*, in two regions in Africa, the strip of coast including Liberia and Sierra Leone and a strip commencing at the Atlantic Ocean and extending about two-thirds of the way to the Indian Ocean, chiefly north of the Equator (see the map in Short, 1991, p. 27). Perhaps such an organism was spread to mankind from the chimpanzee in a manner that can only be guessed at, perhaps a bite, an infection from uncooked chimpanzee meat, or some close sexual contact.

A new phenomenon has appeared—social resistance to the investigation of disease. In attempts to control other case to case (infected to susceptible) infectious diseases, particularly sexually transmitted disease, it has been customary to attack the routes of transmission, but with HIV infections and/or AIDS, libertarian and legal bodies have interfered with a proper epidemiological approach. Yet the ordering of the Wassermann test for syphilis was routine in the 1930s, and from the experience of clinicians at that time it now appears that few political difficulties arose from the practice. Merson (1993) reviews this problem and recommends that people should try to minimize the number of sexual partners during their life; he uses the word *fidelity* at one point. The paper is part of a symposium on AIDS.

# 12
# Mathematics and Epidemiology

## 12.1 Farr's Epidemiology

William Farr, around the time of his appointment as an official statistician in 1837, was aware of possible changes in epidemiological thought. As we have seen in §8.7, a disease of the silkworm and a human skin disease had been shown to be due to fungi; further, yeasts had been demonstrated to cause fermentation of cane sugar, although this interpretation had been bitterly opposed by Liebig, as mentioned in §8.5. Henle had written his essay but the techniques for detecting microbes had not yet been developed to give experimental verification to his ideas. [In what follows in this section, all page numbers refer to Farr (1885/1975).]

In Farr's remarks in the Second Annual Report to the Registrar General in 1840 (p. 317), he introduces a mathematical theory of epidemiology. Farr was expressing his views on "Medical Notes and Reflections" of a Dr. Holland, a review of *Pathologische Untersuchungen, von Dr. Henle, Berlin, 1840*, in the *British and Foreign Medical Review*, April, 1840. Farr in this note accepted the theory of the yeast fermentation of sugars, but presumably not its rejection by Liebig. Farr wrote of Henle's model:

The germs are innumerable, and spread with the greatest rapidity. In mixtures, certain genera of infusoria appear, and then give place to new genera. Individual cases of disease may be caused by one generation of parasites; an epidemic by successive generations. Each epidemic disease has its specific animal contagion, its specific genera of infusoria. Henle has proved the existence of this cause, and the truth of the theory in every way but one; he has never seen the epidemic infusoria. The omission is, no doubt, important; and the more so on the part of Henle, who is justly considered one of the best microscopic observers in Germany. The infusorial hypothesis does not satisfactorily explain the cause of epidemics; it accounts for them by the creation of animalcules, but does not show why the animalcules are created at distant times in swarms. (pp. 317–318)

Henle did look for the "epidemic infusoria" but was unsuccessful because of the absence of effective staining methods, such as were later devised by

Carl Weigert in 1871, which could have distinguished the bacteria from the tissues; moreover, it happened that Henle's choice of the scales in scarlatina was particularly unfortunate.

Farr in the Fourth Annual Report looks forward to further advances in the study of the pathogenesis of disease. He writes:

Liebig, Dumas, and the chemists of this country, will, we sanguinely hope, not rest satisfied with what has been done, but continue to prosecute their labours with ardour and success; and, from the study of the series of transformations of nitrogenous compounds, proceed to investigate the transformations of the blood, tissues, and secretions which accompany the production of varioline, typhine, and the other zymotic principles. (p. 250)

Farr is thus not pressing the claims of bacterial agents; no part of Liebig's general theory is against the possibility of epidemic events; indeed, the modern theory of viruses is that they are not organismal and can be regarded as enzyme-like chemical compounds perpetuating themselves in living tissue and so in a sense fulfilling the Liebig hypothesis. It is remarkable that Greenwood nowhere draws attention to Farr's adherence to the Liebigian theory and, indeed, few medical historians have mentioned its importance in the thought of Farr, Snow, and Semmelweis.

Farr states that "epidemics appear to be generated at intervals in unhealthy places, spread, go through a regular course, and decline" (p. 317). If the latent cause of epidemics could not be discovered, the mode of action could be investigated and laws by which they rose and fell could be determined. Thus there had been an epidemic with deaths observed in successive periods as follows:

| 1 | 2 | 3 | 4 | 5 | 6 | 7 | 8 | 9 | 10 |
|---|---|---|---|---|---|---|---|---|---|
| 2,513, | 3,289 | 4,242, | 4,365, | 4,087, | 3,767, | 3,416, | 2,743, | 2,019, | 1,631 |

Farr noted that 3,289/2,513 and 4,242/3,289 were approximately equal to 1.30 but this ratio falls to 1.06, approximately equal to 4,365/4,242. Farr carried out an interpolation on the figures with the aid of a polynomial of the third degree, thus obtaining a series,

| 1 | 2 | 3 | 4 | 5 | 6 | 7 | 8 | 9 | 10 |
|---|---|---|---|---|---|---|---|---|---|
| 2,513 | 3,289, | 4,242, | 4,364, | 4,147, | 3,767, | 3,272, | 2,716, | 2,156, | 1,634 |

giving an admirable fit. Farr fits other series in the same manner between crest and trough.

Farr was prepared to use his theory of epidemics for forecasting. In 1865 there was cattle plague in England; 9,597 new cases occurred in the four weeks ending 4 November, then 18,817, 33,835 and 47,191 in the three 4-week periods ending on 27 January, 1866. Farr extrapolated by his methods to obtain 43,182, 21,927, 5,226, 494, and 16 new cases in the succeeding five 4-week periods, as against 57,004, 27,958, 15,856, 14,734, and 5,000 approximately observed. The epidemic subsided rather more slowly than Farr forecast but the forecast was better than that of the Rt. Hon. Robert Lowe who said, "there is no reason why the same terrible law of increase which has prevailed hitherto should not prevail henceforth" (see Greenwood, 1936b, p. 116). Farr is then at pains to show that the state of the population can affect epidemic events for,

The mere aggregation of people together in close apartments generates or diffuses the zymotic matter. Thus, place lying-in women in close proximity to each other, or mix them up with the patients of a general hospital and they die of puerperal fever; place many wounded men in a ward where cleanliness is neglected, and erysipelas, pyaemia, gangrene spring up; imprison men within narrow walls, or crowd them in rooms and typhus breaks out. The general and special hospitals of the country have been, until quite recently, erected without any special reference to the dangers accruing from the assemblage of great masses of sick people within the walls of one building, so that the efforts of the most skilful medical officers are frequently defeated; but a better system of hospital construction, with more cubic space, is likely to prevail, with due provision for effective changes of air, and then the evils of agglomeration will be mitigated. It is only recently that the subject has attracted the attention of surgeons, who will no doubt anxiously watch the results of the new arrangements. Sir Henry Thompson and Sir James Simpson will, we may hope, continue their researches so as to determine accurately the mortality after the various kinds of amputation in hospitals and in private houses.

To limit the operation of zymotic diseases overcrowding in towns must be absolutely prohibited: the mere accumulation of masses of living people within narrow limits either generates or insures the diffusion of epidemic disease. The plague which almost destroyed Athens was aggravated by the policy of Pericles when he brought the outlying country population within the walls. (p. 321)

These last two paragraphs make excellent sense and perhaps moderns would read them with a feeling that surely a microbial theory is implied; but there is no mention of it. Perhaps, it would be easier to believe that subtle chemical changes would occur in the crowd rather than that carriers had brought in "infusoria."

Arthur Ransome (1834–1922), a physician, later Professor of Public Health at Owens College, Manchester, was interested in finding interrelations between the times of epidemics and various meteorological events (see Hilts, 1980). He knew of Farr's views on the waveform of the epidemic, indeed, Ransome (1868) wrote: "I believe that Dr. Farr, in his letter upon the Cattle Plague, was one of the first to point out the regularity of the course of most epidemic diseases." He also noted that cycles of infection could occur, in particular, a two-year cycle of whooping cough (pertussis). It should be noted that epidemiologists use the word *cycle* in a broad sense to mean "recurrent"; they seem not necessarily to imply that the distances between successive peaks are equal. Ransome (1881–1882) later noted cycles in the smallpox death rates for Sweden. He suggested that density-of-susceptibles would explain such cycles in the following:

Probably all the facts would be accounted for if we suppose that these disorders can only become epidemic when the proximity between susceptible persons becomes sufficiently close for the infection to pass freely from one to another.

Exanthematous diseases rarely attack the same individual twice in his lifetime.

When, therefore, an epidemic has, by either a fatal or nonfatal attack, cleared away nearly all the susceptible persons up to a certain age, then it must necessarily wait a certain number of years before the requisite nearness of susceptible individuals has again been secured.

There must in the interval be a gradual re-stocking the nation with material fit for the epidemic to feed upon, and it can only spread when the requisite proximity is attained, when the meshes of the network of communication are sufficiently close for it to include all susceptible persons in one grand haul. (Ransome, 1880, cited in Hilts 1980, pp. 47, 48)

Ransome further claimed that such reasoning would explain the differences in cycle length of certain infections in Sweden and England and Wales. As for the shape of the wave, he wrote (1881–1882): "I am inclined to think that any attempt at an explanation of the epidemic wave should have to rely mainly upon a knowledge of the changes in the degrees of virulence of the epidemic poison."

The hypothesis that the density of susceptibles could explain the periodicity of measles was put forward by A. Campbell Munro (1890–1891) but it appears that none of these early workers were satisfied that such a hypothesis was a satisfactory explanation of either the epidemic wave or the periodicity; for example, B.A. Whitelegge (1852–1933) believed that the hypothesis was adequate explanation for "minor" outbreaks of measles but that a qualitative change in the measles virus must be invoked to explain the "major" epidemics (see Whitelegge, 1892–1893).

## 12.2 Measles in London

Many of the investigations into the mathematics of epidemics have taken measles as their paradigm, usually in London or other great city. Annual counts of measles deaths have been available, with a few short breaks, in London since 1629 (see Lancaster, 1990, Table 11.7.1, p. 141). In the 1670s the annual deaths were 295, 7, 118, 15, 795, 1, 83, 87, 93, 117; there was thus a tendency toward highs in the even and lows in the odd years. It is remarkable that later probabilists or statisticians have failed to comment on the single death in 1675; for measles in all likelihood had faded out in that year because one death suggests only a few dozen cases with the high case fatality of those times, and it seems unlikely that so few could have perpetuated the epidemic throughout a whole year, about 25 incubation periods. Presumably in 1676 measles would have been reintroduced from the surrounding country or from another city. Now measles has a seasonal distribution. Percy Stocks (1889–1974) observed that the epidemic periods never began earlier than

TABLE 12.2.1. Measles deaths in London and Birmingham.

| Year* | 0 | 1 | 2 | 3 | 4 | 5 | 6 | 7 | 8 | 9 |
|---|---|---|---|---|---|---|---|---|---|---|
| | | | | | London | | | | | |
| 191– | 3,274 | 1,151 | 2,306 | 958 | 2,676 | 712 | 1,819 | 1,935 | 303 | 1,113 |
| 192– | 157 | 1,656 | 154 | 1,539 | 117 | 1,170 | 53 | 1,480 | 112 | 1,113 |
| 193– | 56 | 885 | 62 | 906 | 10 | 596 | 15 | 245 | 2 | 5 |
| | | | | | Birmingham | | | | | |
| 191– | 328 | 124 | 826 | 174 | 555 | 87 | 356 | 47 | 189 | 81 |
| 192– | 225 | 62 | 194 | 71 | 104 | 62 | 146 | 32 | 194 | 25 |
| 193– | 199 | 22 | 110 | 25 | 49 | 35 | 74 | 7 | 17 | 2 |

* 1910 in the table covers the period 1 October 1910 to 30 September 1911 and so on. Data from Stocks (1942, p. 265). Copyright © *Journal of the Royal Statistical Society* 1942; 105:259–291. Reprinted with permission.

1 October nor later than 31 December (see Stocks, 1942); he therefore defined his years of experience as beginning on 1 October. His graph of annual deaths drawn with this convention displayed a definite sawtooth appearance (Table 12.2.1). Stocks (1942) showed that similar appearances could be observed also in Birmingham, Liverpool, Manchester, Leeds, and Sheffield; the highs of different cities did not necessarily coincide. Thus London and Birmingham had their highs in different years; similar, but not so definite, findings were made for New York City.

Before going on to the mathematical analyses, it is appropriate to list some of the epidemiologic factors mentioned as important by William Heaton Hamer (1862–1936) (Hamer, 1906):

1. Measles shows a persistence of type and so is easily recognizable.
2. Measles is an obligatory parasite of man [but see §11.11].
3. A second attack is unknown or rare.
4. The case fatality rate is approximately 2.5%.
5. About 2,500 births or (allowing for infant deaths) 2,200 new susceptibles appear in the London population each week.
6. Measles has an incubation period of about a fortnight.

It may be added that most modern epidemiologists believe that the epidemiologic behaviour observed in the past in the great cities is explicable as follows. Under conditions of low immigration rates, excluding, for example, the experience of measles in Glasgow in the early nineteenth century, the existence of endemic measles ensures that almost all adults and children above the age of 5 years have already experienced an attack of measles. Measles is then a childhood disease. The susceptible children can be divided into toddlers and infants. Infants are not mobile and would be exposed to infection from toddlers, for their parents are immune; at the end of an epidemic year, there are few toddlers or infants not already infected and so infection rates fall; then in the following year, the infants of the low infective year become the susceptible toddlers of the next year; they become infected and spread the infection to other toddlers and to infants and the cycle is repeated.

Hamer (1906) writes:

The persistence of form of the London measles wave is not a little remarkable. Growth of population and alterations of its age constitution, varying customs, and social conditions have all left it almost undisturbed. Growth of population might, at first sight, have been expected to exert considerable influence. In island communities measles is introduced, prevails, dies out, and it may be many years before it again gains a hold; in Sweden and in country districts of England the disease presents "multiannual fluctuations"; in large towns it has assumed a biennial type. In London and New York we might expect greater frequency of recurrence still.... Records of case mortality in measles, when available, are apt to relate to special classes of population and hence the influences exerted by age, social circumstances, &c., incessantly intrude. In the Norse epidemics, as Dr. Whitelegge has observed, the records relate to a few large towns and a number of country districts; it may be that varying degrees of prevalence in town and country, with accompanying age differences, account in some part for the alterations of case mortality observed. (p. 735)

## 12.3 Hamer, Soper, and Brownlee

Hamer (1906) described the progress of a measles epidemic in a great city, London, supposing that there was a constant inflow of susceptibles by birth and their removal by death or immunity. From Hamer's description, Herbert Edward Soper set up what is now known as the Hamer-Soper model (Soper, 1929); in it, the unit of time can be taken to be the length of the incubation period, infection is assumed to be instantaneous at contact, and an infected person becomes infective immediately at the end of the incubation period. We may follow the discussion of Wilson and Worcester (1945a,b). The authors point out that almost all workers have assumed that the rate of new infections, $C$, is proportional to $r$ times the product of the numbers of susceptibles, $S$, and the infectives, $I$, namely

$$C = rSI. \tag{12.3.1}$$

It is convenient to write $B = S + I$; it is supposed in the simplest model that there are no deaths. There is an assumption, explicit or implicit, that the infectives are mixing homogeneously with the susceptibles; further, infection is assumed to be instantaneous. Wilson and Worcester (1945b) point out that the real utility of the assumption for the explanation of the course of an epidemic is that it gives a curve that approximates to the course of an epidemic if the parameter $r$ is estimated appropriately. It can, indeed, be shown that $r$ cannot be chosen independently of the size of the population. However, $C$ is proportional to the differential coefficient of $S$, so that

$$\frac{dI}{dt} = C = rIS, \qquad S = B - I, \tag{12.3.2}$$

where $B$ is the total of susceptibles at the beginning. It follows that

$$\frac{dI}{dt} = rI(B - I), \tag{12.3.3}$$

the differential equation for the logistic function.

$$I = \frac{B}{1 + e^{rB(t - t_0)}}. \tag{12.3.4}$$

Susceptibles thus decline from $B$ at time $-\infty$ to zero at time $+\infty$. All the susceptibles are exhausted during the epidemic. If we consider the time taken for the epidemic to pass from 99% susceptible to 1% susceptible, equation (12.3.4) enables us to declare that doubling the population $B$ would halve the time. This sort of difficulty appears in every form of hypothesis that introduces the term $rIS$ as in (12.3.2).

It is evident that quite elementary hypotheses lead to difficult mathematics even in the deterministic models. John Brownlee (1868–1927) in Brownlee (1918) used the periodogram technique on the weekly measles notifications for London and believed that there were four principal periods of lengths 87, 98, $109\frac{1}{2}$, and 114 weeks. He believed that corresponding to each period there was a special strain of measles virus; that there was some cross-immunity between the strains but that it was by no means rare for the same patient to be attacked by the strains of two separate periods within such a short time as a month. Now few clinicians would agree with such a statement. Brownlee also believed that he had found a strain of measles with an 87-week cycle on the north side of the Thames and another with a 97-week cycle on the south side. He thus subjected himself to the witticism that 87-week and 97-week measles were separated by Waterloo Bridge.

Perhaps Brownlee (1907) should be cited as an example of the difficulties of finding the "cause" of epidemics. He states several conclusions: (a) there is a high grade of infectivity at the commencement of the epidemic, (b) the sudden increase in infectivity means that there is a life cycle of the parasite (it is difficult to see how the whole of the infective agents in the various infected persons could act as a unit), (c) seasonal effects may be of some importance but they are not all-important, (d) the epidemic does not end because of the exhaustion of the susceptibles, (e) different infective agents may give rise to different forms of epidemic wave [items (c), (d), and (e) seem to be unobjectionable conclusions], and (f) the epidemic curves for some diseases can be fitted with Pearson type curves. (The form of the epidemic curve can be fitted by a variety of curves. There is no harm in doing so but no conclusion should be drawn that the particular form of curve is of general relevance.)

## 12.4 Deterministic Models

Several authors, among them Ronald Ross and T.B. Robertson, have fitted logistic curves to an epidemic. The model is that there is a fixed population with no deaths and all $n$ members are initially susceptible to the disease and all equally infective to other members of the population. A case of the disease is introduced at time $t = 0$. The rate of change of the number infected is proportional to the product of those already infective, $x$, and those not yet infected, $y = (n - x + 1)$.

$$\frac{dx}{dt} = -\beta xy = -\beta x(n + 1 - x). \tag{12.4.1}$$

By a change of scale $\beta \to 1$, the equation can be modified to

$$\frac{dx}{dt} = -x(n + 1 - x) \tag{12.4.2}$$

and the boundary condition is $x \to n + 1$ as $t \to \infty$.

The solution is given by

$$x = \frac{n(n + 1)}{n + e^{-(n+1)t}}. \tag{12.4.3}$$

All susceptibles $y$ are exhausted as $t \to \infty$.

Kermack and McKendrick (1927) introduced a deterministic model in which there is a community of total size $n$ of whom $x$ are susceptibles, $y$ are infectives in circulation, and $z$ are individuals who are isolated, dead, recovered, or immune; $x + y + z = n$. Let $\beta$ be the infection rate and $\gamma$ the removal rate. Then a set of differential equations follows, only two of which are independent,

$$\frac{dx}{dt} = -\beta xy, \tag{12.4.4}$$

$$\frac{dy}{dt} = \beta xy - \gamma y, \tag{12.4.5}$$

$$\frac{dz}{dt} = \gamma y. \tag{12.4.6}$$

The epidemic starts at time $t = 0$ when susceptibles, infectives, and the residue are, respectively, $x_0$, $y_0$, and zero. It follows that the epidemic cannot begin unless in (12.4.5) $\beta x_0 y_0 - \gamma y_0 > 0$ and hence $(\beta x_0 - \gamma) > 0$ or $x_0 - \rho > 0$, where $\rho = \gamma/\beta$.

It is found that in the model the total size of the epidemic is $2v$, where

$$v = x_0 - \rho, \qquad x_0 = \rho + v, \qquad \rho = \gamma/\beta. \tag{12.4.7}$$

This is the threshold theorem of Kermack and McKendrick (1927).

## 12.5 Stochastic Models

Epidemiologic observations, for example in Europe during the nineteenth century, tended to deal with epidemics involving many persons in large populations. It was therefore possible for demographers at a time of stationary death rates to think of epidemics in England and Wales as proceeding as regularly as "stars in their courses" as one put it. Many other physical analogies were used—waves of infection were observed to be approaching a great city, and the epidemic proceeded as a bimolecular reaction; the "forces" acting were sought. So the first models of an epidemic were deterministic and were treated with the aid of differential equations; but although the general mathematical form resembled to some extent those observed, there were features contrary to the observations; thus in the logistic curve solution of equation (12.4.3), all the susceptibles were exhausted; further, the models were inapplicable to epidemics in small populations.

Anderson Gray McKendrick (1876–1943) seems to have been the first to suggest stochastic models (see McKendrick, 1926, 1940), but his work was not immediately followed up. It is instructive to follow another line, the course of infection within a family or small herd. For epidemics within families, by which we mean any set of individuals, or herd, in a community, usually small, the simplest model is by Greenwood (1931): (a) $m_1$ of the family of $M$ is/are initially infected, (b) there is an incubation period of fixed length for all cases, (c) infection occurs at the end of the incubation period and is instantaneous, (d) there is a probability $p$ that any one of other members of the family would be infected and that these infections would be mutually independent. The distribution of the number of cases in the second generation, $m_2$, is thus a binomial with parameters $(M - m_1)$ and $p$. The process is continued; $m_{s+1}$ is a binomial variable with parameters, $M - m_1 - m_2 - \cdots - m_s$ and $p$. A series of random variables $m_1, m_2, \ldots$ is thus obtained, terminating if $m_{k+1} = 0$, or when the population has been exhausted. The probability of a given sequence is readily obtained and the process is given the name *chain binomial*. The expectations of the $\{m_j\}$ form a decreasing sequence. If $m_{k+1} = 0, m_1 + m_2 + \cdots + m_k \leqslant n$. The probability of such a sequence is readily computed. It leads to no theoretical difficulties but the probability of infection passing from the $s$th generation to the $(s + 1)$th being independent of $m_s$ seems unrealistic.

A more elaborate model from unpublished lecture notes of Lowell Jacob Reed (1886–1966) and Wade Hampton Frost (1880–1938) (see Abbey, 1952), specifies that after the $s$th generation any susceptible is exposed with probability $p$ to each of the $m_s$ members of that generation, so that his probability of not being infected is $q^{m_s}$, $q = 1 - p$. The epidemic begins with $m_1 = 1, 2, 3, \ldots$ but usually $m_1 = 1$. So $m_{s+1}$ is now a binomial variable with parameters $n - m_1 - m_2 - \cdots - m_s$ and $1 - q^{m_s}$. This model gives results quite different from the Greenwood chain binomial model. There is a tendency for the $m$'s to increase and then to decline.

Let us now consider the difficulties in selecting a parameter $p$ for different sizes of herd. To fix ideas, let $M = 5$ and $m_1 = 1$. If $p < 0.1$ there is a probability greater than 0.5 that the epidemic will terminate forthwith and sequences such as $\{1, 0, 0, 0, 0\}$, $\{1, 1, 0, 0, 0\}$, and $\{1, 1, 1, 0, 0\}$ will occur much more frequently than series in which 2 or 3 appear; so the choice would be for $p \geqslant 0.1$. Now let $M = 1000$ and, for convenience, $p = 0.1$, as large values would have the same effect; further let $m_1 = 1$. Then $m_2$ is $O(Mp)$ and $m_3$ is $O(M^2p^2)$ where $O(.)$ means "of the order of." The successive parameters are $1 - q^{m_1}, 1 - q^{m_2}, 1 - q^{m_3}, \ldots$ But $m_2$ is $O(Mp) = O(100)$, $q = 0.9$ and $q^{m_2} = 0.9^{m_2} = O(0.9^{100}) = O(10^{-4})$. The probability parameter of the binomial will thus be approximately 0.9999 and so all the susceptibles will usually be exhausted after three generations.

The setting up of a complete stochastic model is too long to be considered here; but in §12.8 we consider some more objections to mathematical modelling of epidemics, whether stochastic or not.

## 12.6  Diseases to be Modelled

Let us examine the list of infectious diseases, which might be considered as great pestilences under appropriate conditions, and see whether any unifying principle is apparent: cholera, enteric fever, gastroenteritis, tuberculosis, bubonic plague, leprosy, diphtheria, whooping cough, acute poliomyelitis, smallpox, measles, yellow fever, influenza, typhus, malaria, American trypanosomiasis, African trypanosomiasis, relapsing fever, and syphilis, to which might be added some diseases due to helminths.

### Cholera

This disease is spread by water as a rule; here the model of the disease being spread independently from those with active disease to the susceptibles is false. Often one or perhaps a few persons will have been responsible for the infection of many by pollution of the water supply (see §8.9).

### Enteric Fever

It may be that similar remarks could be made here as for cholera. There are famous epidemics in which many cases have been infected from a single carrier, perhaps, excreting the bacilli from a chronic infection of the gallbladder. In more recent times, Wilson (1984) has detailed the investigation of an epidemic widespread in the English "home counties," which originated from a farmer who carried the bacillus.

### Dysentery and Gastroenteritis

Similar considerations apply again in these infections.

### Tuberculosis

Here it might be thought that many open cases of the disease would raise the level of the density of the tubercle bacillus in the air surrounding the suscepti-

bles and that new cases might well be proportional to the product (susceptibles) × (open cases). On the other hand, epidemiological research has shown
the importance of infections within the family and in the workshops, mines,
and other occupations, with their special hazards such as silicosis in miners
and dusts in certain trades.

## Leprosy

Infections within the family are important. Irgens (1981), writing on the incidence in Norway, points out that leprosy passes with some difficulty from
case to case but passage is especially favored by conditions of high humidity
and low hygiene, such as obtained in the fisheries of Norway.

We could continue through the great pestilences or other diseases with
such an analysis but we have sufficiently demonstrated that the infectives at
any time may have special properties and some well deserve the title of "hard
core."

## 12.7  Periodicity of Infectious Disease

It is difficult to state precisely what authors mean when they write that an
endemic disease has periodicity. Perhaps this means that the disease is not
always apparent in the herd or that the average number of appearances
or epidemics per unit of time or the mean time between epidemic peaks is
observed. In physics, the definition of periodicity is based on such notions
as the length of day, the distances between high tides in the oceans, length of
waves crossing the Pacific Ocean from below Tasmania and being recorded
in California, and so on. In these cases there are known forces in action
and their nature allows some assessment of the cause of the periodicity. In
epidemiology, Brownlee (1918) believed that he had established the existence
of strains of measles virus that were the cause of the differing periodicity
(see §12.3).

Other mathematical epidemiologists have not gone so far as to infer such
properties in the agent itself. Sometimes they have agreed that there are
annual fluctuations in the conditions for the passage of the infective organisms; indeed, there can be no objection to admitting to an annual cycle of
infections. It has seemed unreasonable, however, for the mathematicians to
detect cycle lengths that are not integer multiples of a year. For example, if
the cycle length for measles in London is taken to be 78 weeks, the finding
must be rejected because this would imply that as many epidemics were
initiated in early March as in early October, which would contradict common observations. More generally, the cycle length cannot be any length that
is not an integer multiple of one year. Suppose, indeed, that we assume the
cycle length to be $97\frac{1}{2}$ weeks, which is approximately $1.875 = 15/8$ years. This

means that the epidemic peak (or alternatively the beginning of the epidemic) will be one-eighth of a year earlier each cycle, which would again be inconsistent with the data for measles and probably for almost any disease.

## 12.8  Critique of Mathematical Epidemiology

Farr considered an epidemic to be an orderly process that does not depend on the capricious activity of some "epidemic tendency" (see §12.1). The next general step was to consider the annual measles deaths in London; from this series, Hamer (1906) concluded that the number of susceptibles in the population was the important statistic. Ronald Ross in these early years introduced differential equations to the study of epidemics, but few solutions resulted from his work. We have mentioned in §12.3 the work of Soper in formalizing the theory of Hamer. McKendrick (1926, 1940) introduced both deterministic and stochastic theories.

The mathematical theorists then settled on the set of equations (12.4.4) through (12.4.6), and worked out the consequences mathematically, without bringing the mathematical theory into line with the properties of actual infective agents. We do not wish to belittle the skill of the mathematicians who solved these equations. We consider only the question as to their relevance to epidemiology.

1. *Geography.* This is considered to be irrelevant to the problem, whether the population is grouped in an institution, in a city, in the country, in a region with both urban and rural communities, and so on.
2. *Age, sex, and individuality.* These also are considered irrelevant. The differential equations, by giving the infectives and susceptibles no individuality, overlook the fact that in many infections there are hard core individuals, for example, the presence of the promiscuous in the spread of the venereal diseases.
3. *Constancy of the parameters in the model where an infinity is assumed.* These are fixed throughout over time and over the members of the community. There is no change in the susceptibility parameter, $\beta$, the resistance of the hosts (see §§11.8 and 11.10). Nor is there any change in the virulence of the parasite. Seasonality is ignored.
4. *Dependence of parameter $\beta$ on $n = x + y + z$ in the application of the theory to a finite population.* The great difficulty is the homogeneous mixing condition where the rate of new cases is the product of the number of susceptibles with the number of infectives, such as is implied by equation (12.4.2). As Bailey (1975, p. 22) notes, the mixing condition is "most nearly realized in small household groups," although one has heard of mothers who will not allow sick members of the family near an infant. Certainly, the condition is "at variance with the observed facts of social behavior in a large town" or, moreover, city. It is sometimes stated that the calcula-

tions of the homogeneous mixing hypothesis can be justified by taking a summation of subepidemics in small geographic units; but this has never been shown to result in the consequences of equations (12.4.4) through (12.4.6). An inconvenient consequence of the homogeneous mixing hypothesis is that "the epidemic sweeps through a single group more rapidly as $n$ increases. In fact, the mean of the normalized epidemic curve $\bar{\tau} \to 0$ as $n \to \infty$" as in Bailey (1975, p. 75). The same point is made in §12.5.

5. *Implicit hypothesis of identity of host classes.* In equation (12.4.4), $z$ is the count of immunes and dead, but the effect of a death of a susceptible host is not equivalent to the passage of a susceptible host into the immune class, for this latter event raises the mean proportion of those immune in the neighbourhood of each susceptible; he/she only meets in the real world a limited number of individuals. Therefore, to mimic reality in the model, the parameter $\beta$ of equation (12.4.4) cannot be kept constant. This difference would be especially notable in families or small groups.

There are other factors to be considered that are not readily brought into the differential equation. Many of the common viral and bacterial diseases have a pronounced seasonal variation, for example, measles. Yorke, Nathanson, et al. (1979) have shown that this has a significant effect on the perpetuation of a disease, in particular malaria, especially if the incubation period is short. They have suggested that some of the damping effect given by Soper's equations of §12.3 can be removed if such seasonality effects are introduced into the model and computing, rather than differential equation, methods are used to estimate the deaths by epoch. This approach is moreover practical for it shows that the time for most effective intervention in control of an epidemic is when the transmissibility rates are low.

In conclusion, we lean to the view that, although much difficult mathematical work has been done on the differential equations, the more biologically and heuristically oriented work of authors such as London and Yorke (1973), Yorke, Nathanson, et al. (1979), and Matumoto (1969) is to be preferred, if the aim is to understand or to control epidemics.

Finally, no disease has yet been found that obeys the differential equations. Mathematical epidemiology has been ignored in the 8th edition of Topley and Wilson (1990).

# 13
# Epidemiology of Noninfectious Diseases

## 13.1 Avitaminoses

From the time of Hippocrates until the 1880s (see §8.2), epidemics had the meaning of diseases falling upon the people and affecting them simultaneously. No restriction to infectious diseases was implied because the idea of contagion was not accepted. In the years 1875 to 1925, many of the epidemics were shown to be caused by living agents, including bacteria, viruses, and parasites. "Epidemic" became identified with infectious diseases. There are, however, noninfectious diseases that appear as epidemics, for example, avitaminoses, environmental poisonings, and lung cancers in miners, and it seems quite proper to use the word epidemiology for their study. As the quotation from Burnet (1941) in §8.6 shows, there would have been great difficulty in the Middle Ages, say, in differentiating between diseases that were infectious in the modern sense and those that were not, even if the diseases had been clearly identifiable. These difficulties have persisted into recent times as is recounted below in the discussion on pellagra.

We could have begun with an early worker, Paracelsus, whom we mention in §16.3 for his epidemiological study of silicosis and tuberculosis as occupational hazards and for his definition of new syndromes. Alternatively, we could have begun with James Lind and his discovery of the etiology of scurvy, yet it appeared preferable to give Lind due credit for the first clinical trial in §17.3 rather than here.

## Pellagra

The search for the cause of this disease has many lessons for the epidemiologist for there was much information, demographic or epidemiologic, available that for a long time did not lead to the discovery of the cause. Pellagra was recognized as a distinct disease by Gaspar Roque Francisco Narciso Casal Julian* (1680–1759), but the first published description was given by François Thiéry (ca 1719–1775). Pellagra is characterized by a dermatitis, a "raw beef" appearance of the tongue and psychiatric symptoms ranging from mild confusion to dementia; the dermatitis was known as *mal de la rosa*, the redness being due to light sensitivity of the roughened skin.

The disease was unknown in Europe until the introduction of maize; by 1800, pellagra was common in southern Europe, the Balkans, the Ukraine, and northern Africa, especially Egypt. Its frequency in South Africa rapidly increased after rinderpest in 1897 had killed most of the cattle in some areas, leading to a change in diet from milk foods to maize. Salas (1863) made the following comments: (a) there was a correlation in time between the introduction of maize into the diet and the appearance of pellagra, (b) a specific agent was indicated, (c) only *verdet* (fungus) could explain the production, and (d) such a specific agent could be expected to produce a specific disease. Sambon (1910) gave nonconclusive evidence for the disease being carried by *Simulium* flies (blood sucking); his points were (a) the spring occurrence of new cases and geographical distribution suggested a cause like tertian malaria, (b) *Simulium* and pellagra had the same geographical distribution, (c) pellagra occurred with the consumption of maize, no matter what its origin. F. Lussana and C. Frua believed in 1856 that pellagra was due to an imbalance between "plastic" and "respiratory" material in the maize of various localities. A puzzle was that pellagra did not occur among the Amerinds although maize constituted a high proportion of their diet; Salas, already cited, had the answer. *Verdet* (fungal infection), he thought, was the cause and the disease was unknown in Mexico because the alkali cooking killed the fungus. [For these authors, see Carpenter (1981, pp. 19–24, 28–30, and 13–18, respectively), where English translations of the original articles are given.]

Many articles, available in Carpenter (1981) and elsewhere, show that there were numerous theories on the disease, sometimes quite consistent with the data, which did not lead to a proper appreciation of its cause nor suggest a specific therapy. Joseph Goldberger* (1874–1929) was requested by the United States Public Health Service to investigate the problem. His results are reported in Goldberger (1914, 1916), Goldberger, Waring, and Willets (1915), Goldberger and Wheeler (1920), and Goldberger, Wheeler, and Sydenstricker (1920). In the first of these articles, Goldberger stated that the diet was unbalanced and that the proportion of maize should be reduced, while the amount of fresh animal foods including meats, egg, and milk should be increased; he did not believe that maize was essential to the production of pellagra. In the third article, he showed that pellagra could be prevented in two orphanages by supplementing the diet by milk, eggs, legumes, and meat. A similar success was achieved in the Georgia State Sanitarium. In the second article, Goldberger showed, by feeding and other experiments, that the opinion widely held in 1916 of the transmissibility of pellagra was false. In the fourth article, it was shown that pellagra could be induced in convict volunteers by giving them a diet not unlike that common in the poorer areas of the southern U.S. In the fifth article, the incidence of pellagra was studied in villages and related to economic and social circumstances; from this study

Edgar Sydenstricker (1881–1936) is said to have developed a lasting interest in the importance of community health.

Spies, Cooper, and Blankenhorn (1938) showed that the disease could be cured by the administration of nicotinic acid. Nicotinic acid and its amide, nicotinamide, are usually termed "niacin" in American usage. Elvehjem, Madden, et al. (1937) showed that nicotinamide was identical with the "pellagra preventing factor" isolated from protein free yeast extracts by Goldberger and Sebrell (1930). Goldberger, Wheeler, et al. (1928) first realized in 1915 that a disease, known variously as black tongue in America and *Typhus der Hunde* in Germany might well be canine pellagra, after reading Chittenden's text on human nutrition; Chittenden and Underhill (1917) indeed drew this conclusion.

## 13.2 Genetic Diseases

Some examples of the human inheritance of deleterious genes have been given: In §5.1, polydactyly, brachydactyly, Huntington's chorea, porphyria variegata, and albinism among the dominant autosomal inheritances; in §5.2, alkaptonuria among the recessive autosomal inheritances; and in §5.3, hemophilia among the sex-linked inheritances. In §13.3 we discuss work indicating that the cause of breast cancers is partly genetic, and in §13.4 the causes of skin cancer and melanoma are partly genetic because skin color is an important factor in the production of each disease and is hereditary.

## 13.3 Neoplastic Diseases

Statistical reasoning has played a large part in determining the importance of the neoplasms, in identifying cancers associated with occupations, in researches into etiology, and in testing the efficacy of treatment, among other functions. We commence with some historical notes about the discoveries of some cancers in early years.

### Breast Cancer in Nuns

Bernardino Ramazzini (1633–1714) of Padua, in his *De morbis artificum diatriba* (1713), reported that breast cancer was relatively common in nuns. He attributed the disease to celibacy, but later opinion was that it was due to childlessness and it was largely postmenopausal. Domenico Antonio Rigoni-Stern (1810–1855) in 1842 analyzed the deaths in Verona between 1760 and 1839, taking note of age at death; he found that cancers were five times more common among nuns than among other women, largely accounted for by cancer of the breast. [See Mustacchi (1961) and Scotto and Bailar (1969) for translations of the discussions of Ramazzini and Rigoni-Stern on cancer of the breast.]

## Cancer of the Scrotum in Chimney Sweeps

An early intimation that cancer could have environmental causation was given by Percivall Pott (1714–1788), who observed a high frequency of cancer of the scrotum in chimney sweeps (Pott, 1775). Experimental verification of carcinogens in soot did not come until Passey (1922), although in the meantime it was observed that skin cancer could be induced by a variety of coal tar products acting in a number of industries. Pott's line of reasoning was that such cancers occurred with extreme rarity in other occupations, and that so many of them appeared in such a restricted population suggested a special relation to the industry and the importance of soot, a coal derivative. His work was still an inspiration to workers in this century [see Potter (1963) and Butlin (1892)].

## Breast Cancer (Female)

Penrose, Mackenzie, and Karn (1948) compared the deaths of the relatives of persons dying from breast cancer with the numbers expected, on the basis of them having the same rate as held in the general population of England and Wales. Of 510 patients dying with breast cancer, 25 mothers had died of breast cancer as against an expected 11.12, and 23 sisters had died as against 6.97 expected, there having been 406 mothers and 307 sisters available for consideration. In contrast to these findings, there was no such excess of other forms of malignancies. There was thus a statistically significant and specific familial concentration of breast cancer among the close relatives.

It is clear from the occupational cancers reported in the nineteenth century that there were many carcinogens, either produced or concentrated by the processes of the Industrial Revolution, which were new to human experience. [For occupational cancers of the lung, see Hueper (1966).] In this century, many other striking relations between cancer rates and occupation have been found (see Decouflé, 1982).

## 13.4 Sunlight and Cancers

We may begin with a consideration of the intensity of the sun's radiation as received at sea level (Lancaster, 1956):

The sun radiates electromagnetic energy over a wide band of frequency or wavelength, as if it were a hot, dark body at a temperature of about 5800°K—that is, 5800° on the absolute temperature scale. However, only [since World War II], (Tousey, 1953) has it been possible, by means of rockets carrying spectrophotometers, to obtain measurements of the intensity of the shorter wave-lengths.

The intensity of sunlight is usually measured as that falling on a unit horizontal area. Without absorption by the atmosphere, the intensity would be given by the formula

$$I_\psi = I_0 \cos \psi \qquad [13.4.1]$$

where $I_0$ is the intensity when the sun is at the zenith and $\psi$ is the angle between the direction of sun and the vertical. However, the existence of an atmosphere brings modifications to this formula in several ways.

First, ozone has an absorption band in the ultra-violet end of the spectrum. It may be taken that the ozone in the atmosphere is equivalent to a layer of ozone a quarter of a centimeter thick at normal temperature and pressure, and that most of the ozone is at great heights—the average height of the ozone is perhaps 22 kilometers [approximately 13.6 miles] (Mitra, 1947). Now the absorptive effect of this layer will be greatly enhanced if the sun is not near the zenith, for the effective thickness with the sun at an angle of $\psi$ from the zenith will be proportional to the secant—that is, the reciprocal of the cosine of $\psi$. This is of great importance, because the logarithm of the intensity of light received after passing through an absorptive layer is equal to the logarithm of the initial intensity less a constant number of times the thickness. The practical effect of this is that little ultra-violet light is received directly from the sun when it is more than about 45° from the zenith. Second, there is scattering of light by the molecules of the air and by fine particles. The amount of scattering is proportional to the fourth power of the frequency, so that the light of the ultra-violet end of the spectrum undergoes more scattering than the red. The blue colour of the sky is, indeed, due to this effect. The scattering effect enables some ultra-violet light to pass through the ozone layer vertically or nearly so, and hence complete absorption does not occur even when the sun is low. Indeed, graphs can be drawn showing the amount of ultra-violet light received on a horizontal surface from the sun directly and from the sky for varying heights of the sun. This radiation from the sky is well known to surfers, who may receive an erythema dose without having been directly exposed to much of the direct radiation of the sun. Third, dust in the atmosphere may also absorb ultra-violet light; this is usually unimportant. Fourth, heavy cloud may absorb or reflect upwards much ultra-violet light; but, in general, countries with wet climates do not receive much less ultra-violet light than those with dry climates. (p. 1056)[1]

As a rule, the physical causes of the change in ultraviolet dosage over latitude have been inadequately discussed.

## Nonmelanoma Cancer of the Skin

The chronic effects of sun acting on the skin are senile or actinic elastosis, a fragmentation of the elastic tissue; this is associated with wrinkling, yellowing, freckling, irregular depigmentation, and telangiectatic blood vessels.

Carl Thiersch (1822–1895) in 1865 and Paul Gerson Unna (1850–1929) in 1894 described the high frequency of skin changes leading to cancer in sailors exposed to the sun. Later studies showed that there was a greater risk of nonmelanoma skin cancers among outdoor workers. Hyde (1906) noted that

---

[1] Copyright © *The Medical Journal of Australia* 1956; 1:1082–1087. Reprinted with permission.

such skin cancers were more common among those with fair complexions engaged in outdoor occupations. Over the years, opinions were formed on the statistical evidence that skin cancer occurred in the lightly pigmented races; it therefore appears among the Celtic peoples, especially when they migrate to tropical or semitropical countries; skin cancer appears on parts of the body directly exposed to sunlight and so outdoor workers are at risk. Albinism, xeroderma pigmentosum, and some other genetic diseases are associated with relatively high rates because of defective melanin in the skin.

## Australia and Skin Cancers

Australia has the highest rates of incidence of nonmelanoma cancers of the skin among the countries of the world. Fair-skinned European ethnic groups have settled in this country and worked for over 200 years without developing any tradition of self-protection against sunlight, and even showing a tendency to do away with clothes at work and sport. Sport, especially surfing, became increasingly popular, adding much to the annual dosage. Yet little was done to examine what preventive measures should be taken. Recently, some epidemiological groups have been collaborating with Professor B.K. Armstrong in Perth, Western Australia and Dr. Adele Green in Brisbane, Queensland.

## Animal Models of the Cancers and Melanoma

George William Marshall Findlay (1893–1952) in 1928 showed that animals exposed to ultraviolet radiation developed skin cancers. In human work it is now usually assumed that the wavelengths causing cancer are those that produce sunburning, namely 290 to 320 nm. (A nanometer is one-billionth of a meter and is equal to 10 Angstrom units.) [See Scotto, Fears, and Fraumeni (1982, pp. 256–257) for further experimental work, especially by H. Blum.] Induction of melanoma in animals was first successfully performed by Epstein (1978).

## Melanoma

It is a curious fact that, although the clinicians knew that melanoma tended to occur in the fair-skinned and its incidence was higher in the tropics, they were reluctant to incriminate insolation as the cause of the geographical distribution of melanoma. An exception was McGovern (1952), who pointed out that lesions tended to occur on areas of the skin exposed to sunlight. Lancaster (1956) saw that comparisons could be made by latitude in Australia from entries in the official statistics. Indeed, the crude death rates from melanoma for males during the years 1953–1955 were 28, 13, 17, 15, 8, 7 per million per annum in Queensland, Western Australia, New South Wales, South Australia, Victoria, and Tasmania, respectively, at latitudes 27°, 32°,

34°, 35°, 38°, and 43° south, respectively. Lancaster and Nelson (1957), in a clinical survey, showed that the evidence for ultraviolet light as the carcinogenic factor for melanoma was as good as that for nonmelanotic skin cancers. [See Lee (1982) and Scotto, Fears, and Fraumeni (1982) for further details of the epidemiology.] With larger population, more detailed geographic comparisons can be made in the United States.

## 13.5  Ionizing Radiation

Although it has been customary not to include ultraviolet light in ionizing radiation, opinion now favors its inclusion.

Let us consider first a classic example of lung cancer in uranium miners. The high rates of sickness in the regions of Schneeberg and Joachimsthal had long been known, for example, to Paracelsus in 1531 and to Agricola in 1556. Härting and Hesse (1879), investigating the cause of sickness of the miners, found that 75% of the deaths of 150 miners were due to malignant diseases of the lung, very high rates as compared with the general population (see Lorenz, 1944). This was notable as the first time an internal cancer had been shown to be caused by exposure to an external or physical carcinogen, probably in this case uranium disintegration products. High rates of lung cancer have been reported in other surveys in Czechoslovakia, the United States, and Canada (see Boice and Land, 1982).

There are difficulties in the gathering of the data and in the interpretation of what may be called the dose-response curve for the causation of lung cancer by uranium and its emanations. The dose will be the number of man-days spent by a target group of persons in a mine, and the response will be the incidence of malignant neoplasms in the same group, followed over a number of years. Upton (1975) made the following points: (a) individual doses were not known and the dosage to a group could only be estimated by infrequent measurements in over 2,500 mines; (b) since the principal dosage of ionizing radiation was in the form of radon gas, there was much variation among the mines because of ventilation, ore quality, or other factors; (c) the absorbed radiation could be delivered by aerosols, dust or particulates, or free in the air; (d) actual dosages received might not be estimated well by "working level months"; (e) the dose to individual cells would depend on their type, whether they were overlaid by other cells, the clearance rates from above the cell, and so on; (f) the dose responsible for the neoplasm cannot be distinguished from the later dosage; (g) dosages in mines, not worked for uranium, might not be determined or, in other words, it was difficult to establish a set of controls; (h) it would be difficult to establish the relative risks attached to the various degradation products of uranium, namely gamma, beta, and alpha rays; (i) smoking by the miners would influence the dose-response curve because of the adjuvant properties of tobacco products.

Another important incident was the occurrence of osteogenic sarcoma in the painters of watch dials in Orange, New Jersey (see Martland and Humphries, 1929); the sources were uranium and other radionuclides.

Natural exposure is only rarely of importance, where there is an outcropping of rock containing dangerous minerals. There appears to be doubt about the effects of cosmic radiation received in travel in high-flying airplanes.

There were occupational hazards in the early days of Roentgen ray diagnosis. Appropriate changes in the machines and shielding of workers and patients have rendered these hazards unimportant. [See Boice and Land (1982) and Lancaster (1990) for references to other hazards in diagnosis and therapy.]

## 13.6  Tobacco

Much work has been done in the years since World War II on the importance of tobacco as an air pollutant.

Tobacco is a product of species of *Nicotiana*, usually *tabacum* but also *rustica*, the latter grown in India and the Soviet Union. Its introduction into European society met with some objections and King James I of England issued a condemnation of its use in 1604. John Hill* (*ca* 1707–1775) believed that the appearance of cancer of the nose in two patients was due to the taking of snuff. Redmond (1970) believes that Hill has a clear priority for pointing to tobacco or other carcinogen as a cause of cancer. His mode of reasoning would resemble that of Percivall Pott in §13.3. Other such relations of tobacco with cancers were reported in the nineteenth century, but techniques were not sufficiently developed to permit proper tests of the hypothesis.

Although our interest in tobacco in this chapter is because of its etiological importance for cancer of the lung, we must note the statement of the United States Surgeon-General (1979, §1, p. 12):

Although mortality ratios are particularly high among cigarette smokers for such diseases as lung cancer, chronic obstructive lung disease, and cancer of the larynx, coronary heart disease is the chief contributor to the excess mortality among cigarette smokers. Lung cancer and chronic obstructive lung disease, in that order, follow after coronary heart disease in accounting for the excess mortality. Pipe and cigar smoking are associated with elevated mortality ratios for cancers of the upper respiratory tract, including cancer of the oral cavity, the larynx, and the esophagus.

[For a bibliography, see United States Office on Smoking and Health (1983).]

## 13.7  Tobacco and Lung Cancer

1. *Clinical intimations.* Clinicians were beginning to realize that cancer of the lung was on the increase before World War II. Thus Dr. Cotter Harvey, an internist at the Royal Prince Alfred Hospital, Sydney, searched

the records for fatalities from cancer of the lung; he found 7, 7, 11, 27, and 28 deaths assigned to lung cancer in the five 5-year epochs commencing at 1910 in a hospital of about 400 beds; in the latest epoch, 1930 to 1934, he found that cancer of the lung constituted 6% to 8% of all cancers in the hospital (Harvey, 1936). Symptoms of mediastinal tumor were often present.

2. *Retrospective clinical surveys.* Others had noticed similar increases. Indeed, Franz Hermann Mueller (1939) at the University Polyclinic at Köln, Germany, carried out a clinical survey; he found that there were 3 (14) nonsmokers, 27 (41) light smokers, and 56 (31) heavy smokers out of 86 (86) patients with (without) cancer of the lung. It is evident that there is a correlation between smoking and cancer of the lung; there are indeed many more heavy smokers among those with cancer of the lung than the controls (that is, without cancer of the lung).

William Richard Shaboe Doll (b. 1912) and A. Bradford Hill, surveyed the results of a questionnaire on 649 cases of cancer of the lung and an equal number of controls (Doll and Hill, 1950). Terms were defined carefully, for example, smoker and nonsmoker. A repeat questionnaire showed that those questioned were consistent in their answers. The respondents were classified into five classes by the amount smoked per diem. For the males, a $\chi^2$ of 36.95 with 4 degrees of freedom was obtained from the $2 \times 5$ table. It is an unfortunate fact that they gave data showing that "inhaling" and the risk of lung cancer were not positively correlated. Later work has suggested that "inhalation" needs to be carefully defined if such comparisons as inhalers versus others are to be made. English workers have tended to overstress the paper of Doll and Hill (1950) at the expense of the American authors, for example, Wynder and Graham (1950) and Levin, Goldstein, and Gerhardt (1950). Items 4 through 10 are in the spirit of Wynder (1972), and the remainder are our comments on Fisher (1959) (see also Wynder, 1955).

3. *Repetition of the Doll and Hill style surveys.* Fisher (1959) noted that there had been at least 19 such surveys and that these added nothing to the evidence. It might well be so, but for public health purposes it would be useful to show that the tobacco–lung cancer problem may exist in the country surveyed. The country concerned might have a different degree of urbanization, as in item 6, below; there may be the possibility of making comparisons between subpopulations as in item 7.

4. *Prospective studies.* Two large surveys have been carried out by Hammond (1966, 1972) for a voluntary sample of the total population in the United States, 1 million persons for 6 years, and by the Royal College of Physicians (1971) for a voluntary sample of physicians, 34,000 for 20 years. These surveys gave a strikingly favorable report for the tobacco–lung cancer hypothesis. Note that the data on religious groups in item 7 are really prospective studies.

5. *Official statistics.* Questions have often been raised about whether the increases in the death rates from cancer of the lung, as revealed by the official statistics, are not spurious; but no substantial evidence has ever been demonstrated that deaths have been transferred over to carcinoma of the lung from gastric or other carcinoma. The reality of the increase could be tested, as by Kennaway and Waller (1953/1954), for the existence of an increase in the rates commencing at young ages, that is, by a cohort analysis.

6. *Urban–rural differences.* There is evidence that urban death rates from lung cancer are, perhaps, 10% higher than the rural rates.

7. *Comparison between subpopulations.* Wynder (1972) states that the lung cancer sex ratio is correlated with long-term smoking habits. Lung cancer is rare among nonsmoking Seventh Day Adventists [see Wynder, Lemon, and Bross (1959) and Phillips, Kuzma, and Lotz (1980)]. Other such surveys have been made on religious groups, in which the use of tobacco has been proscribed [see Lyon, Gardner, and West (1980) for comparison of Mormons in Utah with other religious groups].

8. *Refinements of the variables.* In surveys with matched controls, the comparison between the various smoking classes can be made more precise by considering the Kreyberg Group I (Kreyberg, 1962), an extremely rare cancer in nonsmokers but the most numerous type seen in heavy cigarette smokers.

9. *Industrial hazards.* Substantial risks occur in workers in the uranium, chromate, asbestos, and other industries. Cigarette smoking has been shown to be synergistic with some hazards.

10. *Animal experiments and chemical analysis.* By forced smoking methods on animals, it has been shown that there are carcinogenic substances in cigarette smoke. This has been confirmed by chemical methods. Some of this work was published after 1959 and cannot be used against the arguments of Fisher (1959).

11. *Fisher's objections.* R.A. Fisher felt for some years that Doll and Hill (1950) had not made a satisfactory case against cigarette smoking as a cause of lung cancer. He gathered together his letters to journals on this topic and added some philosophizing. Echoes of his work can still be heard. The following items examine the various lines of argument he used (Fisher, 1959, pp. 11–25).

12. *Doctrine of fear.* Fisher regards it as important not to implant fear in the minds of the general populace, since the hypothesis has not yet been verified. Unfortunately it is difficult to implant fear, as is obvious from the reaction of the general populace to the lung cancer and AIDS problems.

13. *The use of precise statistical methodology.* Fisher (1959, p. 13) uses the forecasting of the appearance of the next card as an easily understood example of the necessity of randomization; in agricultural trials, one avoids conscious choice, with its biases, by the use of random sampling numbers. Errors have been made in interpreting the fall of cards by not using

methods approved by the agricultural statisticians. Fisher says that correlation is not causation; proper randomization must be used. He absolves Doll and Hill (1950) from blame in not doing a prospective survey. Note that there is no criticism on the inhalation table; he needs this result for his own purposes.

14. *Chemical difficulties.* Fisher (1959) counters suggestions by Doll and Hill (1950) and others that more smoking leads to a greater risk of lung cancer by saying that, in smoking by cigarette, cigar, or pipe, tobacco is burned; however, one might add that the same chemical changes at different temperatures may produce a different mix of end products. Already in the 1920s it was known that there were carcinogens in soot, but the early investigations on the tars produced in smoking had been inconclusive. Certainly by 1959, there was a good deal known about the tars; moreover, the temperature of combustion was critical and was higher in the cigarette than in pipe or cigar.

15. *Difficulties of classifying smoking classes.* One can only be surprised at this point; errors of classification usually act the other way, making the null hypothesis more difficult to overthrow.

16. *Inhalation.* It has always been our belief that Doll and Hill (1950) erred in bringing in a discussion on inhalation without making suitable criteria for detecting inhalers. Fisher makes much of this oversight. He thinks that if further research were done, a negative correlation between the amount of tobacco smoked and inhalation would destroy the evidence for positive association between smoking and lung cancer.

It is impossible to let this incident or story pass without comment. First, we have initially the potential damage caused by Doll and Hill (1950) raising the question of inhalation; little did they realize how a hostile critic could magnify the importance of such a side issue. Second, we have Fisher's antigovernment predisposition; he disliked the interference of a government in almost any field and allowed his judgment to be distorted. As he well knew, the final judgment should be made not by repeated references to such side issues as inhalation but by chemical analyses of tobacco, analyzing the effect of the temperature at which the tobacco was burned on the concentrations of carcinogens, by animal experiments, by comparisons between subpopulations such as religious groups, and much other work that had become available before 1959. Third, probabilistic arguments fare badly when used in an "adversary" situation, and the truth can be distorted as in Fisher (1959, pp. 44–45). Fourth, we must criticize some of Fisher's disciples, as Gamow would say; surely they accepted too uncritically the views of the master! [For a contrary opinion see Bennett (1991)].

## 13.8 Etiology and Statistics

Long after a disease has been established as an entity by statistical methods and found to be associated with certain causative factors, it is sometimes denied that the association is genuine and that the case has been proved. A good example is the way public debate is carried on about the cause of the epidemic of lung cancer (see §13.7).

With the infective diseases, it was the proof of contagion that was important; we may recall the statistical proof of contagion of puerperal sepsis by A. Gordon, T. Nunneley, O.W. Holmes, and others in §9.1, in the years 1795 to 1855. The proof remained statistical until the development of bacteriology, when L. Pasteur and others found the causal organisms. Belief in the contagion hypothesis was thus greatly increased by Pasteur's (1878) demonstration of the streptococcus; thus the mode of infection was identified. Later, in the 1930s, with the typing of the streptococci, there was even more convincing evidence, when it could be shown that streptococci of a given Lancefield type could be identified in the cases of an epidemic, whereas they were not usually found in control patients. Once again the argument is statistical but it was now almost impossible to believe that the particular microbe was not the cause of the disease. Similarly, when bacteriological methods could be applied, it was found that diphtheria bacilli were present in the throats of those having certain signs and symptoms, especially those associated with toxic paralyses. The disease was now defined as having those same signs and symptoms but requiring that the causal agent should be found in each case. There could now be no reference to merely statistical proof of the cause of the disease; that the disease was associated with the causative organism, *Corynebacterium diphtheriae*, was now a tautology. Other diseases could be cited such as amebiasis due to *Entamoeba histolytica*, typhoid fever due to *Salmonella typhi*, and so on. When the cause is not microbial, there is not the same ready consensus of opinion.

# 14
# Metrical Characterizations of Individuals and Populations

## 14.1 Human Growth

### Proportions

The great painters and sculptors of the Renaissance would be expected to have made observations on the human body (that is, in anthropometry), but they had greater interest in form than in absolute size. Leone Battista Alberti* (1404–1472) seems to have been the first since ancient times to measure people, according to Tanner (1981); but he was more interested in proportions than in size; his ideal man should be 6 feet tall, the upper edge of his pelvic symphysis should be 3 feet from the ground, his external auditory meatus $5\frac{1}{10}$ feet; the distance between his hip joints $1\frac{1}{10}$ feet. Leonardo da Vinci* (1452–1519) carried on the traditions begun by Alberti but made no measurements. He introduced the "golden section" as the name for the "divine proportion" of his friend Luca Pacioli* (*ca* 1445–*ca* 1517). Albrecht Dürer* (1471–1528), after a visit to Pacioli to learn of perspective, returned to write a book on the geometry of drawing figures. Johann Sigismund Elsholtz (1623–1688) introduced the term *anthropometry* and gave instructions for the use of a measuring rod but seems not to have amassed any data. Gérard Audran (1640–1703) carried out measurements on the statues of antiquity. Perhaps the work of the artists, with their theoretical emphasis on proportions, should be assigned to the prehistory of anthropometry.

### Embryo and Fetus

Modern anthropometry is especially concerned with growth. According to Tanner (1981), Georges Louis le Clerc de Buffon* (1707–1788), assisted perhaps by Louis Jean Marie Daubenton* (1716–1800), was the first to measure embryos and fetuses (see Gilbert, 1991).

### Birth Weights and Lengths

Obstetricians appear to have had fanciful views on the weights of infants at birth, for example, 12 lbs or more, 16 or 17 lbs, and other weights clearly not obtained by actual measurement. Johann Georg Roederer (1726–1763) in 1754 made accurate measurements on the weights and lengths of babies in 232 confinements between 1751 and 1762 with a view to estimating the degree of immaturity in the baby; weights ranged from $5\frac{1}{2}$ lbs to just over 8 lbs (2.5 to 4.0 kg approximately) and lengths from 18 inches (43 cm) to 23 inches (55 cm); these results are quite in line with modern estimates.

## Growth in Stature

Johann Gottfried Schadow (1764–1850) in 1834 to 1835 measured the heights of babies and children and claimed, Tanner (1981) says correctly, that he was the first to do so. He gives the tables of proportions at 0, 4, 8, 12, 18, 24, 30, 48, and 54 months and at years 5, 6, 7, ..., 15, and 17; his illustrations show a gradually decreasing velocity in the rate of growth up to about 11 years, then it jumps to 3 inches per year for the next three years. Angerstein (1865) made measurements of 1,000 to 1,200 boys over a period of five years; he thought that the maximum growth rate occurred at age 13.0 years.

## Tables of Heights

Christian Friedrich Jampert (1727–1758) in 1754 is credited by Tanner (1981) as the first to publish a real table of measurements of the growth of the human. He has heights for each age of life up to 25, which were selected as most typical of the age that he had in his sample.

## Heights by Age

The first longitudinal study of height through life was made at the request of Buffon by Philibert Guéneau de Montbeillard (1720–1785) on his son over the years 1759 to 1777 at six-month intervals, given as Fig. 5.3 on p. 104 of Tanner (1981), first as a cumulative curve, the graph of height, and second as a graph of the increments of height (see Buffon, 1774–1789). In this second graph, the process of growth, the height velocity, is more evident; it is highest in infancy, falls rather rapidly to about the age of 4 years and then more slowly to attain a minimum at age 11 years; the velocity then rises to a maximum at age $14\frac{1}{2}$ years and falls sharply, becoming negligible after 18 years (see also Thompson, 1942, pp. 99–100). Of course, it was known that persons differed in the values of their heights and height velocities. In modern times it has become customary to draw accompanying lines at arbitrary percentile points to indicate the spread of the observations at the various ages. An interesting graph is given by Tanner (1981) showing the growth of students at the Karlsschule in Stuttgart from 1772 to 1794, the records of whom were only recently discovered by Uhland (1953). [See also Antonelli (1985), Bogin (1988), and Rosenbaum (1988).]

Systematic sampling errors may occur, as detailed by Davenport (1926). A table of increments of growth of over 2 million persons by strict year of age was available to Thorburn Brailsford Robertson (1884–1930) in his study of human growth and senescence. He fitted a logistic curve to the annual increments and maintained that he had discovered a spurt of growth between infancy and adolescence. Davenport (1926) notes:

There are certain features of the secondary maximum of Fig. 4 that arrest attention—principally that it affects only the sixth year; there is no premonition of it in the fifth year, nor after-effect in the seventh ... The reason for the break is apparently that 6 years is the first of the school years. Most of the measurements have been made on school children. Not all children go to school in their sixth year. Those who go may well be selected for a development that is above the average. (pp. 213–214)[1]

## Age at Puberty

The growth curves of heights and weights are greatly influenced by the onset of puberty. For females, Hippolyt Guarinoni (1571–1654) and Jampert agree that the menarche falls between 12 and 20 years; it appears earlier in well-nourished girls living a sedentary life and later in peasant girls who do hard work and have scanty food; the early maturers only cease menstruating in their forty-ninth year, whereas the later maturers may cease in their thirty-eighth year. According to Osiander (1795), Marseille girls mature at 12, contrasting with Strasbourg girls at 14. For males, an indirect way of measuring the age at puberty is to note the change in voice; S.F. Daw (1970) has examined the records of the Thomasschule at Leipzig in the years of employment there of Johann Sebastian Bach (1685–1750); all the singers were males, most boarding at the school. The boys with transitional (alto) voices were of ages $16\frac{1}{2}$ to 17 years, compared with modern United Kingdom averages of 13 to $13\frac{1}{2}$ years. During the War of Austrian Succession (1740–1748), the average age of the altos rose by over half a year, puberty probably being delayed by the poverty and undernutrition due to the effects of the war.

## Pelvic Measurements

Gustav Adolph Michaelis (1798–1848), during his work as Director of the Kiel University Hospital of Gynecology in 1840 to 1848, carried out routine pelvic measurements of every pregnant woman entering the hospital. His classification of the bony pelvis is based on the five measuring points that are still used today according to Semm and Weichert-von Hassel (1985); he was able to use also an external feature, the so-called Michaelis rhomboid, as a prognostic sign for labor.

[See Eveleth and Tanner (1991) for worldwide variation in growth.]

---

[1] Copyright © *American Journal of Physical Anthropology* 1926; 9:205–226. Reprinted by permission of Wiley-Liss, a division of John Wiley and Sons, Inc.

## 14.2 Mathematics of Growth

There have been attempts to give mathematical formulations to the phenomena of growth, for example by Thompson (1942) on length of the whole body or of its appendages, considered either singly or in sets of lengths whereby shape would also be considered. In general, the length of the body, $l = l(x)$ say, is a never decreasing function of the age $x$ over a span of the life history. Such functions, $l(x)$, are, except for a multiplier, $a$, (statistical) distribution functions. Moreover, weighted sums of such functions have the same property; but we must avoid considering these sums as that of a single individual; for example, taking heights of boys, the time of maximal growth of individuals is much less than that of the mean (see Merrill, 1931).

### Allometry

In a mathematical model of growth, $F$, the growth curve can be taken to be absolutely continuous or differentiable a certain number of times. It is usual to set

$$F' \equiv \frac{\mathrm{d}}{\mathrm{d}x} F(x) = f \equiv f(x) \tag{14.2.1}$$

so that the curve of the rate of growth, except for a multiplier (equal to the final value), is a density function in statistical terms. Following Thompson (1942), workers have attempted to obtain an estimate of $F$, the growth curve, or of $f$, the rate of growth, as simple functions. Alternatively, a simple formula is often assumed, for example

$$y = bx^\alpha \tag{14.2.2}$$

or equivalently

$$\log y = \log b + \alpha \log x. \tag{14.2.3}$$

In 1927, Paul Georges Teissier* (1900–1972) began his studies with work on similitude in Teissier and Lambert (1927) and soon decided that quantitative methods were necessary and so adopted biometric and statistical methods. He and Julian Sorel Huxley (1887–1975) agreed in Teissier and Huxley (1936) that equation (14.2.1) should be called the "allometric" equation (see also Reeve and Huxley, 1945).

### Estimation of the Parameters of Growth

We have seen that

$$f(\text{Mean of Observations}) = \text{Mean of the } f(x_j) \tag{14.2.4}$$

holds only in special cases. If there is not homogeneity among the animals, the growth equation of an individual cannot be deduced from the growth equation of the means of the individuals. Nevertheless, such deductions are often made as an act of faith. See Clark and Medawar (1945), in which Peter Brian Medawar (1915–1987) gives a table by Gray (1929) showing how difficult it is to obtain a unique answer as to the form of the growth curves because the curves for functions $g(x)$ and $h(x)$ may resemble each other over a long range of the time but not for every point. The models may not give specific answers to all questions yet they help to give a general idea.

Comparisons between rates of growth of the whole animal against an organ, or between different organs, can be made by plotting logarithm of, say, total length against length of a femur, for the differential coefficients are of the form $y^{-1} \, dy/dt$. Two such curves are compared in the biography of Teissier by C. Petit (DSB, XVIII, pp. 901–904). Teissier went on to carry out a survey that gives a mathematical expression for allelic and genotypic frequencies through generations. Teissier also investigated the importance of polymorphisms for changes of genetic frequencies.

## Similitude

To show the relevance of mathematical thinking to biology, we mention the principle of similitude, now an unfashionable research topic in general biology although still always in the mind of the civil or mechanical engineer or biogeographer. There is a large literature on the principle, see for example Chapter 2 of Thompson (1942), with Galilei (1638) as the first great classic (see Medawar, 1962). We give some very simple examples of the theory. Two objects are said to be geometrically similar if the distance between two points in the one bears a constant ratio to the distance between the two corresponding points in the other.

We follow Thompson (1942) in taking a simple example; for a ball, the radius, area, and volume are $r$, $4\pi r^2$, and $\frac{4}{3}\pi r^3$, respectively. Let us study the changes in the other variables with powers of $r$; let $L$ be a linear expression, $S$ a quadratic expression, and $V$ a cubic expression, then $S$ varies as $L^2$ and $V$ as $L^3$.

$$S = kL^2, \qquad V = k'L^3 \tag{14.2.5}$$

where $k$ and $k'$ are suitable constants.

$$\frac{V}{S} = \frac{k'L^3}{kL^2} = KL \tag{14.2.6}$$

where $K = k'/k$.

Similarly, $S = k''L^2$ for a constant, $k''$. From such equations, we deduce that if we have two bodies, geometrically similar, and the length of the second is double that of the first, then its surface is quadrupled and its volume is multiplied by 8. This is the simplest application of the principle of similitude. Another example is when two similar weights are suspended by two similar wires

$$\frac{\text{Force}}{\text{Area}} = \frac{\text{Weight}}{\text{Cross-Sectional Area of Wires}}$$

$$= \frac{k'L^3}{kL^2} = KL \tag{14.2.7}$$

and so the weight per cross-sectional area becomes greater as $L$ increases.

There are many biological consequences of such computations. With fixed materials in its composition, a spider cannot be increased indefinitely in size; an ox, several times the length of a real ox, could not exist for the same reason that, of two similar bridges made of the same metal, the smaller bridge is more stable.

George-Louis Lesage* (1724–1803) deduced that the larger ratio of surface to mass in a small animal would lead to dehydration if its skin possessed the same porosity as a human skin. Big men do not win marathons. Trees cannot grow nor towers be built, dependent on the materials used, beyond certain limits.

## Bergmann's Rule

Carl Georg Lukas Christian Bergmann (1814–1865), the son of an important Hanoverian official, graduated from Göttingen and served there as Privatdozent and extraordinary professor; he then accepted a call as full professor to the less prestigious university of Rostock. In 1852, Bergmann and Karl Georg Friedrich Rudolf Leuckart* (1822–1890) wrote a text on the anatomical-physiological survey of the animal kingdom. Incidentally, it was Leuckart who first coined the word *polymorphism*.

We may follow Coleman (1979) in noting the distribution of homoiothermic, the so-called warm-blooded, animals particularly in the temperate to Arctic regions of Europe. Since these animals maintain constant temperature against a cooler or colder environment, they must be able to control the passage of heat either through their skins or by their respiratory system. *Bergmann's rule* states that within a given group of animals, the larger forms tend to occupy the cooler portion of the animals' overall range, the smaller forms the warmer portion. For since the amount of energy goes up as the weight of the animal, it increases as $L^3$, and the loss of heat as the surface area, and so as $L^2$, larger animals will be at an advantage over smaller animals in the colder zones. Coleman (1979) cites an example from Allee and Schmidt (1951, pp. 462–469) where the American mole has average skull lengths of 30.8 mm in Florida, 31.8 mm in North Carolina, 34.1 mm in Maryland, and 35.5 mm in Connecticut, giving a total increase of 11%. Nevertheless, other examples are available where the rule does not work. The theme has become of importance as an ecogeographical generalization. Moreover, since it could only be properly discussed by a consideration of the whole animal, and even species, it served against an excessive degree of reductionism in zoology (see §2.2), as practised by Ludwig and others. Bergmann's rule was later found relevant by Charles Robert Richet* (1850–1935).

## Surface Area

Dubois and Dubois (1916) gave a formula for the surface area of the body,

$$S = \text{Surface Area} = W^{0.425} \times H^{0.725} \times 0.007184 \qquad (14.2.8)$$

where $W$ is the weight in kilograms and $H$ is the height in centimetres. In a specific case, $H$ is 173 cm and $W = 70$ kg. Note that the dimensions are consistent for $3 \times 0.425 + 1 \times 0.725 = 2$.

There are now electronic means for computing the surface area directly.

[For the mechanics of mammalian terrestrial locomotion see Biewener (1990). For the interpretation of animal form, see Coleman (1967) and Gould (1970).]

## 14.3 Craniology and Craniometry

Franz Joseph Gall* (1758–1828) was born at Tiefenbronn near Pforzheim in the Grand-Duchy of Baden. After primary education at schools in Baden and Bruchsal, he began his medical course at Strasbourg in 1777, where he came under the influence of Jean Hermann (1738–1800), an anatomist and naturalist; in 1781, he moved to Vienna where in 1785 he graduated with an M.D. degree. In Vienna, he had a large and successful practice, with many eminent patients. In 1791, he wrote a "philosophy of medicine." His theses were, according to Ackerknecht and Vallois (1956), (a) moral and intellectual qualities are innate (that is, given by the Creator of the World); (b) the functioning of the qualities depends on organic supports so that no faculty manifests itself without a material condition; (c) the brain is the organ of all the faculties, of all the tendencies, and of all feelings; and (d) the brain is composed of as many organs as there are faculties, tendencies, and feelings. Gall developed these ideas in public lectures and demonstrations but was forbidden to continue them by a personal letter from the Emperor, as they were considered materialistic and so subversive to State and Church. Gall was joined by Johann Christian Spurzheim* (1776–1832) as a collaborator in 1800. Gall and Spurzheim toured the German states, Switzerland, Scandinavia, and France. Gall arrived in Paris in 1807 and died there in 1828, having become a French citizen in 1819.

Gall and Spurzheim made many advances in neuroanatomy. The significance of some of their discoveries was not appreciated for many years. According to Young (*DSB*, V, pp. 250–256) Gall conceived of the nervous system as a "hierarchically ordered series of separate but interrelated ganglia designed on a uniform plan." He made numerous discoveries such as the origins of the first eight cranial nerves. Marie-Jean-Pierre Flourens* (1794–1867) called Gall the "author of the true anatomy of the brain." He originated the modern doctrine of cerebral localization of functions, established psychology as a biological science and contributed to the naturalistic approach to the study of man.

Phrenology came to overshadow Gall's more solid scientific work. If the brain consisted of a number of ganglia joined together by nervous fibres, then perhaps there would be merit in determining the sizes of the ganglia; in the absence of direct observation it could be supposed that the size of the ganglia was reflected in the size of the "bumps" on the skull. One could begin from the reverse direction. When a bump is discovered, what faculty is represented by the presumed underlying ganglion? Spurzheim elaborated on the theme but unfortunately some of the work of these founders was uncritical and localization of the "faculties" was finally rejected as a scientific theory. However, an interest in craniology had been generated. It is often said that the only one of Gall's localizations to be confirmed is now known as Broca's convolution, in the left inferior frontal gyrus, a disease of which is associated with aphasia.

[See also Temkin (1953); for biodiversity, see §3.1.]

## 14.4  Adolphe Quetelet

Lambert-Adolphe-Jacques Quetelet* (1796–1874) was born in Ghent, Belgium and died in Brussels. As a student, his chief interest was in mathematics but after a visit to Paris at the request of the government to study the problems of an astronomical observatory, his mind turned toward statistics.

In an early application of the methods learned in France, Quetelet (1827) criticized past estimates of the population of Belgium and proposed adopting Laplace's method of counting births throughout the kingdom in a year, then multiplying the birth number by a factor obtained by sampling chosen districts for both births and deaths and so obtaining an average of births per year to population. He finally recommended a complete census to take place in 1829 as a result of criticism by Baron de Keverberg, who suggested that lack of homogeneity made the Laplace method impracticable. Justification of its use would have been necessary, perhaps, by preliminary sample surveys followed by appropriate planning; but these procedures were matters for the future.

In the years 1825–1835, Quetelet collected many sets of data on a great variety of topics. Methods of estimation characterizing a population had been evolved by others. At first a subsample that appeared representative was chosen, and then a few measurements were made on its members. More details were required and so, under the influence of Quetelet, distributions were recorded, and the arithmetic means were computed. If populations are to be compared, the arithmetic mean is usually the statistic or parameter of choice.

Quetelet published a number of papers on social statistics. There had been no tradition of annotating the official social statistics. In this work, Quetelet carefully planned the tables, thought out the meaning of the figures, and attempted to relate them to causes and influences. His work on the construction of tables of observed distributions was widely followed, although he did not develop a theory of goodness of fit to a given distribution nor make any formal test of significance between distributions.

Quetelet's work in constructing tables of observed distributions was followed by many, including Henry Thomas Buckle (1821–1862) and Florence Nightingale in England, and Adolph Heinrich Gotthilf Wagner (1835–1917) in Germany.

Quetelet (1835/1842) was impressed with the regularity of the numbers of crimes as recorded each year. One supposes in our time that such figures result from a mixture of Poisson processes and only small groups of persons to be counted are involved in any particular incident. Quetelet supposed that the regularity was due to some special causes and that they could be eradicated by administrative or other action.

Quetelet's use of mathematics was less valuable. Quetelet (1845), considering anthropometry at a time when he had few predecessors, introduced the expression *homme moyen* for the first time; he believed that every nation had its *average man* or *type* and that even the whole world had its type (see §19.4).

## 14.5  Correlation in Biology

Originally "correlation" was a biological principle. Georges Cuvier* (1769–1832) made two morphological generalizations: first, the principle of the correlation of parts, and second, the principle of the subordination of characters. The first states that each organ of the body is functionally related to every other organ; this is justified by the consideration that such a relation is necessary for the efficient working of the body; an example given is that the lengths of the humerus and the femur must vary together the same way within and between species. Cuvier believed that from an examination of a small piece of bone, one could reconstruct the whole animal. In his time the phyla had not yet been determined, but the principle showed why there could be no close relations between such differing groups of animals. The second of Cuvier's principles enabled him to establish rules to recognize and rank the higher taxa of animals [see Coleman (1964); see also §19.3].

In his textbook on biometrics and variability, the Russian geneticist Juriĭ Aleksandrovič Filipčenko* (1882–1930) asserted that traits varying within a species are different in kind from those characterizing the higher taxa and moreover the latter exhibit "less variation" and "appear significantly earlier during individual development" (Filipčenko, 1923).

# 14.6 Characterizations of Human Populations

By a characterization of a population we mean a description of it by the frequencies of the various characters or properties of interest. We consider first the early attempts to classify mankind.

According to (M.) G. Retzius (1909), Carl Linnaeus (von Linné)* (1707–1778) in the various editions of *Systema Naturae* was the first to classify man; his races were *Americanus rufus, Europaeus alba, Asiaticus luridus, Afer niger,* and a miscellaneous category of varieties not seen by him; he acknowledged only one variety in Europe. Johann Friedrich Blumenbach* (1752–1840) described European, Mongolian, Ethiopian, American, and Malayan varieties; he transferred Lapps and Finns from the European variety to the Mongolian. He gave many details of head shapes, for example, and believed that there was uniformity within the varieties, although he gave no head measurements. It is a remarkable fact that Blumenbach rarely made a measurement and so did not use the ratios between length and breadth of the skull.

A change in outlook came with Anders Adolf Retzius* (1796–1860), who studied prehistoric graves in Scandinavia and found great variety in the shapes of the skulls. In 1840, Anders Retzius defined an index 1,000:$x$, where $x$ = maximum length of cranium/maximum breadth of cranium. It has since been used in the form.

$$\text{Cephalic Index} = \frac{\text{Maximum Width of Cranium}}{\text{Maximum Length of Cranium}} \times 1000,$$

with values found to be 773 in Swedes and 888 in Slavs [see A. Retzius (1845) and (M.) G. Retzius (1909)]. Thus Retzius gave the impetus to the measurements of the cranium. Both brachycephaly and dolichocephaly have been found all over the world, except in Africa. Anders Retzius thus disposed of the theory of uniformity of the Europeans.

Pierre Paul Broca* (1824–1880), a versatile researcher in anatomy, surgery, pathology, and anthropology, had followed several lines initiated by Gall. In 1847, Broca served on a commission to report on excavations in the cemetery of the Celestins in Paris. This work was congenial to him because of his broad interests in mathematics and anatomy, especially craniometry, for the more accurate study of which he is said to have invented at least 27 instruments.

Magnus Gustav Retzius* (1842–1919) carried on the family traditions in the study of anatomy and anthropology including a correlation of talent and brain structure. (M.) G. Retzius (1909) states that in the classification of Europeans five principal characters were used:

1. length and breadth of the head and the cephalic index,
2. the form of the face,
3. stature or body length,

4. head hair color,
5. color of the iris.

He says that the following varieties may be admitted: (a) Northern European (dolichocephalic, blue eyed, tall) variety, that is, nordic; (b) Middle European (brachycephalic, dark haired, dark eyed, short statured) or alpine variety; (c) Mediterranean (dolichocephalic, dark eyed, short statured) variety.

Deniker (1900/1924) points out that the cephalic index does not give a sharp answer to the question of grouping as there is overlap between the varieties; he gives a detailed explanation.

The classification given above for Europe has been extended to other regions, and no system of characters has threatened to replace it. There are problems. Usually not enough observations are available and the variance of the sample has not been estimated. Much was hoped of the blood groups but there are only two loci effective and the different loci do not give consistent results because of different environmental stresses.

Laughlin and Osborne (1967) give a collection of excellent popularizing accounts in their *Human Variation and Origins* of the works of anthropology, genetics, and so on. Oxnard (1984) gives a discussion, largely anatomical, of the place of man among the primates. Young (1971) gives a broader view of man in nature, with many references. Variation is a leading theme of Mayr (1982). For biodiversity, see Ehrlich and Wilson (1991) and other authors in the same issue of *Science,* and also a letter from van Vark and Bilsborough (1991) who used multivariate analysis to show that more variation occurs in prehistoric West Asian skulls than in the whole modern world, citing Howells (1973) and van Vark and Howells (1984) for methodology. [See also Hillis and Moritz (1990) for molecular systematics and the immunological approach of G.H.F. Nuttall in §3.1.]

[See also J. Bertillon (1883) for the life of his father, Louis Adolphe Bertillon (1821–1883).]

## 14.7  Identification of the Individual

There is medicolegal interest in the identification of individuals, for example, criminals, heirs, bodies, etc. The first systematization of the methods was created in 1879 by Alphonse Bertillon (1853–1914), chief of the Judicial Identification Service of France (see S. Bertillon, 1941). The system (of *bertillonage* as it came to be called) made use of (a) bodily measurements such as height, weight, various head measurements, span of arms, etc., (b) descriptive (qualitative) features such as color of eye and hair, form of ear; (c) special bodily marks such as moles, scars, vaccination scars, tattoos; and (d) photographs. The general principle was to classify by (a) and (b) and then to attempt to identify the individual by the more specialized tests of (c) and (d). In anthropometric work it was assumed that the relations between variables could be measured by the correlation coefficient. Bertillon indeed claimed that the chosen variables were practically independent; but Galton was not of that opinion; he was worried that the slope of the regression line depended on the scale of the variables. In 1889, he realized that each character should be measured in its own variability with, originally, the semi-interquartile range as the unit and, after Pearson's entry into the field, the standard deviation, as it would later be called. Further, since the characters studied were continuous, there were always likely to be misclassifications of measurements near a dividing line between classes.

Fingerprints had been used in India by William James Herschel (1833–1917) to detect impersonation and false repudiation of signatures. He observed that there was no material change in the fingerprint after twenty years of observation. Henry Faulds (1843–1930), a Scottish medical missionary in Japan, had his interest directed to the subject by impressions on ancient pottery. He studied prints on recent pottery and made inked "nature-prints" of many Japanese and a few English persons. He concluded that such studies might well prove useful in comparative primatology, ethnology, archeology, and the demonstration of blood relationship. (See the introduction of H. Cummins to the 1965 reprint of Galton's *Finger Prints*.)

Francis Galton had been invited to give a lecture on *bertillonage* to the Royal Institution. He read the available literature and visited Bertillon in Paris to get details. He recalled that "thumb marks" had been used for identification and so wrote to *Nature* asking for further information. He included remarks on fingerprints in his lecture and went on in later years to devise a classification. Galton (1908) remarks that he has established that (a) the pattern of fingerprint is constant throughout life, (b) the number of distinguishable patterns is very large, and (c) it is possible to "lexiconise" the fingerprints and so to produce a suitable dictionary or its equivalent against which unknowns can be tested.

*Bertillonage* did not work well in India because accuracy could not be guaranteed with untrained personnel. Galton (1908) remarks that the correlations between variables measured are not negliglible and that the number of distinguishable patterns is thereby diminished.

The various blood group systems have been used for identification. Few of them have frequencies near to 0.5, which is the ideal for identification, and there is no prospect of them being of value in the general identity case. Nevertheless they are very useful in medicolegal paternity cases (see Chapter 25 of Race and Sanger, 1975).

# 15
# Quantitative Diagnostic and Physiological Methods

## 15.1 Introduction and Vision

In this chapter, complementary to Chapter 14, the emphasis is not on the properties of the human or animal body but on the methods of studying them.

Mayr (1982) and Crombie (1967) have pointed out that the historians of science have allowed an excessive enthusiasm for mathematical physics and astronomy to lead them to the neglect of biological topics. At a late stage, I have changed my priorities from Harvey to Johannes Kepler* (1571–1630), who solved one of the problems of vision by a reduction. By his time accurate accounts of the anatomy of the eye were available, and so Kepler (1604) decided to attend to the purely physical conception of light. He was familiar with the operation of the *camera obscura*, the localization of images in mirrors as well as the measurement of refraction. Light, he said, radiates from any luminous source as a two-dimensional surface. For a given ray, the intensity of light remains constant, so that when impinging on a spherical surface, the same amount of light spreads over a larger area and the amount of light falling on unit area declines as the inverse of the square of the distance from the source of light. He had thus discovered the fundamental law of photometry. From some of the rays proceeding in all directions from an illuminated object and passing through the lens, an inverted and reversed image is formed as in the *camera obscura*.

Crombie (1967) gave such a picture of a man looking at an image formed by removing the back of the retina. At the retina, the light rays ended and were succeeded by a different kind of motion (we can see now that the scene on the retina is transferred by the optic nerve to the brain) but Kepler left this motion to be described by the natural philosophers rather than confuse his readers with a discussion.

Christoph Scheiner* (1573–1650) carried out the above experiment by cutting out the back of the retina of freshly dead animals and humans. René du Perron Descartes made several corrections assuming that the lens was lenticulate in form and making use of the sine law of refraction; Scheiner and Descartes gave the explanation of accommodation by changes of shape rather than of position of the lens. We leave to the reader a discussion of three problems in Crombie (1967):

1. How is a body considered to be without sensation stimulated to coordinated behavior?
2. How does the animate body receive and differentiate information through its senses?
3. How does the animate body coordinate the information received through the different senses?

For further information, see *Physiological Optics* (1924–1925), an English translation of the treatise by Hermann von Helmholtz* (1821–1894), the great exponent of applying mathematics and physics to physiology in general; in particular, he invented the ophthalmoscope [see also Koenigsberger (1906) and Kandel, Schwartz, and Jessell (1991)].

Helmholtz and Ernst Wilhelm von Bruecke* (1819–1892), Emil Heinrich Du Bois-Raymond* (1818–1896), and Carl Friedrich Wilhelm Ludwig* (1816–1895) were influential in the removal of vitalism as an explanatory theory of biology; thus they insisted that the life processes were to be explained by physicochemical laws, as we have just seen. To do so they have often studied an organ as a unit and not the whole body; this is a *reduction*.

## 15.2 Pulse and Circulation

The pulse has been studied since ancient times. Thus we find references to it in the Hippocratic corpus. Praxagoras of Cos* (b. *ca* 340 B.C.), a student of Diocles, belonged to the school of Dogmatists. He observed that the beats of the heart and pulse were synchronous. His student, Herophilos* (b. *ca* 330 B.C.), counted the pulse with the aid of a water clock (clepsydra). He taught that the cardinal properties of the pulse were frequency, rhythm, size, and strength.

Another example of quantitative methods is provided by Bryan Robinson (1680–1754), who, according to Plackett (1988), took a number of readings of the pulse rate with the aid of a watch that ran for a minute; he found that the pulse rates per minute $n$ were inversely proportional to the length $h$ of the animal,

$$n = 1606.6 \, h^{-0.75} \tag{15.2.1}$$

where $h$ is measured in inches.

A great classic of the application of mathematical or quantitative reasoning to medicine is by William Harvey* (1578–1657) on the circulation of the blood, *De Motu Cordis*, published in 1628. Note must be taken of a predecessor, Mattheo Realdo Colombo* (*ca* 1510–1559; *Isis* has *ca* 1516–*ca* 1577). Colombo discovered the pulmonary circuit, the passage of blood through the pulmonary artery, the lungs, and to the left ventricle, by means of vivisectional observations. [See J.J. Bylebyl for biographical details of both Colombo (*DSB*, III, pp. 354–357) and Harvey (*DSB*, VI, pp. 150–162), and for some discussion on priorities.]

Harvey used mathematical types of reasoning when necessary in his study of the dynamics of the circulation. He observed that in the living, blood was expelled from the heart at systole into the aorta and hence into the arteries; he also proved that the valves in the veins would ensure the passage of venous blood toward the heart, but it was not altogether clear that some blood might be transmitted out from the heart by the veins or back to the heart from the aorta. Harvey made a rough quantitative assessment of the amount of blood that was delivered from the left ventricle of the heart into the arteries and found it was too great to be retained in the arteries and that it must in some way find its way into the veins and be returned by them to the heart; he began to consider therefore that the flow of blood was in a circle and afterwards confirmed that this was true. Harvey showed that blood in the veins can be pushed toward the heart but not in the reverse direction. Further, if a segment of vein in the arm is emptied by pressure from two fingers, the segment cannot be filled by release of the finger proximal to the heart, but can be filled indefinitely often by releasing the distal finger; and so the total blood entering from the distal end exceeds the total body blood volume, which of course requires circulation of the blood. This is a simple mathematical argument. Later Harvey's findings were to be verified by the discovery of the arteriovenous connection through the microscopic observations of Marcello Malpighi and of Leeuwenhoek. It is of interest that Graunt (1662) justified his interest in the numbers of deaths by referring to successful use of numerical reasoning by Harvey. [For a history of the timing of the pulse, see Rosenbloom (1922).]

## 15.3  Blood Pressure and the Work of the Heart

Giovanni Alfonso Borelli* (1608–1679), the leading exponent of the iatro-mathematical school, attempted to solve the problem of whether man or other living thing could or should be considered as a machine. Of the internal organs, the most obvious one is the heart, which drives the circulation, but here he was in a territory with the physical concepts not well defined.

James Keill* (1673–1719) made some calculations of the absolute velocity of the blood travelling through the aorta and smaller vessels; he recognized the relation between the speed of the blood and the cross-sectional area, which increased as the vessels branched from the aorta to the arteries, arterioles, and capillaries. Keill was not a good experimenter but the problem was taken up by one who *was* very effective, Stephen Hales* (1677–1761), who in his *Haemastaticks* (1733) measured the blood pressure of a live mare, tied supine, by the insertion of a bronze cannula into one of her femoral arteries and its connection to a glass tube nine feet high; the blood rose in the glass more than 8 feet (2.43 m). During exsanguination of the mare, the blood pressure fell. In later experiments he estimated the cardiac output as the product of the pulse rate and the internal volume of the left ventricle, calculated postmortem from a cast of the ventricle. He noted that pulse was faster in small than in large animals and that the blood pressure increased with the size of the animal.

Daniel Bernoulli* (1700–1782) was better informed about the physical side. He could see that there were two important factors, the volume of blood pumped and the energy per unit volume of the blood, implied by its motion or equivalently by the maximum height to which it could be driven by the heart. So Bernoulli's preliminary calculations, announced in an unpublished oration of 1737, were wide of the mark because of a lack of good experimental data, which he subsequently obtained from the *Haemastaticks* of Hales. Bernoulli handed the problem to Daniel Passavant (1722–1799), the results of whose calculations were published in a dissertation of 1748 and were close to the modern accepted values. The works of Bernoulli and Passavant were not taken up by others and fell into obscurity [see Spiess and Verzár (1941) and Huber (1959)].

Jean Léonard Marie Poiseuille* (1797–1869) used a mercury manometer to measure the blood pressure; the vertical tube was connected to a needle inserted into an artery by a tube containing potassium carbonate to prevent coagulation. He was able to show that the blood pressure rises and falls on expiration and inspiration. Ludwig later improved on the model by adding a float that wrote on a rotating drum. The models above were not suitable for clinical work because they measured the arterial pressure directly; but it was realized that it was sufficient to measure the pressure that would collapse the artery. So the present sphygmomanometers follow a design by Samuel Siegfried von Basch (1837–1905); such a machine was greeted by the *British Medical Journal* in 1874 with the comment that "by such methods we pauperize our senses and weaken clinical acuity."

As a result of the use of the instruments, a new disease was named "hypertension" or high blood pressure, with a note whether it was "essential" or secondary to some disease such as chronic nephritis.

We digress now to discuss Daniel Bernoulli, who was the first to be employed as a biostatistician in the modern sense of the word. He was taught mathematics by his father Johann Bernoulli I and older brother Nikolaus Bernoulli II. He studied medicine in 1718 and 1719 and obtained his doctorate with a dissertation *De Respiratione* (1721). He wrote on the mechanical aspects of physiology. In his dissertation, as a typical iatrophysicist under the decisive influence of Borelli and Johann Bernoulli, he furnished a comprehensive review of the mechanics of breathing. Failing to find a suitable post in Switzerland, he accepted a position in St. Petersburg in 1727, where he worked on a number of physiological topics, so that it comes as a surprise that his name is not mentioned more frequently in histories of medicine. For example, he described the anatomy of the entrance of the optic nerve at the bulbus or blind spot. He published a purely mechanical theory of muscular contraction. In the same spirit, Bernoulli determined the maximum work that a man can perform, in *Hydrodynamica* (sec. 9) and a prize-winning treatise of 1753. (In this context, Bernoulli meant by "maximum work" the quantity that a man could do over a sustained period of time, for example, a working day.) Bernoulli (1760, 1766) made a calculation on the advantage of inoculating against smallpox to rid the world of the disease. His great work is the *Hydrodynamica*, number 31 in his bibliography given by H. Straub (*DSB*, II, p. 45), which has little of direct physiological application. He worked at the highest levels in many fields and was for some years a colleague of L. Euler in St. Petersburg. He made many advances in pure and applied mathematics and in mathematical statistics.

Adolf Eugen Fick* (1829–1901) may be taken as a prime example of a mathematician making progress in biology with the aid of mathematics. For the purposes of this section, we note that he developed the principle later termed "Fick's law." Since all the blood passes through the heart, we can estimate the amount of oxygen passing through the heart over any interval, a minute say. The oxygen taken up over any interval is equal to the difference between the arterial and venous concentrations entering and leaving the heart, multiplied by the volume of blood passing through the heart. The pulse rate can be estimated. The volume pumped through at each stroke is the only unknown and can be calculated from a simple fraction. K.E. Rothschuh, Fick's biographer (*DSB*, IV, pp. 614–617), mentions many physiological problems that he treated by a great diversity of mathematical and experimental techniques.

## 15.4 Thermometry

The priority for the invention of the thermometer is still a matter of dispute. Santorio at some time between 1602 and 1612 added a scale to Galileo's thermoscope to make it a thermometer with end points at the temperature of snow and of a candle flame; he used it in health and disease, but it was not a satisfactory instrument because the reading was affected by the barometric pressure (see Middleton, 1966). Gabriel Daniel Fahrenheit* (1686–1736) invented the mercury thermometer in 1710. About this time, Hermann Boerhaave* (1668–1738) and later Gerard van Swieten* (1700–1772) began to apply the thermometer to clinical practice, but not until the work of Wunderlich (1871) could it be said that it had been received into general medical practice. In 1835, the normal temperature was estimated to be 37°C (= 98.6° Fahrenheit). The construction of easily portable thermometers in 1868 in both England and Germany can be taken as the beginning of the modern era of clinical use. According to Wunderlich, Anton de Haen (1704–1776), the first clinical teacher at Vienna and in all Germany, greatly extended the practical application of thermometry in disease. To that time, fever had been defined as being recognizable by a rapid pulse; de Haen showed that elevated temperature was the important sign of fever.

The instrument was still clumsy and readings were not yet made on a properly calibrated scale. Indeed, de Haen was accustomed to leaving the thermometer in situ for $7\frac{1}{2}$ minutes and then adding $1°$ or $2°F$ to the temperature registered to allow for increase that a further period would have ensured. Yet with such methodology he was able to obtain much information, in particular, the contrast between subjective feelings of temperature and the objective readings. Wunderlich (1871) mentions Ch. Martin (*ca* 1740) as providing the first accurate observations on temperature in healthy man and animals. Charles Blagden* (1748–1820) established the constancy of the internal temperature with greatly increased ambient temperature, and others confirmed this was true also with lowered ambient temperature, a feature of the *milieu intérieur*. Although the level of temperature is an important feature of the *milieu intérieur* and more readily estimated than other factors, at a time when Lavoisier and Laplace were ascribing the formation of heat in the animal body to the chemical combination of oxygen with hydrogen and carbon, a deeper knowledge of biochemistry showed that other factors were important. For example, Walter Bradford Cannon* (1871–1945) and his colleagues were able to show that the autonomic nervous system maintains a uniform condition in the body fluids, for example, blood glucose levels. He reviewed progress in Cannon (1929), but later developments can be obtained in Mrosovsky (1990).

Milne-Edwards and Villermé (1829) had written on the influence of cold on infant mortality. For light relief we find that, as Lesky (1976, p. 26) has reported, after de Haen had introduced the thermometer into the Vienna Clinic, Valentin von Hildenbrand (1763–1818) "continued to use the back of his hand for taking the patient's temperature and considered that counting the pulse by means of a watch was a "ceremony" which "had better be left to the quacks."[1]

## 15.5 Percussion and Auscultation: Josef Škoda

Josef Škoda* (1805–1881) was born in Pilsen, Bohemia, now Plzen, Czech Republic, the son of a locksmith; after a sickly childhood he entered high school at the age of 12, and graduated from it in 1825. He proceeded as a poor student to the Faculty of Medicine in the University of Vienna and was fortunate to be offered accommodation in the house of a lady, a manufacturer of silk materials. He earned his maintenance by tutoring and, in addition to his medical studies, attended lectures in physics and mathematics. After

---

[1] From Lesky, E. *The Vienna Medical School of the 19th century*. Copyright © Johns Hopkins University Press, Baltimore, 1976. Reprinted with permission.

graduation, he returned to work in Kouřim and Plzen (1831–1832), where he became cholera specialist in the cholera epidemic of 1832; here he realized that more could be done by preventive measures than by clinical medicine.

Škoda returned to Vienna as an unpaid assistant in the General Hospital. Here he was anxious to correlate physical signs with the autopsy findings of Rokitansky. He began to use percussion and auscultation, although there had been no other Viennese physician interested in these methods, introduced into medicine by such physicians as Corvisart, Laënnec, and Gaspard Bayle in France and Charles Williams, Robert Graves, and William Stokes in Great Britain. The methods were not greatly used in general practice because they were difficult to understand. Škoda, by the use of physical theory and experimental work, was able to classify the percussion sounds by four criteria: from full to empty, from clear to muffled, from tympanous to nontympanous, from high to deep. His name survives in "skodaic resonance," the tympanous percussion in the presence of serous pleurisy. Škoda rebutted the thesis of Pierre Adolphe Piorry (1794–1879), the inventor of the pleximeter, that every organ had its special sound, but many at first failed to agree.

In the theory of auscultation he distinguished the normal heart sounds from murmurs by an examination of both healthy and ill patients (Škoda, 1839, English translation 1853). Thus, his work had already raised the standards of diagnosis in cardiovascular and respiratory medicine. Škoda achieved a spectacular victory by diagnosing the cause of the French ambassador's "liver affliction" as aneurism of the abdominal aorta, the diagnosis being later confirmed at autopsy. Such events brought him a reputation for lightning diagnosis. The improved methods enabled him to carry out a pericardial puncture in 1840.

Škoda introduced the method of diagnosis by exclusion into his clinic, differential diagnosis as we would say; given the physical signs, the question was asked what disease could have caused them. Reasons for exclusion of various causes must be given, leaving finally a single cause diagnosed.

For a time Škoda had left the General Hospital to work as a physician to the poor in a suburb of Vienna. In February 1840, Ludwig von Türkheim (1777–1846) established a separate station for the treatment of chest disease under Škoda and in 1841 raised him to the position of a chief physician. After some difficulties with the authorities, Škoda became professor of internal medicine in 1846. At this time he was acknowledged as the leader of the progressives in the General Hospital.

Škoda was clearly a contagionist. Thus the treatment of diseases of the skin was put under his control in the department of diseases of the chest; he persuaded Ferdinand von Hebra (1816–1880), who had graduated in 1841, to care for them. Although he had had no special training beforehand, Hebra's work in the field was the foundation of the modern specialty of dermatology.

Škoda was convinced that cholera and enteric fevers were contagious and water borne; he repeatedly urged the provision of a high-altitude water supply conduit for Vienna. Indeed, this was a great success for in 1871 there were 1,530 cases of enteric and in 1879, 180 cases admitted to the General Hospital, the latter after the installation of the pure water supply. Lesky (1976) provides us with other examples of his important work in social hygiene and epidemiology.

The charge of "therapeutic nihilism" was often levelled against Škoda. According to Willius and Dry (1948), Škoda had begun practice using the traditional remedies of venesection, blistering, emetics, and mercury, but he came to doubt their value and he reexamined their efficacy. Indiscriminate venesection and many of the old therapies were out. Some new medicinals, such as chloral hydrate and salicylic acid, were introduced. It is said (for example, Lesky, 1976) that he carried out serial tests on patients in "accordance with the numerical-statistical method" of Louis and the French observers. Such analyses, casting doubts on the usefulness of traditional remedies, were certain to evoke criticism. A crisis in the Vienna medical school arose when Joseph Dietl (1804–1878), one of Škoda's students, condemned much of contemporary therapy in 1845. The conservatives reacted against Dietl by attacks on his master, Škoda. Fortunately, Protomedicus I.F. von Nadherny (1789–1867) of Prague, who was drafting a report for the Vienna Hospital on staff appointments, pointed out that Škoda's aim was to advance the therapeutic art and he needed to deepen the knowledge of disease; in any case, he had been successful in the treatment of thoracic effusions, typhoid fever, gout, and other diseases.

Balfour (1847) defended Škoda with the following comparison of the statistics of pneumonia in Edinburgh, where it was treated vigorously, and in Škoda's clinic, where it was treated conservatively: In 41 months, 392 patients with pneumonia were admitted to the wards of Škoda with 53 deaths, a case fatality rate of 13.5%; whereas in 63 months, 253 patients with pneumonia were admitted to the Edinburgh Infirmary with 91 deaths, a case fatality rate of 36.0%

According to Lesky (1976, p. 124), Škoda remarked on 14 March 1871, at his farewell address: "The development of the physical method of examination has given us a solid basis for the understanding of many diseases; it is now the turn of chemistry and microscopy and thus we shall at last acquire the necessary insight into the pathological changes of our organism to enable us to establish a useful science of medicine."[2]

## 15.6  Cell Counting

The taking of red or white cell counts has been a familiar clinical test since about 1880, at which time precision engineering, including very accurately spaced gratings incised on the floor of the well of the hemocytometer, had

---

[2] From Lesky, E. *The Vienna Medical School of the 19th Century.* Copyright © Johns Hopkins University Press, Baltimore, 1976. Reprinted with permission.

become available. Ernst Abbe* (1840–1905) secured a school and university training as the result of a scholarship provided by the employers of his father, a spinning-mill worker. After graduating from Eisenach Gymnasium in 1857, he studied physics in Jena and Göttingen, receiving his doctorate in 1861. In 1866, while a lecturer at Jena, he cooperated with Carl Zeiss in the production of improved lenses and his was long the standard account of the microscopic lens. He became a partner with Zeiss in the firm and finally owner. He used his fortune to create the Carl Zeiss Foundation. His greatest achievement was the construction of the apochromatic microscope. He also constructed the hemocytometer, of which the chief features are the constant depth of the chamber and the accurately ruled grid. Plum (1936) has described the evolution of the blood counting apparatus and Eisenhart and Wilson (1943) have described the applications of the same methods to the counting of bacteria.

Although Siméon-Denis Poisson* (1781–1840) had in 1837 first used the distribution that goes by his name, Abbe (1879) derived the same law for the distribution of the cells over the squares of a hemocytometer chamber, but did not give any experimental verification of the theory as he was more interested in the accuracy of his pipettes and hemocytometer chamber from the volumetric point of view. Lyon and Thoma (1881) supplied an experimental verification, but the current statistical techniques were not sufficient to make it convincing. These experimental results were probably not known to Bortkiewicz (1898), whose much-quoted example of the number of fatalities from the kick of a horse in the different corps of the Prussian army is rather artificial and has come in for much criticism from modern authors (see Quine and Seneta, 1987).

The counting technique was used by William Sealy Gosset* (1876–1937) in "Student" (1907) to estimate the error of his cell counts and to determine the most efficient dilution for obtaining pure cultures from a single yeast cell; if the dilution is too high many tubes are wasted and if the dilution is too low there will be cultures springing from more than one cell and so not being necessarily pure (see also E.S. Pearson, 1990). Incidentally, Joseph Lister* (1827–1912) used a dilution method to obtain pure cultures before solid media had been introduced into bacteriology (Lister, 1878).

The importance of Student's work was noted by a well-known physician, Goodall (1908), in a very favorable review notice but for many years the possibilities of a proper statistical analysis of blood counting were overlooked. Nevertheless, the fundamental importance of the Poisson distribution became generally recognized, especially in the field of counting in bacteriology, largely as a result of Student's paper. Indeed, R.A. Fisher rated the Poisson as the most important discrete distribution in biological work. Fisher, Thornton, and MacKenzie (1922) and Fisher (1925) supplied useful methods by which the consistency of a small number of parallel counts could be tested, which are, essentially, new applications of the $\chi^2$ test for goodness of fit. At the same time they introduced the concept of "statistical control" of laboratory work. These methods of control appear never to have been applied before to the counting of cells. Besides being a convenient check on current laboratory work, Fisher showed that the methods could be used to test the consistency of published data and gave examples of agreement in the count of parallel bacterial plates that were so close that the results could only have occurred with excessive rarity under conditions of strictly random sampling and were, therefore, suspect. Later, Joseph Berkson (1899–1982) of the Mayo Clinic, Rochester, Minnesota, and his co-workers showed by photographic methods that many individuals systematically count too low and attempted to have the over-strict criteria of a "good count" modified, so that technicians might avoid the temptation to count erroneously in order to bring the dispersion within preconceived bounds (Berkson, Magath, and Hurn, 1935, 1940; Magath, Berkson, and Hurn, 1936).

If laboratory conditions are good and red cell counts are made over the 400 ruled squares of the hemocytometer, then the resulting empirical distribution approximates closely to the theoretical Poisson distribution, with parameter equal to the mean of the observed distribution, if that mean is less than, say, 4 per small ruled square; at levels above this, and markedly for mean counts of 8 per small ruled square, the variance falls, as noticed by Berkson and his co-workers. Lancaster (1950a) pointed out that, although nearly theoretical conditions might hold in the fluid initially, after settling there is crowding on the hemocytometer surface because the total area of the cells is not negligible with respect to the area of the small squares, although the total volume of the cells is so with respect to the volume before settling; if therefore too many red cells are falling onto a small square, some will slip, because they fall from a red cell already on the glass onto the next square. This effect causes a positive correlation between counts on neighbouring squares and a diminution of the variance, as noted by Berkson and his co-workers. It has always seemed a pity to the author that students, introduced to their first chance of observing errors in a natural experiment, have been forced to "cook" their results to meet unreasonable requirements of persons ignorant of the sampling theory.

It is appropriate to mention here other applications of the Poisson distribution.

The random variable $X$ is said to have the Poisson distribution with parameter $\lambda$ if

$$P(X = x) = e^{-\lambda}\frac{\lambda^x}{x!}, \qquad x = 0, 1, 2, \ldots; \qquad (15.6.1)$$

the frequencies in (15.6.1) are nonnegative, and their sum is unity. Equation (15.6.1) may be written as a probability generating function $Q(t)$ and then

$$Q(t) = \sum t^x e^{-\lambda}\lambda^x/x! = e^{\lambda(t-1)}; \qquad (15.6.2)$$

it follows that the distribution has the additive property that sums of independent Poisson variables are again Poisson. Counts of nuclear disintegrations (not in atomic bombs), of bacteria in water supplies or soil are other applications [see Greenwood and Yule (1917) and Lilienfeld (1977)].

Note that for $\lambda > 11$, say, the distribution function of $(X - \lambda)\lambda^{-1/2}$ is well approximated by that of the standard normal distribution.

[For a bibliography of the Poisson distribution, see Haight (1967).]

## 15.7 Mechanical Aids in Diagnosis and in Physiology

The history of the mathematization of biology might well begin with Nikolaus of Kues (Nicolaus Cusanus)* (ca 1401–1464). In his *Idiota de staticis experimentis* of 1450, he showed how to determine physical parameters through the use of apparatus such as scales or a water clock; he tells us how to estimate the humidity of the air by measuring the weight of wool, for example. "Static," as in the title above, refers to weighing in particular, but more generally it refers to a quantification of the experimental work. It is derived from the same Latin root *status* as are state and statecraft. These uses are independent of statics from the same root, meaning the mechanics of stationary objects (see also the *Shorter Oxford English Dictionary*).

Others in the faculty of medicine at Harvey's university, namely Padua, were following the quantitative trend in experimental medicine. Thus Santorio di Capodistria* (1561–1636) made many observations on his own weight in a specially designed chair and established that *perspiratio insensibilis*, known from ancient times, was greater in weight than the sum of all other bodily excretions combined and that it was not constant, depending on external temperature, sleep, and fever, among other factors. He built this knowledge into a system of medicine that asserted the primacy of these measurements for hygiene, diagnostics, and therapeutics. Notwithstanding his own many accomplishments, he seems not to have appreciated the importance of Harvey's discovery of the circulation of the blood. Nevertheless, Harvey and Santorio have been praised for their introduction of quantitative experimentation into the biological sciences, particularly medicine (see Renbourn, 1960).

The use of mechanical aids in the examination of the patient or of the experimental subjects was greatly extended by Ludwig, whose guiding principles were (a) all actions might be the result of simple attractions and repulsions (Lesky, 1976), and (b) all that was weighable or measurable should be weighed or measured, in the tradition of Galileo. Visitors to his laboratory from abroad were numerous; his example surely was great in the mathematization of medicine and biology.

Ludwig graduated from Marburg in 1840 and in 1842 qualified to teach at that university with a dissertation on the mechanism of renal function; in 1849 he became professor of anatomy and physiology at Zürich and in 1855 professor in the same subjects at the Austrian military-medical academy, the Josephinum, in Vienna. He introduced automatic recording in physiology by means of the kymograph in 1847 and in 1851 discovered the secretory nerves. His early work showed that the secretion of the salivary glands was dependent on nerve stimulation and not blood supply. He devised the mercury blood pump, which enabled him to separate the gases from a given quantity of blood directly in vivo in 1859; later, among other inventions, there were the steam gauge in 1867 and the method of maintaining circulation in isolated organs in 1865. He designed numerous instruments, such as a multiplier and a slide magnetic electric engine, kymographs, myographs, compasses, ophthalmometers, scales, air pumps, induction apparatus, spirometers, and gasometers [see Schröer (1967), Bonner (1963), and Rothschuh (1973)].

One of Ludwig's students, Christian Harald Lauritz Peter Emil Bohr* (1855–1911), studied especially the exchange of gases in the lungs with novel methods and theory (Harald and Niels were his sons). He, like Ludwig, was a capable designer of instruments.

Ludwig has often been criticized as a reductionist: the discussion often takes on a theological slant and at least follows on ideas held by vitalists in earlier times. [See Ayala and Dobzhansky (1972), especially the articles by Goodfield on pp. 65–86 and W.H.T. Thorpe on pp. 109–138; see also Mayr (1982, pp. 56–63).]

Ludwig brought graphical methods to a high degree of efficiency. According to Tilling (1975), who performed a thorough search of old journals, credit should go to Johann Heinrich Lambert* (1728–1777) for the first graphs in experimental work (see also Royston, 1956).

## 15.8 Electronic Devices

In 1903, Willem Einthoven* (1860–1927) adapted the string galvanometer to record the action currents produced by the muscles of the heart. He showed that the resulting curve recorded could be analyzed into deflections denoted by the letters $P$ through $T$, which could be related to events in the heart; aberrations of these deflections could be interpreted as signs of disease. This was the first of numerous electronic devices for use in clinical medicine and physiology (see §14.2).

## 15.9 Hearing

There are many other branches of the science of medicine where the fundamental sciences have played an important part in explaining function. A short account of the biophysics of hearing has been given by Dittrich (and his collaborator R.C. Extermann) (1963), but it seemed not to lend itself to our exposition, for the main discussion depends on Fourier transform methods. S.R. Ellis in his biography of Georg von Békésy* (1899–1972) remarks that Békésy may have been the first to recognize that the modes of vibration of the basilar membrane can be imitated by a mechanical model. The most famous of his models is "a slotted brass tube filled with water and covered with a membrane varying in thickness. When the water in the tube is vibrated at different frequencies, the movement of the point of maximum vibration can be readily felt along the slot."[3] Békésy won the Nobel Prize in 1961 for clarifying "the physical mechanisms of stimulation in the cochlea of the middle ear," work that he had published in 1928 [see also the biography by Ratliff (1976)].

## 15.10 Nutrition

A fundamental step in the analysis of nutrition is to determine the metabolism of the different elements, especially hydrogen, carbon, nitrogen, and oxygen. In this field, Carl von Voit* (1831–1908) was the acknowledged leader after about 1860. A good account is given of his work (*DSB*, XIV, pp. 63–67) by F.L. Holmes and in Holmes (1984), including Voit's technical methods and formulation of hypotheses, but little on statistical methodology. The nitrogen values could not be understood without a proper knowledge of those of the other three—a good example of the difficulties caused by reductionism [see also Bylebyl (1977) for the history of metabolic enquiries].

According to Florkin (1972, pp. 215–249), about the middle of the nineteenth century, the concept of conservation of energy became generally accepted. Moreover, calculations had become common in industrial and business transactions, for example, in accountancy. Florkin's Table IV (p. 242), credited to M. Rubner in 1901, gives food, number of days of observation in the calorimeter, heat calculated from the metabolism, heat directly determined by day, and difference as a percentage. Such studies have shown that the amount of heat produced in animals under various conditions can be accounted for by physical and chemical laws and is not varied by "vital forces."

---

[3] *Dictionary of Scientific Biography*, vol. XVII, p. 64a. Charles Scribner's Sons, New York. Copyright © American Council of Learned Societies, 1990. Reprinted with permission.

# 16
# Classification of Diseases

## 16.1 Clinical Observation

Little is gained by considering classification of diseases as a part of a larger topic classification, generalizing the Linnean methods. There are two main justifications for this separation. The Linnean system is unnatural, for the classification into classes and orders by the numbers of stamens and of pistils does not always succeed in bringing together closely related taxa. Further, in biology there is a need to harmonize the classification with the notion of evolution.

A fundamental difficulty with the classification of the taxa (diseases) is uniqueness. Are diseases or syndromes unique? Opinion was confused or divided on this point. Perhaps, it was only in the infective diseases that it could be said that there was uniqueness. We have cited Fuller (1730) in §8.6 as claiming such uniqueness for certain infective diseases. Some diseases were progressive in the sense that the main emphasis of the disease changed; some authorities would claim that a new disease had developed.

The important ancient classifications of disease are due to Hippocrates, Aulus Cornelius Celsus* (fl. Rome, *ca* 25 A.D.), and Galen. Hippocrates used diagnostic terminology to describe the forms of disease that he had observed, such as fever, consumption, and asthma.

With the renaissance of learning, we find new authors writing in the same manner and finding new syndromes. For example, Guillaume de Baillou (also Ballon and Ballonius)* (*ca* 1538–1616) attempted to follow the example of Hippocrates and establish syndromes and disease entities, but perhaps he gives too much detail. Franciscus de la Boë (Sylvius)* (1614–1672) was the first to describe the tubercles in phthisis (*DSB*, XIII, pp. 222–223). Thomas Sydenham gave a classic description of gout, differentiating it from other rheumatic or joint diseases and emphasizing its development in time. The concept of disease as a process could not be understood until pathology had developed. Sydenham believed in the healing power of nature, nature's orderly production of diseases by species, and the possibility of deriving treatment from the observable phenomena of a disease.

## 16.2  Diseases in the Hospitals

There were many attempts by physicians to name and classify diseases before the studies that brought pathology and clinical findings together.

The greatest work of Philippe Pinel* (1745–1826) was the *Nosographie Philosophique* of 1798. He introduced statistical *tables synoptiques* showing the frequencies of certain diseases, their development, and prognosis. He is said to have conducted tests of the efficacy of drugs and used drugs only conservatively; he was against polypharmacy, purges, and blood-letting. He created an inoculation clinic at the Salpêtrière in 1799 and the first vaccination in Paris was given at that clinic in April 1800. From our point of view, he was among the first to attempt proper recording and analysis of hospital statistics and he was an early researcher into testing the effectiveness of medicines or treatments. Further, Bichat admitted that he owed the idea for the *Traité des Membranes* to Pinel's *Nosographie Philosophique*, which had proved a felicitous connection between the structure and various affections of the membranes (see the footnote in Goldschmid, 1953, p. 113). Pinel (1798) in his *Nosographie Philosophique*, an artificial nosology based on a minute description of the diseases, divided diseases into five classes–fevers, phlegmasias, hemorrhages, neuroses, and diseases caused by organic lesions [see Chabbert (*DSB*, X, p. 613) for a description of his classification]. To our eyes, there are some strange conjunctions of diseases in the few examples given, as is suggested in the quotation from Burnet given in §8.6.

Pinel (1807) classifies 1,002 patients, admitted over a period of $3\frac{3}{4}$ years, by year of admission, broad clinical diagnosis, clinical findings at admission and outcome of hospital stay; he gives proportions of cures superior to those of other institutions, so justifying his therapy.

As an example of a natural nosology, the reader is referred to the discussion of Goldschmid (1953) on Jean Louis Alibert (1768–1837).

Pierre Charles Alexandre Louis (1787–1872) was born in Aï in Marne, the son of a vineyard proprietor. After graduation in medicine at Paris in 1813, he travelled to Russia and lived there until 1820. In that year he was shocked by his inability to combat an epidemic of diphtheria. In the account of Jackson (1836), Louis returned to France, dissatisfied with the state of the science of medicine; he spent seven years collecting observations on patients, avoiding particular doctrines and not making special selections of patients, at La Charité under Auguste-François Chomel (1788–1858). Here he made far more careful and detailed histories of the patients than had been the custom of his predecessors; Louis (1829/1836, 1825/1844) gave detailed discussions on the enteric fevers and tuberculosis.

Louis followed up the notion that disease was a process; he collected and analyzed detailed observations on over 2,000 hospital patients. Louis's great achievement was to realize that questions of diagnosis, prognosis, and the results of treatment could only be answered in a statistical sense; although every case has individual properties, it is possible to subdivide cases into classes by signs and symptoms and to give reasonable statements about diagnosis, prognosis, and therapy, although in most cases the statements would only be true in a proportion of cases. The weakness in this argument is that an enormous number of patients would have to be studied; thus if Louis's 2,000 cases are classified into, say, ten established syndromes and a remainder of many poorly represented or unclassified syndromes, then except in a few diseases there will be less than 200 cases to work on. If these 200 are further divided into subclasses by sex and, say, four age intervals, there will be even less material to work on; for example, in typhoid fever, a common cause of death, he constructs what is in effect a fourfold table by length of stay in Paris before entry to hospital and by result, death or survival. His number 129 is not large enough to obtain a significant answer to the hypothesis that the disease is more lethal for newcomers to the city; in other words the test with 129 patients would have little power.

In 1832, Louis was elected *Président perpétuel* of the *Société Médicale d'Observation*, a society lasting until 1870. [See Armitage (1983) for an appreciation and Steiner (1939) for his influence on his American medical students. See also Greenwood (1936b), Rosen (1968), and Bariéty (1972) for further biographical details and Lancaster (1982b) for further references.]

Failure of the project of Louis and of the *Société Médicale d'Observation* was perhaps due to the difficulty of defining diseases using signs and symptoms as the criteria for classification; from knowledge of the inconstancy of appearance of a given sign in such diseases as syphilis or tuberculosis, it is easy to see that we cannot easily define diseases by means of multiple twofold classifications giving results such as AbCDeFgh.... It was only after etiology became established by a variety of disciplines, particularly pathology and microbiology, that it was possible to classify diseases, as we explain in the succeeding sections (see Rosen, 1955).

## 16.3 Iatrochemistry

Since chemistry was one of the earliest of the sciences to develop in the new intellectual life of the fifteenth century, its application to physiology was responsible for the development of iatrochemistry, the use of chemical substances in therapy, and for the construction of related theories of disease by Paracelsus* (*ca* 1493–1541). In therapy, he imposed strict controls on the use of the newly introduced chemical substances; from our present point of view, he was the first to identify silicosis and tuberculosis as occupational hazards. He recognized endemic goiter as being due to the mineral content of drinking water. In silicosis, he identified a definite chemical compound as a cause of disease.

Sylvius believed that all vital processes could be reduced to chemical processes as then understood. Johannes Baptista van Helmont* (1579–1644), according to Pagel (*DSB*, VI, pp. 253–259), introduced an ontological concept of each disease as a specific entity; further, some diseases are determined by specific pathogenic agents acting on the body to yield local changes. The works of the iatrochemical school, represented by these two authors, soon degenerated into theoretical and deductive methods of looking at disease, but perhaps some of their ideas should be regarded as advances, pointing to the need to look for causes and objective evidence of them.

A modern version of the iatrochemical methods is provided by the criteria for the diagnosis of diabetes mellitus and of another category, impaired glucose tolerance (IGT) as set out by Keen (1982); there is good agreement by authorities on the level of fasting blood glucose, namely 140 mg per 100 ml as the critical level. IGT is a necessary category because some normal persons have blood glucose levels overlapping the set limits. It appears that both insulin-dependent diabetes mellitus (IDDM), in which there is a predominantly juvenile onset, and non–insulin-dependent diabetes mellitus (NIDDM), in which onset is usually above the age of 40 years, can be the primary or secondary disease. The cause of IDDM is genetic and an example of a polymorphism (see §5.6). [For a symposium, see Koebberling and Tattersall (1982).]

## 16.4  Physical Methods

Physical methods of diagnosis have already been discussed in Chapter 15. A good amount of the history of physical diagnosis is available in Chapter 6 of Mettler (1947). Neither Škoda nor Austin Flint (1812–1886) believed that many diseases are associated with pathognomonic (that is, peculiar to a fixed disease) signs or symptoms, such as Corrigan's sign in aortic aneurysm or the signs of exophthalmic goiter (Mettler, 1947, p. 309).

A simple example is given by hypertension. The blood pressure of a patient is taken under certain standard conditions. If the systolic blood pressure exceeds 160 mm of mercury, the patient is said to be hypertensive. If the hypertension is secondary to some known cause, it is said to be *secondary* hypertension. If no such cause can be found, the hypertension is said to be *essential*. Another example is diabetes (see §16.3).

## 16.5  Pathology

We have seen that there had been great difficulties in characterizing individual diseases by clinical methods. Studies of the function of the body in health and disease, now gathered under the headings of anatomy, physiology, and pathology (or form, function, and disease), were required. Of these subjects, pathology (including bacteriology) has the most direct interest for classification.

Giovanni Battista Morgagni* (1682–1771) studied anatomy, gross and microscopic, in health and disease. By viewing the body as a mechanism he was able to associate some diseases with particular pathological findings. His views were set out in the book, *De Sedibus et Causis Morborum per Anatomen Indagatis*, published in 1761 in Venice, in which he reasoned that a breakdown in some part of the organism must be the seat and cause of the clinical manifestations of the disease. He also believed that external causes such as environment and occupation could be important. Morgagni noticed parallels existing between anatomical lesion and clinical symptom and made them serve as a basis of his *historiae anatomico-medicae* in the *De Sedibus*, which excelled past collections of case histories. Morgagni thus can be regarded as the founder of pathological anatomy.

Matthew Baillie* (1761–1823) in 1795 published *Morbid Anatomy of Some of the Most Important Parts of the Human Body*, the first English-language text on pathology and the first systematic study of pathology in any language. He, too, hoped to correlate the altered structure of morbid anatomy with the physical signs. Perhaps, this would have given clinicians an indication of what signs to look for, for example, consolidated lung in lobar pneumonia would cause dullness in percussion. Marie-François-Xavier Bichat* (1771–1802) drew attention not to the topography of the organs but to the "membranes" in them, that is, the tissues. He believed that tissues of the same type would be prone to the same kind of lesions in whatever organ they were situated. Laënnec, a skilled anatomist and pathologist and colleague of Bichat, related the clinical findings in chest disease to the pathological changes by the development of mediate auscultation, that is, by stethoscope in 1818.

Karl von Rokitansky continued the systematization of the macroscopic appearances of diseases with the publication of the *Handbuch* (later *Lehrbuch*) *der pathologischen Anatomie* in several volumes over the years 1842 to 1846. This book was based on many thousands of autopsies by the author, but there were two weaknesses. First, the importance of cell structure had not yet been recognized; Rokitansky rarely used the microscope for the purpose of description and diagnosis. Second, his interpretations of the appearances were made under the humoral theory, that tissues or cells could develop out of the unspecialized body fluids (see §3.3). A more rational pathogenesis, that is, view of the development of disease, could only come after new discoveries in botany and zoology of the importance of the cell. General anatomy and the pathology of tissues were transformed in the nineteenth century by the use of the microscope, allowing the development of histology and cytology and the study of morbid processes in the cells, cellular pathology (see §3.2).

In 1855, the difficulties of histological examination were still great. Although leukemia had been first correctly interpreted with the aid of the microscope in 1845, methods for the fixing and cutting of sections and for staining were not available until Joseph von Gerlach (1820–1896) in 1847 noticed that carmine stained the nuclei of cells. He used carmine in ammoniacal solution to study the cells of the brain in 1858; such was "one of the most valuable additions to our means of investigation that has ever been discovered" according to another histologist in 1865. According to W.D. Foster (1961), among the staining fluids available by 1880 were carmine, osmic acid, indigo, silver nitrate, eosin, and logwood (now known as hematoxylin). Of the aniline dyes, eosin was the only one much used by the earlier workers. Staining with a double stain, so that cell structures could be more readily distinguished, became common and eosin-hematoxylin has been used ever since. By 1900, all the common histological techniques now in use had been developed. For the effective use of staining, methods of support for the tisssues were developed, so that thin slices could be cut (sectioned) suitable for microscopic examination. The freezing of tissues was used in 1843; they were supported externally in 1853 and then in 1869 paraffin embedding was introduced.

The sequence of events in inflammation was elucidated in 1857 by Joseph Lister and by Julius Cohnheim (1839–1884). Although these advances gave insight into the development of diseases in the body and the reaction of the body, they sometimes were little help in finding the correct cause of the disease. Some disease processes, such as diabetes, do not leave evident pathological changes; they must be examined by other methods such as biochemistry.

From the point of view of the classification of disease lesions, we have come to the modern period with Virchow; but for a classification of diseases it was necessary to study microbiology and parasitology (see Chapter 8).

[For the history of clinical pathology, see Foster (1961), Bracegirdle (1978), and Virchow (1959).]

## 16.6  Bacteriology and Parasitology

With the aid of these sciences, specificity is assured not by clinical signs but by diagnosis of the particular agent of disease [see Lancaster (1990, Ch. 2) and §§8.4–8.8 in this book].

## 16.7  Nosology

As the notion of distinct disease entities became established, a nomenclature became essential. Here we might well begin with Thomas Sydenham, who rejected "speculative hypotheses and philosophical systems," that is, the dominant iatrochemical theories, and insisted that "clincial reasoning should be founded on direct bedside observation" (Feinstein, 1967). Sydenham introduced two concepts: (a) the coincidence of several individual signs and symptoms, that is a *syndrome* (*cluster* might also be used nowadays in statistical writings); and (b) disease as a process, which Feinstein introduces as temporal correlation, meaning the evolution of the clinical course or its natural history, e.g., syphilis and tuberculosis. (See notes on Louis in §16.2 and Sylvius in §16.1.)

The art or science of nosology (including nosography) has thus begun with the notion that diseases are specific identities and that each must be given a constant name. Of course, a nomenclature must be exhaustive, that is, include all diseases. Further, it is desirable that the taxa should be mutually exclusive. Goldschmid (1953) has pointed out the relation between botanical and nosological classifications. In his Table I are given the botanical systems due to Linnaeus (2), de Jussieu (2), Erasmus Darwin, de Candolle (2), Eichler, and finally Engler and Prantl, over the years 1735–1887. Systems of skin diseases have a longer history, commencing with Girolamo Mercuriale (1530–1606) in 1570 and ending with that of Ferdinand von Hebra, sixteen in all. Some authors used an "artificial" system requiring the appearance of certain specified signs; others used a "natural" system based on a total knowledge of the whole course of the disease. It has often been pointed out by authors that all systems must be artificial in the usual sense of the word, since nature is not artificial. Goldschmid (1953, Table III) gives a list of nineteen "systems of diseases in general" beginning with Jean François Fernel* (*ca* 1497–1558) in 1554 and ending with Conrad Heinrich Fuchs (1803–1855) in 1844. Goldschmid believes that the nosologies reached a climax with the work of Pinel.

## 16.8  Farr and his Eclectic Classification of Diseases

William Farr was born on 30 November 1807, the son of a small farmer, at Kenley, a small, ancient, and remote village of Shropshire. Farr's educational opportunities as a boy were limited, although he was able to read much. He also learned some Latin and Hebrew and later Italian. On 10 July 1837, Farr was appointed to the General Register Office as Compiler of Abstracts at a salary of £350 per annum. He thus ended his brief career as a medical practitioner and devoted the next 40 years almost exclusively to the task of developing a national system of vital statistics. Farr died on 14 April 1883, a little over three years after his retirement [see Humphreys (1885) for a longer account].

[For the development of Farr and his work, see Farr (1885/1975). In the following discussion, the page numbers refer to this Memorial Volume.]

In the First Annual Report of the Registrar General of England and Wales, 1839, Farr notes that the *Nosologia Methodica* of François Boissier de Sauvages de la Croix (1706–1767) was the first important work of its kind (p. 231); his successors in general had added little new to it, with the exception of William Cullen* (1710–1790), a teacher and writer whose popularity helped his nosology to supersede that of de Sauvages. Farr cites Pinel, Richerand, Bichat, Parr, Young, and Mason Good as authors of modern nosologies. Farr notes that:

The advantages of a uniform statistical nomenclature, however imperfect, are so obvious, that it is surprising no attention has been paid to its enforcement in Bills of Mortality. Each disease has, in many instances, been denoted by three or four terms, and each term has been applied to as many different diseases; vague, inconvenient names have been employed, or complications have been registered instead of primary diseases. The nomenclature is of as much importance in this department of inquiry as weights and measures in the physical sciences, and should be settled without delay. (p. 234)

In the Fourth Annual Report, 1842, Farr (pp. 234–250) points out that the alphabetical list contains more names than the nosology and cites with appreciation Cullen's remark that species (that is, individual diseases) are created by nature, genera by the human mind. Farr says that it is not his object to propose new names for the species except insofar as new discoveries in pathology will be made. He proceeds to give a general overview of medical theory. He reveals himself as a humoralist, for humoralism was the prevailing doctrine. "All the *tissues* are formed out of the blood, and they form the parts, organs, and systems of which the aggregate is the organization" (p. 235).

Farr remarks that diseases are recognized by changes in the anatomical structures, production of unusual discharges, or other signs; of outstanding importance is inflammation, the symptoms of which are redness, swelling, heat, and pain. Farr notes that, if the epidemic diseases are excluded, the great bulk of the other diagnoses given in the register of deaths show a reference to the organ affected. This policy is parallel to the specialism of medical officers and of medical texts. French authors, indeed, have followed such classifications by organ site (p. 237).

Farr then turns to the class of (our) epidemic diseases, to which he refers as zymotic diseases:

The blood, which pervades the whole system, is the primary seat of zymotic diseases; but this does not diminish the importance of the local phenomena with which they commence, proceed, or terminate; for they affect (as poisons do) particular organs more extensively and frequently than others, give rise to specific pathological formations or secretions, and derive their character from the lesions and affected organs. The heat disengaged in these diseases suggested the term fever, derived from *ferveo*, as *fermentum* is from *fervimentum*. Some zymotic diseases recur, others only happen once in life, or if they happen twice, it is the exception ... The tendency of zymotic diseases to increase and decline in activity, is one of their most remarkable properties; and the suddenness of their outbreaks, with the great mortality of which they were the cause, excited at an early period the attention and solicitude of mankind. This tendency is indicated by the terms epidemic and endemic; the latter serving to designate diseases which are excited by miasmata, and prevail in proportion to the quantity of miasm developed; the former, epidemic, denoting the diseases transmitted by man to man, independently of locality, or only dependent on locality, temperature, and moisture as adventitious circumstances. For statistical purposes, the epidemic, endemic, and contagious diseases have been classed under one head, as they may all be excited by organic matter in a state of pathological transformation. (pp. 247–248)

It must be noted that "pathological transformation" is a reference to the Liebigian hypothesis (see §8.5).

In the Sixteenth Annual Report of 1856, Farr (p. 253) gives details of his fivefold classification of diseases:

*Class I*: Epidemic, endemic, and contagious diseases. *Zymotici.*
*Class II*: Constitutional diseases. *Cachectici.* [Note that Farr includes tuber-culosis here rather than among the *zymotici*.]
*Class III*: Local diseases. *Monorganici.*
*Class IV*: Developmental diseases. *Metamorphici.*
*Class V*: Violent deaths or diseases. *Thanatici.*

A report by Farr on this classification, which had already been in use for some years, was in 1856 submitted by invitation to the International Statistical Congress in Brussels. In the Thirty-eighth Annual Report of 1877, Farr notes:

I was requested by the International Statistical Congress, in conjunction with Dr. Marc D'Espine, to frame a project based on this resolution, passed at Brussels: "Il y a lieu de former une nomenclature uniforme des causes de décès applicable à tous les pays".... Profiting by experience and by criticism, I carefully revised the English classification, and submitted it in proof to the most eminent physicians, surgeons and statists of England, Scotland and Ireland, to whom I was indebted for many valuable suggestions. The causes of death were thus definitively divided into five classes; and the classes were further subdivided into twenty-three new orders. (p. 261)

## 16.9  International Statistical Classification of Diseases

Farr's nomenclature based on eclectic criteria prevailed over that of Jacob Marc d'Espine (1806–1860) based on general pathological criteria that were becoming obsolescent. Farr's classification was revised in 1864, 1874, 1880, and 1886, and became the basis of the International List of Causes of Death. As a result of a request in 1891 by the International Statistical Institute, Dr. Jacques Bertillon (1851–1922) and a committee prepared a classification of causes of death, which was adopted at the meeting of the Institute at Chicago in 1893. It represented a synthesis of English, German, and Swiss classifica-tions used by the city of Paris and was based on the principle, introduced by Farr, of distinguishing between general diseases and those localized to a particular organ or anatomical site. The classification of 161 titles could be abridged to either 44 or 99 titles. The Bertillon Classification of Causes of Death, as it was first called, received general approval and was adopted by several countries as well as by many cities. The Institute at its meeting in 1899 recommended that since the classification had been adopted by all the North and South American states, it should be adopted in principle and without revision by all the statistical institutions of Europe. The Institute further approved generally the system of decennial revisions proposed by the American Public Health Association in 1898 and it urged all statistical offices to adhere to the general agreement on classification.

The Sixth Decennial Revision Conference of 1948 marked the beginning of a new era in international vital and health statistics. It approved a comprehensive list suitable for mortality and morbidity purposes and agreed on international rules for selecting the underlying cause of death. It also recommended the adoption of a comprehensive program of international cooperation in the field of vital and health statistics.

## 16.10  Biological Classifications Before Linnaeus

As Mayr (1982) points out, the history of taxonomy starts with Aristotle and his main work in descriptive zoology, *Historia animalium* (see §1.3). At first, his method consisted in applying a series of dichotomies to the superset, as we might call it, for example, the kingdom of plants; later, he saw that the system of dichotomies was not the best. Finally, he made a series of comparisons: comparative anatomy, reproductive biology, and behavior. Thus he was able to group 580 kinds of animals into assemblages like "birds" and "fishes." This was later to be of great importance in the revival of learning, the Renaissance. That a first-rate philosopher had interested himself in biology was a great encouragement to other scientists. In the meantime, the level of natural history fell into decline. Mayr praises Dioscorides* (floruit 50–70 A.D.) for his observations made on plants, the basis of a *Materia Medica*.

There was later a return to observation and so a return to nature in the sixteenth century with the three German fathers of botany, Jerome Bock* (1498–1554), Otto Brunfels* (*ca* 1489–1534), and Leonhard Fuchs* (1501–1566). Not only did these workers describe plants but they also had artist assistants or colleagues who drew them from nature and who did not embellish them with allegorical figures or distortions.

Andrea Cesalpino* (1519–1603) extended the works of these authors; he was the first to elaborate a system of the plants based on a unified and coherent group of notions according to K. Mägdefrau (*DSB*, XV, pp. 80–81). He designed a rudimentary classification of plants into *arbores, frutices, suffrutices* (shrubby herbs), and *herbae*.

John Ray* (1627–1705) can be cited as classifying plants and animals by a "natural" system (see Raven, 1950). He believed that breeding true from seed was a necessary test of a natural species. He had believed in the fixity of species; indeed, many case histories had shown that limited transmutation was possible. Ray (1686, transl. E. Silk in Beddall, 1957), cited from Mayr (1982), defined species thus:

> In order that an inventory of plants may be begun and a classification of them correctly established, we must try to discover criteria of some sort for distinguishing what are called "species." After a long and considerable investigation, no surer criterion for determining species has occurred to me than the distinguishing features that

perpetuate themselves in propagation from seed. Thus, no matter what variations occur in the individuals or the species, if they spring from the seed of one and the same plant, they are accidental variations and not such as to distinguish a species ... Animals likewise that differ specifically preserve their distinct species permanently; one species never springs from the seed of another nor vice versa. (p. 256)

We mention briefly some other authors. The notions of genus were elaborated by Joseph Pitton de Tournefort* (1656–1708) as a collection of species. His aim was to exhibit each species as a member of a genus, defined by character and thus by diagnosis. His genus was to be a "cluster of species," independent of the observer and identifiable. Another worker of interest is Augustus Quirinus Rivinus (family previously Bachmann)* (1652–1723), who used a binomial system for the names of species.

Michel Adanson* (1727–1806) disliked the notion of systems and proposed a natural classification based on a number of characters. It has been established that his views were shared by many Parisian botanists, including Bernard de Jussieu* (1699–1777). With the cooperation of the head gardener of the royal garden at Trianon, Jussieu managed to arrange part of it as a "botany school" to illustrate his views on a natural system of classification.

G. Ledyard Stebbins, Jr. notes that Jens Christian Clausen* (1891–1969) found that "a few characteristics used by taxonomists to differentiate species are inherited in simple Mendelian fashion and vary within populations, so that they are not diagnostic of biological species."[1]

## 16.11  Linnean System

The Linnean system has served as a paradigm to many other classifications since 1760. It came under much criticism from continental authors, French and German among others, because it seemed unnatural. Its features do not lend themselves to generalization to fields other than botany. This system is both a nomenclature and a classification, whereas the nomenclature follows the classification of the plants into genera.

Carl Linnaeus (or von Linné) was introduced to botany at an early age by his father, Nils Linnaeus, a country parson (Lindroth, *DSB* VIII, pp. 374–381). Carl had wide interests, for example, entomology, archeology, agriculture, and geography, but above all botany and classification. While a student at Uppsala University and after 1728, Linnaeus became interested in the sexuality of plants. The existence of a sexual system in animals had long been established but knowledge of such sexuality in plants was vague. Rudolf Jakob Camerarius* (1665–1721) in 1694 summarized the literature on the sex of plants and made many original observations of his own. He described the

---

[1] *Dictionary of Scientific Biography*, vol. XVII, p. 169a. Charles Scribner's Sons, New York. Copyright © American Council of Learned Societies, 1990. Reprinted with permission.

"structure of flowers, the male and female organs of plants ..." (see Stubbe, 1972, pp. 92–94). Linnaeus soon convinced himself of the truth of the matter and it later became the basis of his system. In a manuscript version of his *Hortus Uplandicus*, he announced in 1730 that the stamens and pistils were the sexual organs of plants. He divided plants into 24 classes according to the number of stamens. Each class was then subdivided into orders by the number of pistils. This procedure was attacked on the continent but accepted later in the 1760s with strong support from England. This choice of criterion for a preliminary classification was clearly unnatural and explains why Linnaeus's methods worked well in plants but did not generalize well to other fields of interest.

His views on classification have been set out in *Fundamenta Botanica* (1736), *Philosophia Botanica* (1751), and *Species Plantarum* (1753). Tournefort had established the genus, although it remained for Linnaeus to work with species as a clearly defined concept; he took them to be fixed and unchangeable. He gave precise methods of procedure; every species must be distinguished from all other species in the genus by a short diagnosis of at most twelve words and given a specific name; the whole plant had to be described in terms laid down by Linnaeus. But he found that it was impossible to use merely the specific name so he introduced the binomial nomenclature, following Rivinus as we have noted, of genus name followed by species name, a system adhered to ever since.

The Linnean system is designed for identification of species—a kind of index of the plant world. Since Linnaeus believed in the primacy of the species and their constancy, he did not attempt to derive higher taxa in order to suggest evidence of change or evolution. He originally believed, as set out in the *Philosophia Botanica* of 1751, that all species arose at the creation of the world and were therefore fixed; but observations caused him to change his views and so he could later write that different species of the same genus had descended from a single species. Thus he abandoned the principle of the invariability or fixity of species. It was not the intention of Linnaeus to link his classification to ideas on evolution. His system was successful in that it gave a place to all the plants whether European in origin or introduced as a result of the explorations abroad.

## 16.12 Biological Classification and Mathematics

This section explains why classification in biology is difficult. Almost all biological classsifications have been qualitative; the most successful classification in botany has been that of Linnaeus (see §16.11) but it has always been seen to be artificial. The corresponding theory is mathematical set theory. We try to explain with its aid a translation of a passage from Cuvier's *Règne Animal* given by Coleman (1964):

For a good classification ... we employ an assiduous comparison of creatures directed by the principle of the *subordination of characters*, which itself derives from the conditions of existence. The parts of an animal possessing a mutual fitness, there are some traits of them which exclude others and there are some which require others; when we know such and such traits of an animal we may calculate those which are coexistent with them and those which are incompatible; the parts, properties, or consistent traits which have the greatest number of these incompatible or coexistent relations with other animals, in other words, which exercise the most marked influence on the creature, we call *caractères importants, caractères dominateurs*; the others are the *caractères subordonnés*, and there are thus different degrees of them. (p. 77)

We have taxa (objects to be classified) with their totality forming a superset, which has to be broken down into sets by characters $\mathscr{A}, \mathscr{B}, \mathscr{C}, \ldots$ A character may take $2, 3, 4, \ldots$ different values, for example, A and a or $A_1$ and $A_2$ for two different values, $A_1$, $A_2$, and $A_3$ for three different values and so on. We use the first notation, A and a, below and note that it is easy to generalize to the second notation.

For example, with two characters, $\mathscr{A}$ and $\mathscr{B}$, we have the possibility of four distinct sets:

$$AB, Ab, aB, \text{ and } ab \tag{16.12.1}$$

which can be written

$$\{AB, Ab, aB, ab\} \tag{16.12.2}$$

There are classifications where we can write the sets derived by $n$ characters, each with two distinct values, as formulae such as $AbcDE\ldots$, which are called *hierarchical*; with $n$ characters there are at most $2^n$ possible sets. Such seem not to occur often in biology, even for splitting a genus into species.

*Set-theoretical independence* is defined with respect to the partition of a superset of taxa into sets by given characters as there being no zeroes in these sets; for example, with set-theoretical independence, the sets represented by equations (16.12.1) and (16.12.2) above contain no zeroes; there may be no content in the cell AB. As a rule, set-theoretical independence does not hold for the partition; in biology, the contrary property of *set-theoretical dependence* is well-nigh universal.

Cuvier's statement cited above asserts that there are *caractères importants*, important characters, which by the nature of things determine many other characters and naturally ensure that the taxa having the complementary form(s) of the character do not exist; for example, all vertebrata metabolize glucose, or the set defined by possessing vertebra and not metabolizing glucose is empty. It can happen that, for a set with given important characters, there are characters that are irrelevant to it.

It may be asked, for a given superset of taxa, how many characters are required to partition it so that each subset contains precisely those with the same set of values of the characters?

Alfred Rényi* (1921–1970) has given a theorem that states that if A, B, C,..., N is a finite set of $N$ taxa and if for any pair, $\{B, C\}$ say, there is a test that differentiates B from C, then a full classification is possible with the use of only $(N - 1)$ tests (Rényi, 1965).

So far we have been using tests with binary response, yes or no, A or a, etc. With several responses, $A_1, A_2, \ldots$ some care in statement will be needed. If the responses are all binary, there are two extreme cases, namely a minimum, $|\log_2 N| + 1$ where $2^{k-1} \leqslant N < 2^k$ and a maximum, $N - 1$, tests.

The reader is referred to other works by Rényi (1970a, 1970b) and to some of his 1961 papers, as well as to Jardine and Sibson (1971). The classification of living agents has an analogue in the establishment of discrete distributions by taking subsets of a finite set of points.

# 17
# Numerical Analysis of Clinical Experience

## 17.1 Early Difficulties

In this chapter, we consider some history of numerical analysis applied to clinical observations on individuals rather than to the massed data of the official statistics. The history of such numerical analysis is short; perhaps 1790 could be chosen as a beginning, with the works of Alexander Gordon (§9.1) and Philippe Pinel (§16.2) following soon after. Some reasons for this late development can be given:

*Specificity of disease.* The difficulties of classifying diseases to form homogeneous groups for comparisons could only be solved after the bacteriological revolution initiated by Pasteur and Koch in the last quarter of the nineteenth century; before that time, diseases were not thought of as entities (see §8.6 and Chapter 16).

*Individuality of patients.* There were doubts that patients could be classified in a meaningful way, the academic view being that the individuality (variance) of the patient was of a greater order than that of the disease, for example, Double's view as set forth in the debate in the French Academy of Science (see §§17.2 and 17.10).

*Numbers of patients observable.* For comparisons between treatments of any disease, large numbers of patients are usually necessary, with the size of the necessary experience being dependent on the relative effectiveness of the treatments. The assembly of such large numbers would have been possible in the great hospitals but we have seen that conditions in them were usually chaotic.

*Lack of effective medicines.* In the early nineteenth century, there were few effective remedies; often the supposed remedy was actually harmful.

*Opposition from the academic physicians.* There was a general attitude that "mathematics" did not have anything to give to medicine.

It is easy to see that these factors would usually count against appointments of statisticians unless their work was confined to vital statistics.

In the remainder of this chapter, we will be preparing the way to the study of clinical trials after 1935, the date of the publication of *The Design of Experiments* of R.A. Fisher. It is now widely recognized that clinical trials are special cases of experiment in the sense of that book, with economic and ethical complications. It can be said that man is a large, expensive, and litigious animal. Special properties are required of practitioners of clinical surveys, who tend to become isolated from the natural scientific aspects of medicine. Clinical trials can be defined after the manner of Bull (1959) as "deliberate experiments designed to assess upon patients the value of therapeutic procedures." The aim of clinical trials, so defined, is to obtain as much information from the trial with as little expenditure as possible. The phrase "clinical trials in the strict sense" means that the experiments have been carried out with proper attention to rules of procedures that have been laid down by modern workers and are explained in detail in Chapter 18.

## 17.2  Gavarret as a Modern Medical Statistician

Louis-Denis-Jules Gavarret (1809–1890), physician and physiologist, gave possibly the first formal statement of the principles of medical statistics (Gavarret, 1840). In his Preface, he pointed out that propositions were spoken of as *mathematically* proved and laws as *mathematically* established; and whereas there appeared to be no barrier to the progress of medicine in discovering the seat of disease and diagnosing disease, therapy lagged behind. Some distinguished physicians struggled to use statistics in medicine as the sole method of collecting the experience of the "centuries." Gavarret found that the discussions centered on whether numerical ratios could replace such terms as *often, rarely, in the great majority of cases,* etc. At a special sitting of 5 October 1835 at the French Academy of Sciences, the scope of the discussion was enlarged from the "numerical method" to the use of the calculus of probabilities in therapy. F.J. Double, the rapporteur, was against such applications, but C.L.M.H. Navier argued for them with great lucidity. Gavarret could see that the question was not trifling but indeed exciting. The works of Laplace, moreover, were to demonstrate beautiful solutions to problems in astronomy. Gavarret's interest was further aroused when he noted that D.F.J. Arago had commented how applications of probability theory had given confidence to the astronomers to distinguish real groupings of stars in the year 1767. (See §9.6 for the analysis of contagion.)

Gavarret attended another meeting in April 1837, where the limited notion of numerical ratios replacing such words as *often* and *rarely* was discussed again; but Gavarret was determined to show how rules deduced from the calculus of probability formed an indispensable complement to the experimental method of his day; moreover, in view of the lack of familiarity of medical workers with the higher mathematics, he was determined to give an account shorn of all algebraic formulae. Here he is following the example of Jean-Baptiste Bouillaud (1796–1881) mentioned on p. 189 of that author's work, *Essai sur la Philosophie Médicale*. It is important to remember for the interpretation of Gavarret's judgment on Louis's work on phlebotomy that Bouillaud was a colleague of François Joseph Victor Broussais* (1772–1838), whose excesses in the operation were criticized by Louis.

Gavarret (1840) begins his introduction with two citations, one from S.D. Poisson's *Recherches sur la Probabilité des Jugements*: "*La médecine ne serait ni une science ni un art, si elle n'était pas fondée sur de nombreuses observations, et sur le tact et l'expérience propres du médecin, qui lui font juger de la similitude des cas et apprécier les circonstances exceptionnelles.*" [Medicine would be neither a science nor an art, if it were not founded on numerous observations, and on the doctor's own tact and experience, which help him to judge the similarity of cases and to appreciate the exceptional circumstances.] Since Gavarret (1840) was the most sophisticated of texts, designed as an explanation of applied medical statistics, it is necessary to give some further ideas of its contents. He notes that numerical methods in the descriptions of diseases have been accepted but they have led to inconsistencies because no account has been taken of the limits of the error. Thus he finds that Louis (1835/1836) has founded his conclusions on the efficacy of phlebotomy on too few cases—107 cases of pneumonia, 44 cases of erysipelas of the face, and 23 cases of laryngeal angina (diphtheria). Perhaps, he has erred here because of prejudgment of the phlebotomy issue; but, in any case, there was no general theory that he could call on to suggest appropriate size for such tests. It is possible to obtain, we may note, strikingly low probabilities with quite small numbers of observations. With this reservation, we can applaud Gavarret's (1840, p. 26) general principles:

1. To determine what one understands by similar or comparable facts (*faits semblables ou comparables*) suitable for entry into the composition of a statistic [that is, for the estimation of a parameter].
2. To prove that every conclusion deduced from a small number of facts deserves, in therapeutics, no confidence and that every statistic [that is, estimate], to furnish admissible information, must be based on several hundreds of observations.
3. To demonstrate how laws deduced *a posteriori* are never true except within certain limits of variation (*oscillation*) and to give the means of determining these limits in each particular case.

Gavarret says the first of these considerations has hardly been skimmed (*effleurée*) by the medical statisticians; a glance at their publications will show how little they heed the second when making generalizations; and the most superficial examination will suffice to show that they do not heed the third. He points out that, although the calculus of probabilities will resolve these three problems by giving a methodology, there remain many questions in medicine outside the scope of medical statistics.

Gavarret notes that the empiricists have stressed the importance of individuals but this line of reasoning leads to a denial that experience can teach us anything in the treatment of cases. Nevertheless, as Gavarret points out (p. 49), before there was an adequate system of diagnosis researchers would necessarily be comparing cases of a heterogeneous nature, which would make the drawing of legitimate conclusions very difficult and, indeed, would render many classical aphorisms invalid. He cites the dictum of Morgagni "*Non numerandae sed perpendendae sunt observationes*" [Observations must be not counted but weighed carefully]; but since his time diagnostic skill had improved, so that a term of varied etiology like *ascites* would not be used to form a diagnostic group to be regarded as sufficiently homogeneous. As he has already remarked, "one can count in medicine; it is only a matter of demonstrating how it should be done" (p. 44). His method essentially is to study the application of the binomial distribution to medical problems using the normal approximation of Laplace, which he refers to as the "*loi des grandes nombres*".

In the calculus of probabilities, there are (a) the direct calculus, essentially the working out of hypotheses involving known parameters, and (b) the inverse calculus, founded essentially on the law of large numbers (actually on Laplace's central limit theorem), whereby the probability of an event can be estimated from the data and an estimate of error given. Gavarret gives urn examples of events with constant probability and those with variable probability.

The mathematical side of Gavarret's work is treated in several lengthy appendices. His theory concerns binomial variables with parameters $\mu$, $p$, $q$, say. He writes $\mu$ for the number of trials, $n$ for the number of appearances of the event, and $m$ for the number of non-appearances of the event. He uses a normal distribution with a variance 2 rather than the modern variance of 1. His limits of significance are twice his modulus and so $2\sqrt{2}$ times the modern standard deviation; an observation has a probability of 0.9953 of not exceeding these limits; he might be said to be working at approximately a significance of 0.005. For a binomial experiment he is saying that the probability of the estimate $\hat{p}$ of the unknown parameter being within two moduli (or our $2\sqrt{2}$ standard deviations) of the true result is 0.9953 and that with increasing size of experiment $|p - \hat{p}|$ becomes increasingly small. He develops a similar theory for the difference between two means.

## 17.3 The First Clinical Trial and Scurvy

With the definition in §17.1 of a clinical trial in the strict sense, we can assign the priority for the first clinical trial to James Lind* (1716–1794). There are competitors for this priority. It may well be that further translations of Ibn Sina* (980–1037) (or the Latinized form Avicenna) would cause the priority to be awarded to him. He was the first, perhaps, to think clearly about testing the efficacy of remedies. In his *Canon*, he states that the trial should be of a remedy in its natural state upon uncomplicated disease and that for the test of efficacy, two opposed cases should be observed; studies should also be made of the time of action of the remedy and of the reproducibility of its effects; results in other species may be irrelevant for man. The claims of Ambroise Paré* (ca 1510–1590) must be set aside, as his experiment was unintentional (see §17.6). Indeed, if a certain comparison in a clinical trial has not been specified beforehand, the experimenter cannot claim that its significance is verified by the experiment, but suggestions can be permitted that testing the comparison in a further experiment is worthwhile. The claims for the priority of Leonardo da Vinci can be countered because he did not carry out experiments. It seems now that the efficacy of quinine (Jesuits' bark, etc.) was not adequately tested by clinical trial.

References to scurvy are known in the Ebers Papyrus of about 1500 B.C.. and in the works of Hippocrates, for the mention of pain in the legs, gangrene of the gums, and loss of teeth are diagnostic of the disease. Scurvy was a common disease in Europe during the Middle Ages. In the fourteenth and fifteenth centuries, sea voyages became longer, with long intervals without access to fresh food; as a result, there were huge losses of seamen from scurvy in the voyages of Magellan and of Vasco da Gama, for example [see Watt, Freeman, and Bynum (1981) for many other specific examples].

The treatment of scurvy first recorded in the European literature was due to Jacques Cartier (1491–1557), who learned from the Amerinds of Quebec the art of making an infusion of the swamp spruce and so saved his expedition from disaster. According to Lloyd (1961), Sir Richard Hawkins (ca 1562–1622) had already known of the efficacy of citrus juices in 1593. Captain James Lancaster (d. 1618) in 1600 clearly demonstrated the value of lemon juice on the voyage to India; he had no case of scurvy on his ship, the only one of four ships carrying lemons. In 1753, James Lind, appalled by the loss of 1,051 out of 1,955 men in Anson's voyage around the world in 1740 to 1744, described his own experiments on the prevention and cure of scurvy. In May 1747, he had selected 12 scurvy patients, to all of whom he gave the same basic diet. He added different supplements to six pairs of them as follows: (a) a quart of cider a day; (b) 25 drops of *elixir vitrio* three times daily; (c) spoonfuls of vinegar; (d) seawater; (e) an electuary (a powder mixed with honey) of garlic, mustard seed, horseradish, and other ingredients; (f) two oranges and a lemon daily. Those given the citrus fruits (f) did best and were declared fit for duty in six days; Lind concluded that cider also had some effect but that all the other treatments were useless. Lind obviously should be praised for his experiment. A modern statistician would perhaps have liked to see more cases treated with juices, possibly at the cost of the trial of salt water in pair (d). One supposes that Lind proceeded to treat scurvy with citrus juices wherever possible and, seeing the results continually in favor of his treatment, would regard it as unnecessary and even pedantic to do more trials. Lord Anson, First Lord of the Admiralty, after the publication of Lind's *A Treatise of the Scurvy* in 1753, appointed Lind to the Royal Naval Hospital at Haslar. Of the 5,734 admissions to his care in the first two years of his administration, 1,146 were due to scurvy, so that his views could be confirmed; but his conclusions were not adopted by the Navy [see Meiklejohn (1954) and Lloyd (1961)].

Lloyd (1961) has pointed out how the circumstances of the voyages of James Cook* (1728–1779) made them quite unfit for a test of the efficacy of remedies; first, his men were picked and better victualled than any others on a comparable voyage; they were also better quartered than those in the great battleships. Moreover, Cook insisted on the men taking a varied diet and he made as many calls to port as he could. Cook had the advice of a surgeon, who appears to have nothing to say about fruit juices; special foods were a kind of condensed soup, sauerkraut, or pickled cabbage; fruit juices were only carried in small quantities as a necessity for the sick; 40 bushels of malt were also available. Cook's surgeons in their report stressed the importance of malt as a preventative. We can see the defects of the observations and the contemporary comments. First, there was no control, and second, there may have been little deprivation of vitamin C because of Cook's policy, mentioned above. It is of interest that Joseph Banks* (1743–1820) recorded in his diary that, noting the swelling of his gums and incipient ulcerations inside his mouth, "I then flew to the lemon juice," nearly 6 oz. per day and "in less than a week my gums were as firm as ever." Cook, however, reported on the value of malt on the advice of his surgeons.

That the Admiralty regarded Cook's voyages as efficient tests of the efficacy of treatment by malt was a disaster for the Navy; the surgeons on his voyages had no special knowledge, had been misled by recent literature on the efficacy of malt taken in the form of wort, omitted any reference to fruit juices, and were not interested in the results.

Naval operations during the American revolution forced a review of the effectiveness of malt since in 1780, 2,400 cases, one seventh of the personnel of the fleet, were landed for treatment of scurvy. In 1781, the Commander in Chief of the Channel Fleet reported after a ten-weeks' cruise "the scurvy is making strong strides, so that we cannot keep the sea." In the Navy serving in the West Indies, where fruit was readily available, there were 89 deaths out of 1,844 cases in five months; Gilbert Blane (1749–1834), the chief physician of the fleet, could say, after his doubts about citrus juices had been removed, that the juice worked like a "charm." Finally, the issue of citrus juices became obligatory in the fleet (see Lloyd and Coulter, 1961, pp. 319–326). At the end of the Napoleonic Wars a naval surgeon is reported to have said "It is the opinion of some of the most experienced officers that the blockading system of warfare which annihilated the naval powers of France could never have been carried on unless sea-scurvy had been subdued" (Roddis, 1950, p. 51).

## 17.4 Quinine

The use of quinine in the treatment of malaria can be examined as a case history illustrating the difficulties of assessing the effectiveness of a drug in the prescientific days.

Quinine, by which we mean here Peruvian bark or cinchona or any other medicinal containing quinine as its active principle, was first brought to Europe in 1632, from Peru to Rome in the form of Peruvian bark. Cardinal di Lugo developed a severe attack of malaria some time later, for which he was treated with quinine with dramatic results. Further tests were made at Santo Spirito Hospital in Rome, leading to a very favorable report and a recommendation from the Cardinal that its use be widespread. Quinine became recognized as effective against malaria, but the Peruvian bark was sometimes mixed with the bark of a Peruvian balsam tree reducing, and leading to conflicting ideas on, its therapeutic value.

After its introduction into India in 1657, quinine or its bark was much used. James Lind treated some 200 cases of malaria in India in 1765, losing only two cases, neither of whom had had an adequate course of quinine. He was convinced of the adequacy of the drug, remarking that "in the administration of the bark, the cure of ague may be said to entirely consist." A change in medical opinion was brought about by James Johnson (1777–1845), a naval surgeon arriving in Calcutta in 1804. His first fever patient, which he diagnosed as malaria, failed to respond to cinchona and died on the third day. At his autopsy, Johnson was greatly impressed by the "intense engorgement of his liver." He concluded that evacuations and venesection would have been more logical. His next patient was so treated and survived. Neither case appears to have been malaria according to Russell (1955). Johnson never again treated a malaria patient with cinchona; he used mercury treatment to excessive salivation and calomel purging or blood letting with knife or leech. After a short experience in India, Johnson returned to England and published in 1813 a book, *The Influence of Tropical Climate ... on European Constitutions*, which ran through four editions. The damage was done; cinchona was no longer the drug of choice in India. Edward Hare (1812–1897), arriving in India in 1839, followed the same practice, because quinine was forbidden by all authorities, with disastrous results. Fortunately, he read the old reports of James Lind and began to use cinchona alkaloid cautiously. During nine years he treated nearly 7,000 fever patients with the alkaloid and was able to report a mortality of less than 0.5%. A year's trial at Calcutta Hospital reduced the mortality "tenfold," which restored confidence in cinchona as the specific remedy. Once again Lind's notes on the cure rate of quinine therapy give him an interesting priority. It is curious that Johnson's conclusion based on one case was so easily accepted.

## 17.5 Jennerian Vaccination

Edward Jenner* (1749–1823) was born into a wealthy county family at Berkeley in Gloucestershire. As part of his medical education Jenner studied anatomy under John Hunter, whose assistant he became, arranging zoological specimens brought back from Cook's first voyage of the *Endeavour* in 1770. He returned to practise in Berkeley, where he was often asked to inoculate persons against smallpox. This inoculation was dangerous, as it could cause a severe attack in the recipient or cause other actual cases of smallpox. Jenner noticed that some patients, completely resistant to the inoculation with smallpox, had had attacks of cowpox. Further, he found that milkmen or milkmaids were usually immune to smallpox. He concluded that an attack of cowpox might be useful as a prophylactic against smallpox. He determined that only one of the forms of eruption on the teats of cows was effective in creating a resistance to smallpox. On 14 May 1796, Jenner vaccinated the 8-year-old James Phipps and after a successful "take" inoculated him with smallpox on 1 July 1796; Phipps was found to be immune to smallpox. In June 1798, Jenner was able to publish *An Inquiry into the Causes and Effects of the Variolae Vaccinae*, in which 23 instances of vaccination had conferred a lasting immunity of the subject against smallpox.

This testing of the vaccine can be counted as a clinical trial in the strict sense, with a hypothesis to test, a procedure to be followed, and a criterion of success (immunity to smallpox inoculation).

## 17.6  Early Tests of the Efficacy of Therapy

There was early interest in the test of the effectiveness of therapy. As an example, Antoine-Laurent Lavoisier wrote in his *Méthode de Nomenclature Chimique* (1787):

Le seul moyen de prévenir ces écarts, consiste à supprimer, ou au moins à simplifier, autant qu'il est possible, le raisonnement qui est de nous, & qui peut seul nous égarer, à le mettre continuellement à l'épreuve de l'expérience; à ne conserver que les faits qui sont des vérités données par la nature, & qui ne peuvent nous tromper; à ne chercher la vérité que dans l'enchaînement des expériences & des observations, sur-tout dans l'ordre dans lequel elles sont présentées, de la même manière que les mathématiciens parviennent à la solution d'un problème par le simple arrangement des données, & en réduisant le raisonnement à des opérations si simples, à des jugemens si courts, qu'ils ne perdent jamais de vue l'évidence qui leur sert de guide. [The only way to prevent these errors is to suppress, or at least to simplify, as much as possible, the reasoning which is our own, and which by itself can mislead us, and to put it continually to the test of experience; to keep only the facts that are the truths given by nature, and that cannot deceive us; to seek the truth only in the link between experiments and observations, above all in the order in which they are presented, in the same manner as the mathematicians arrive at a solution of a problem by a simple arrangement of the data, and by reducing the reasoning to operations so simple, and judgments so short, that they never lose sight of the evidence that serves as their guide.] *Oeuvres* 5:358

This statement does not get us very far but it is a first step against notable clinicians drawing conclusions from very scant data, that is, few cases.

There must be some consideration as to what is measured or who are counted. Some of the general conclusions of Louis (1835/1836) may first be mentioned. He would like to have some objective criteria, such as the length of stay in hospital, to measure the effectiveness of treatment; but it will be asked, he adds, how can criteria be used of an unclassified group of patients when it is known that age, sex, physique, and so on are determinants of the outcome; the alternative is to have many more homogeneous groups of patients; but this process cannot be carried too far for then there would be nothing but individualities in medicine. On the other hand, certain medicines applied to individuals of different ages, strengths, temperaments, and so on have almost uniform success, for example, drastic purges in painter's colic, quinine in malaria; indeed, the malady itself seems to efface such differences. Louis believes that if 500 were given one treatment in an epidemic and 500 were given another, and if one treatment was associated with fewer deaths, then we must accept it as the superior treatment. He remarks that

it is impossible to appreciate each case with mathematical exactness, and it is precisely on this account that enumeration becomes necessary; by so doing, the errors, (which are inevitable) being the same in two groups of patients subjected to different treatment, mutually compensate each other, and they may be disregarded without sensibly affecting the exactness of the results. (p. 60)

Here he could have obtained help from contemporary mathematicians such as Poisson and tested the difference between proportions.

Louis (1835/1836, p. 96) writes,

Let those, who engage hereafter in the study of therapeutics, pursue an opposite course to that of their predecessors. Let them not think that they have done anything effectual, when they have only displayed their own theories, or stated what is done by the most celebrated physicians in such or such a case. But let them labor to demonstrate, rigorously, the influence and the degree of influence of any therapeutic agent, on the *duration, progress, and termination of a particular disease*. Let them not forget that nothing is more difficult, than to verify a fact of this nature; that it can be effected only by means of an extensive series of observations, collected with exactness; instead of touching upon a boundless inquiry, let them limit the subject, that they may master it completely, and study it in all its aspects. Let them reflect that while this is the only means of being useful to science and to mankind, it is at the same time the only source of true fame to the student in therapeutics.

It is clear to us now that Pinel and Louis were correct; the writings of previous authors and established doctrines were not sufficient guides for therapy. Careful and detailed observations must be made to classify the diseases and enable the clinician to observe the effects of therapy on patients with given diseases or syndromes. Further, the best therapy could only be selected after competing therapies had been tested on many patients, perhaps even hundreds of them.

The reading of Louis (1835/1836, pp. 1–54) will come as a disappointment to a modern statistician, for example, the table (p. 3), where the individual cases are classified by day of first bleeding and the variables noted are the duration of illness and the number of bleedings. There are 78 cases of pleuropneumonia treated, with 28 deaths; the deaths are omitted from the table, leaving 50 cases. It appears that the antiphlogistic treatment (the blood-letting), when applied in the first two days, may very much abridge the duration of the disease, whereas after that the day seems to be irrelevant to the abridgement. The same conclusion follows if only cases in one age group are considered. It is exceedingly difficult to see what conclusions should have been reached as there has been no plan or design of experiment. Further, as Gavarret (1840) was to remark, the numbers of cases are too low for valid conclusions to be drawn (see §17.2).

The discussion on Louis can be concluded by the remarks of one of his American followers, Jackson (1836), who, after a comparison of Louis's results in Paris with his own in Boston, writes,

In conclusion, many readers may ask if it is thought that the researches, of which this volume contains the results, are to be considered as leading to any positive conclusions. Certainly not. M. Louis has done us great service in stating his own accurate observations. They must have great weight in the minds of reflecting men. We have added all the observations that we have of sufficient accuracy to be compared with his, which will be received for what they are worth. The whole are to be regarded as materials, to which others are solicited to make additions from time to time; that, at length, so many cases, impartially collected, may be brought together, as shall justify entire confidence in the inferences to be made from them. Ten hospitals, under the care of honest physicians, may settle the questions discussed in this work within five years, so that our posterity will not for ages be able to make any material correction in the answers. Seasons and epidemics will vary no doubt; but the general laws will be found the same, and little else would remain for future ages than to settle the allowance to be made for disturbing forces. (pp. 170–171)

Later events have shown that Jackson was overly optimistic and could not see the difficulties of clinical trials, a topic that we can defer to Chapter 18.

[See Bull (1959, p. 231) for a note on the admirable precepts, laid down by Louis (1831/1834) in *Essay on Clinical Instruction*, for collecting clinical data and testing the efficacy of drugs.]

Several unplanned experiments yielded important information and are now classic. At the siege of Turin in 1536, Ambroise Paré's supply of boiling oil for the cauterization of wounds had become exhausted. He, therefore, used a dressing composed of egg yolk, oil of roses, and turpentine. The next day, he found that soldiers treated according to the improvised method were doing better than those treated by the conventional boiling oil treatment. He later experimented with other dressings and concluded that wounds in themselves were not poisonous and did not require cautery, conclusions that he published in a treatise of 1545.

John Hunter* (1728–1793) in the Belle Isle campaign found that patients were no worse off if the projectiles were not deliberately sought and removed from wounds. Of course, this is an observation rather than an experiment. Comparisons were made with other troops who had suffered the standard treatment.

## 17.7  Rejection of Venesection

In the absence of a rational physiology, bloodletting, venesection, or phlebotomy was regarded as a useful, indeed important, branch of therapy. The practice of phlebotomy dates back to ancient Egypt and to Hippocrates, although Erasistratus* (*ca* 304–250 B.C.) rarely used it. Celsus and Galen both practised the procedure; Wangensteen and Wangensteen (1978, pp. 246–250) may be consulted for some historical remarks showing how popular or well established the practice was. But there were opponents. Thus Johannes Baptista van Helmont, in the early seventeenth century, rejected much of traditional therapy, especially bloodletting. His criticism is said to have been the inspiration of a Viennese veterinary surgeon, Johann Wolstein (1738–1820), who published a twenty-year experimental study of bloodletting and concluded that it was of value in very few clinical conditions. Louis (1835/1836) wrote a famous work on venesection, which is difficult to follow but probably is an adequate condemnation of the method (see §17.6.) Škoda is also said to have proved it ineffective.

Later in Edinburgh, Bennett (1865) analyzed 129 cases of pneumonia by sex, age, and severity, the mortality rate, and length of duration of the illness. He concluded that "restorative" methods were just as good as the heroic methods and that tartar emetic should be abandoned. Venesection was justified in various ways. Thus the doctrines of F.J.V. Broussais, that disease was due to gastroenteritis and must be treated by repeated bleedings and debilitating diets, had been immensely popular in the early nineteenth century in France. Leeches began to be imported into France with numbers of (in millions) 0.3, 33.6, 41.7, and 21.9 in the years 1823, 1827, 1833, and 1834, respectively. Applied to the treatment of hemorrhages, cancer, malaria, scurvy, or syphilis, these doctrines had been disastrous. In particular, Broussais's doctrines began to be criticized, even by his students, when the treatment was applied to cases of cholera in the epidemic of 1832 in France. Further, bloodletting was not free of complications, which included anemia, fever, phlebitis, and pyemia.

## 17.8  Medical Theories of Therapy

Before 1860, there had been many schools of therapy, of which we may mention a few in order to understand why clinical trials of therapy, coming to eminently sound conclusions according to modern theory, were not accepted by contemporaries.

1. *Vis mediatrix naturae,* the healing power of nature. This is a force believed in by many schools of medical thought from ancient times onward. A. Paré's motto is famous "*Je le pensai, Dieu le guarist*"—I dressed him, God cured him. The belief is the basis of expectant treatment—symptoms are

to be treated as they arise and the patient is to be given good conditions of food and shelter, and good nursing. Any other treatment must be justified by showing a superiority over the expectant. The following quotation from Lavoisier (1784) cited by Hill (1960, pp. 170–171) is apt:

Since the principle of life in animals is a force which is ever active, which is constantly endeavouring to overcome obstacles, and since nature when left to its own devices cures many diseases by itself, it follows that, when a remedy is applied, it is infinitely difficult to determine what effects are due to nature and what to the remedy. The result presents itself to the wise man merely as a greater or lesser degree of probability, and that probability can be converted into certainty only by a large number of facts of the same kind.

It is remarkable that in the second half of the eighteenth century, dominant personalities could cause vast changes of treatment and of the ways of considering disease in a short time. To Brown and Hahnemann (see below), we could add Johnson, considered in §17.4.

2. *Brunonism*. According to Lesky (1976), the theory of John Brown (1735–1788) was based on the consideration that life was but a condition induced and maintained by stimuli acting on the human body, which thus had the property of excitability. Ill health according to his doctrine resulted when excitability departed too much from what could be called the median degree. Too many or too few stimuli thus brought about disease, sthenia or asthenia, respectively. His theory gave little attention to the humors of the body. In place of purgatives, laxatives, and expectorants, the "emptying" medicines, so-called stimulating medicines were now given— opium, Peruvian bark, camphor, wine, etc.

3. *Allopathy*. The system of heroic symptomatic treatment, allopathy, spread widely under the influence of James Gregory (1753–1821), with bloodletting, emetics, purges, leeching, and so on.

4. *Homeopathy*. A reaction against such monstrosities as allopathy, summed up in the statement that the patient had died cured, was the rise of homeopathy, the exhibition of drugs in small doses. The three principles of this system, enunciated by Christian Friedrich Samuel Hahnemann* (1755–1843) were (a) like is cured by like (*similia similibus curantur*), since the disease depends upon a perversion of the purely spiritual powers and is entirely immaterial it cannot be combatted by physical agencies; (b) its symptoms can be treated by substances whose potency is released by dilution and whose action, originally like the disease, is reversed by dilution; (c) the medicinal treatment must be supported by dietetic and hygienic measures. Homeopathy probably did much less harm than allopathy but degenerated into absurdities when in 1829 drugs could be exhibited in dilutions of one in $10^{60}$, with an almost complete certainty that no molecule of the drug administered would reach the patient. The principal lessons of these aberrations were that some drugs had an effect and that the body was often able to cure its own ills without aid from drugs.

# 17.9 Immunological Control of an Endemic Bacterial Disease

We take diphtheria as our example and discuss it in a way that can be applied to other diseases. A more clinical approach has been adopted in Lancaster (1990).

## History

Bretonneau in 1826 stated that the essential pathological lesion was a pseudomembrane in the pharynx or elsewhere. F.Loeffler in 1884 discovered the bacteria, grew them in culture, and produced lesions in animals resembling those in man. Further, he found that bacteria could be found only at the site of injection and not at the site of the lesions. He speculated that there was a poisonous substance released by the bacteria. Further, he discovered the bacillus in healthy carriers. This reasoning was confirmed by the work of Émile Roux and Alexandre-John-Émile Yersin* (1863–1943). Diphtheria toxoid for active immunization was available from 1928 but was little used until 1941.

## Homogeneity

The species *Corynebacterium diphtheriae* is not homogeneous but consists of several bacterial types, *gravis, intermedius,* and *mitis,* which vary in their virulence (see Anderson, Cooper, et al., 1931). The existence of such inhomogeneity makes it quite misleading to use historical series of cases as controls in any clinical trial of therapy, for it is now known that the ratio of the types varies between epidemics. If possible (or practical) a trial should be split into subtrials, in which the therapy for cases and controls is considered for a single type, *gravis* for example.

## Antitoxic serum

Roux and others in 1892 to 1893 produced such a serum. From the 1890s onward it has been improved and thus its potency would have to be specified in terms of some agreed standard. The subjects of the trial must also be defined as there is possibly a heterogeneity of response to diphtheria infection by age and sex. Such heterogeneities occur in other infections including smallpox, the pneumonias, streptococcal diseases, and possibly others. [See McLeod (1943) and G.S. Wilson and G. Smith (1984) for authoritative notes and reviews of the general problem.]

## Diphtheria in Denmark

Th. Madsen and S. Madsen (1956) gave the experience of Denmark in some detail. Thorvald Johannes Marius Madsen (1870–1957), indeed, had helped to prepare the first antidiphtheritic serum in Denmark, available in the summer of 1895. One has no reason to suspect that he had set out to belittle his own pioneering efforts. Thorvald noted that the case fatality rates in Copenhagen had already fallen before the availability of serum, some months in 1894 passing without a single diphtheritic death. Also, in Sweden, case fatality rates had fallen in Stockholm before serum had been introduced, being 38.3%, 34.7%, 20.3%, and 9.0% in the four quarters of 1894. In Norway also, serotherapy was introduced at a time of rapidly declining death rates. Madsen and Madsen believed that much of the decline in the mortality around 1895 was due to changes in the type of diphtheria bacillus prevalent at the time. S. Madsen stated in Part II of the cited article that "diphtheria was defeated by [active] immunization and not by serum therapy."

## Diphtheria in England and Wales

MacIntyre (1926) commented:

It is just over 30 years since antitoxin was discovered for the treatment of diphtheria, and it may be of interest to take a brief review of the results obtained with it and the progress which has been made in our methods of administration. For the first few years after its introduction in 1894 [in England and Wales], the serum used was often of uncertain strength and the dosage, according to our present standard, was very small; but in spite of this the results obtained were very definite, as shown by the improvement of patients after injection and the marked decrease in the case mortality ... However, after 30 years' experience, during which antitoxin has been extensively used, we find that diphtheria still causes a large number of deaths in this country every year. And in comparing the deaths in 1924 with the number registered from the disease in 1871 we find that the decrease has been very small. Also, if we compare diphtheria in this respect with scarlet fever and typhoid fever, we find that during the past 54 years a much more marked and progressive fall has occurred in the annual deaths from these two diseases than from diphtheria. (p. 855)

There had been no serious controlled clinical trial except Fibiger (1898). The Danish statistician, G. Rasch, cited by Madsen and Madsen (1956), believed that even Fibiger's trial could not be used with confidence to make the required comparison, although the case fatality rates were 3.5% in the treated and 12.2% in the untreated cases (controls).

## 17.10  Individuality of Patients

The medical academics made great play with the individuality of their patients; it is impossible, they would say, to make prognoses on individuals from statistical data. F.J. Double (1837) expressed the idea:

For myself I must say that the more I see of a disease the more does each case appear to me a new and a separate problem ... Individuality is an invariant element in pathology. A disease is not a simple fixed and uniform entity; it is a series of varied and changing actions;... Numerical and statistical calculations, open to many sources of fallacy, are in no degree applicable to therapeutics. (Armitage, 1983, p. 325)[1]

Louis (1837) answered:

The object of medical statistics is the most rigorous determination which is possible of general facts, which in my opinion cannot be arrived at without their assistance. Thus a therapeutic agent cannot be employed with any discrimination or probability of success in a given case, unless its general efficacy, in analogous cases, has been previously ascertained; therefore I conceive that without the aid of statistics nothing like real medical science is possible.... Those who argue against statistics talk a good deal of their study of medical literature; but what is the use of it if it cannot lead to any general views of treatment? Or what is the use of writing, if we can only state particular cases, without drawing any general inferences, or making any advantageous use of the past?[1]

Bernard (1865/1957, p. 137) brought another objection, pointing out that if the case fatality rate from an operation is 40%, "this ratio means literally nothing scientifically and gives us no certainty in performing the next operation ... the law of large numbers never teaches us anything about any particular case." As Armitage (1983) says, this argument of Double against Louis still persists.

Carried to its logical conclusion, amassing cases by clinical experience is no guide to prognosis. Fortunately, there is an escape; the patient can be regarded in some cases as a member of a "recognized subset," defined by age, sex, and certain features of the disease, which may classify the patient as low or high or other risk. However, there usually remains a stochastic element in prognosis, for the outcome of biological events is rarely determined.

## 17.11  Reference Set

In a properly designed experiment, in which treated and controls have been chosen by an appropriate randomization, the probability of every outcome can, in principle, be computed. In considering observations, rather than experiments, difficulties may arise.

We can take the analysis of Daniel Bernoulli (1735), mentioned by Hans Straub (*DSB*, II, p. 45) as problem 24, as the first example of the clustering problem. Bernoulli considered the inclinations to the ecliptic of the five planets; none exceeded $7°30'$, which Bernoulli took to be one-twelfth of a right angle. The probability that all inclinations were less than $7°30'$ was $p^5 = 12^{-5}$ under the null hypothesis that the inclinations were mutually independent (excluding the earth); $p^5$ being a very small number, Bernoulli concluded that the inclinations were not mutually independent.

---

[1] Citations from Double, and Louis taken from Armitage (1983, p. 325). Copyright © *Journal of the Royal Statistical Society* A, 1983; 146:321–334. Reprinted with permission.

A second astronomical example was given by John Michell* (*ca* 1724–1793) in 1767, which was summarized in Todhunter (1865, p. 332) and mentioned favorably by Gavarret (1840). It was required to find whether binary stars or larger groupings of the bright stars such as the Pleiades (Seven Sisters) could be explained as a result of a uniform random scattering of stars over the heavenly vault or of some physical process favoring aggregations. What makes the problem difficult mathematically is that it does not require the calculation of seven stars falling within a circular area of apparent diameter 1°, say, but of them falling within any such region of the same area, somewhere in the heavens.

Perhaps the earliest biological treatment of this problem was by Maupertuis (1752), which we have partly discussed in §4.1. He surely rejected the null hypothesis that polydactyly was not heritable. He writes in his Lettres (1752) No. XIV:

> But if one wished to regard the continuation of polydactyly as an effect of pure chance, it would be necessary to see what the probability is that this accidental variation in a first parent would be repeated in his descendants. After a search which I have made in a city which has one hundred thousand inhabitants, I have found two men who had this singularity. Let us suppose, which is difficult, that three others have escaped me; and that in 20,000 men one can reckon on one six-digited: the probability that his son or daughter will not be born with polydactyly at all is 20,000 to 1; and that his son and his grandson will not be six-digited at all is 20,000 × 20,000 or 400,000,000 to 1; finally the probability that this singularity will not continue during three generations would be 8,000,000,000,000 to 1; a number so great that the certainty of the best demonstrated things of physics does not approach these probabilities.[2]

Maupertuis (1752) was a pioneer in statistical theory, for in the passage quoted above he had constructed a "reference set" on which he could assign a measure (or probability distribution) and thus compute the "probability of the observations" on the Ruhe family data.

A second biological example was provided by the unknown mathematician advising O.W. Holmes (1855) (see §9.1), who used a similar method. Effectively, in the Holmes article, the null hypothesis is that there is no contagion between women at confinement and that the probability of puerperal fever occurring in a confinement is $p$, $p$ being taken larger than the observed probability; thus the probabilities in any set of $n$ confinements are $q^n$, $\binom{n}{1} pq^{n-1}$, $\binom{n}{2} p^2q^{n-2}, \ldots, q = 1 - p$, that there will be $0, 1, 2, \ldots$ cases in the $n$ confinements. To obtain a reference set of number $N$, it is now assumed that it would be appropriate to consider the confinements in the population of England and Wales within the previous 50 years. With such an $N$, it is then shown to be quite improbable that such a sequence or sequences of cases, as has been observed, would have occurred under the null hypothesis.

---

[2] From Glass (1959, p. 72). In Glass, B., Temkin, O., and Straus, W.L., Jr. (Eds.), *Forerunners of Darwin: 1745–1859.* Copyright © Johns Hopkins University Press, Baltimore, 1959. Reprinted with permission.

# 18
# Modern Clinical Trials

## 18.1 Conduct of a Clinical Trial

After Lind's clinical trial, outlined in §17.3, there had been many remarks on specific features of such comparative trials although no general theory had been developed before R.A. Fisher (1935b). He had worked in biological experiments often of an agricultural nature. Here he was fortunate for the results had immediate application with assessable economic value; there were realistic simplifying hypotheses, such as the normality of the errors and the mutual independence of the errors; there were few entrenched doctrines to combat; the price of experiment was low, and there were no ethical problems. Statisticians familiar with the Fisherian theory began to work on medical problems. Possibly the first clinical trial using a proper randomized approach in the selection of controls was that of the Medical Research Council (1948), a test of streptomycin for the treatment of tuberculosis. This article has now been reprinted in Hill (1962)[1] and we summarize here its salient points.

### Choice of Hypothesis

A special committee of the Medical Research Council of England and Wales had begun in 1946 to plan a trial of streptomycin, developed in 1944, as a curative agent in tuberculosis, especially pulmonary tuberculosis. It had been determined that streptomycin inhibited the growth of the tubercle bacillus in vitro and the results of trials of it on the treatment of experimental infection in guinea pigs had been strikingly superior to those of other drugs. Some suggestive but inconclusive results had recently been obtained with human infections but the evidence was strong enough to suggest that a properly designed trial should be undertaken.

---

[1] Citations from Hill, A.B. *Statistical Methods in Clinical and Preventive Medicine.* Copyright © E. & S. Livingstone, Edinburgh, 1962. Reprinted by permission of Churchill Livingstone.

## Type of Case

The first prerequisite was to limit the entry of cases to one type of the disease: (a) the case must be unsuitable for other forms of therapy; (b) the chances of remission must be small, but the form of the lesion should not be such as to offer no prospect of action by any effective chemotherapeutic agent; thus cases of long standing and/or thick-walled cavities were excluded; and (c) the age group must be limited since the total number in the trial could not be large. Thus cases came from a type characterized by "acute progressive bilateral pulmonary tuberculosis of presumably recent origin, bacteriologically proved, unsuitable for collapse therapy, age-group 15 to 25 (later extended to 30)." Some 109 patients were accepted but two patients died in the preliminary observation week. Of the above prerequisites, (a) is an ethical requirement; (b) is practical, since it is not desirable to have a class of treated ($S$) and controls ($C$), each member of which will contribute a death; and (c) attempts to have the variances at time of selection of $S$ and $C$ as small as possible.

## Choice of Treated and Controls

The patients were divided into treated and controls by reference to a prepared series of random sampling numbers drawn up for each sex at each center by the statistician and not known to the investigators. The medical officers at the various hospitals were informed by the organiser that a given patient was $S$ or $C$, and $S$ and $C$ patients were usually not treated in the same ward, the $S$ patients having 2 g streptomycin intramuscularly four times daily and standard treatment, and the $C$ patients having standard treatment only. It was agreed beforehand that either $S$ or $C$ patients were permitted to have collapse therapy if it appeared necessary to the principal clinician in charge at the particular hospital. In the event, such collapse therapy was given to 5 of the 52 $C$ (controls) but to none of the 55 $S$ (streptomycin treated) cases.

## Observation and Treatment Period

Each patient was to remain in bed at the centre for at least six months, and the results were to be assessed on the clinical status at the end of that period. In addition to the usual hospital records, clinical observations were entered on standard record forms designed particularly for this trial; these forms provided for details of history, criteria of acceptance, examination on admission, monthly routine re-examinations with assessment of progress since last examination, observation of toxic reactions, temperature and treatment records, and finally a pathological record form. Instructions on required frequency of examinations were given.

## Results

For the sake of brevity we mention only the survival to 12 months from entry into the trial: Of 55 streptomycin treated (52 controls), 31 (16) improved, 4 (5) were unchanged, 8 (7) deteriorated, and 12 (24) died. Thus 22% died in the $S$ series and 46% in the $C$ series and this was statistically significant, whether it was a one-sided or a two-sided test of significance.

Semmelweis (1861) did not bother to specify precisely when he initiated his therapy. [See §8.11 for a planned veterinary experiment by Theobald Smith.]

## 18.2 Choice of Hypothesis

A clinical trial, as in §18.1, presupposes some knowledge of the disease, the therapy, and the subjects tested. An informal statement might be that a certain drug is effective against a certain disease, for example, quinine against malaria. A more formal statement would be that quinine is not effective against malaria. The aim of the trial is then to upset this formal hypothesis, usually denoted by $H_0$ and termed the null hypothesis.

The criteria for diagnosis should be precisely stated, for example, the proven presence of a bacterial cause.

The possible final results will be cure, death, loss to the survey, and so on. Procedures to be adopted in determining each are to be written out.

There will be a chosen line of therapy, if the trial is to be deliberate, by which all possible contingencies are provided for. The statistical method of analysis must also be specified (see §18.4). The written specification of these points is called the *protocol* and should be prepared before initiating the clinical trial. A change cannot be made in the hypothesis tested during the course of the experiment, but the experimenter can make a note of any such requirement and carry out a new experiment.

## 18.3 Choice of Cases

The choice of cases or subjects (that is, individuals), whether treated or controls, for a clinical trial will usually be determined by the nature of the particular disease being studied. There may be restraints on the type of disease, as in §18.1. It will usually be desirable for the cases to belong to a class relatively homogeneous by age, sex, or other relevant characteristic. It may be desirable to stratify the cases and to specify beforehand how such stratified classes can be pooled.

## 18.4 Statistical Tests of the Hypothesis

The outcome of a clinical trial may be enumerative or metrical. In either case, probabilities will be calculated from the form of the mathematical model chosen and stated in the protocol. The aim will be the calculation of a "dimensionless" quantity, the test function. It is of interest that Bernard (1865/1957) was aware that hypotheses could not be proved by statistical means but that it was the way of science to test falsifiable hypotheses [see Wynn (1982) and Armitage (1983), who find that Bernard's views on this point are consonant with those of Karl Popper and of R.A. Fisher (1935b)]. In most cases there are possible alternative hypotheses, usually written $H_1$. In some cases the form of $H_1$ may determine the most powerful method of testing $H_0$.

In the enumerative type of trial, tests of the differences between binomial variables are relevant and P.S. Laplace and Poisson could be followed after the manner of Gavarret (1840) who gave two examples:

1. $H_0$ specifies the probability $p = 1 - q$ in a Bernoulli trial of $n$ independent events; the event $M$ appears $m$ times. Then

$$P\left\{\left|p - \frac{m}{n}\right| \leqslant u \sqrt{\frac{2m(n-m)}{n^3}}\right\} = 1 - \frac{2}{\sqrt{\pi}} \int_u^\infty e^{-t^2} \, dt. \qquad (18.4.1)$$

The final expression in modern notation is given by

$$P = 1 - \sqrt{\frac{2}{\pi}} \int_{v\sqrt{2}}^\infty e^{-1/2t^2} \, dt, \qquad u = v\sqrt{2}. \qquad (18.4.2)$$

If we follow Gavarret, taking $u = 2$, the size of the test is $1 - 0.9953$; so his test is two-tailed at the size or level of significance of 0.0047 and his significance value of $u$ is 2. In modern notation as in equation (18.4.2), the significance value at the 0.47% level is $2\sqrt{2}$.

2. In a comparative trial of two Bernoulli series, of lengths $n$ and $n_1$, $m$ and $m_1$ events, respectively, occurred. Then Gavarret compares the means and finds that

$$\left|\frac{m}{n} - \frac{m_1}{n_1}\right| \bigg/ 2\sqrt{\left[\frac{2m(n-m)}{n^3} + \frac{2m_1(n_1-m_1)}{n_1^3}\right]} \qquad (18.4.3)$$

is the variable to be tested as in the first example.

Liebermeister (1877), by a different method, gave

$$U^2 = \frac{(n + n_1 - 1)(mn_1 - m_1 n)^2}{mm_1 nn_1} \qquad (18.4.4)$$

which differed very little from Gavarret's test function when a two-tailed test was used.

The design may be such that there are several, rather than two, mutually exclusive outcomes. Here the Pearson $\chi^2$ has been available since 1900 (see §19.5).

With metrical data, there may be tests between sums, for which the $t$-test and the multigroup extension of it, Fisher's $F$-test, have been available from 1908 and 1925, respectively. K. Pearson and others avoided the use of the $t$-test by assuming that the sample was so large that the variance could be computed from the sample sufficiently accurately.

It may be concluded that in medical statistics after the works of Laplace,

Poisson, Gavarret, and Liebermeister, the development of the clinical trial in the nineteenth century would not have been hampered by the lack of mathematical techniques.

The modern theory was set up by Jerzy Neyman* (1894–1981) and Egon Sharpe Pearson (1895–1980). R.A. Fisher never admitted the relevance of their theory but sentences in his *Design of Experiments* show that he was aware of the two kinds of error described (see Neyman and Pearson, 1967).

## 18.5 Size of Trial

For purely statistical reasons a trial should contain a sufficient number of cases. Given $H_0$ and $H_1$, and binomial parameters $p$ and $p^*$, the power of any significance test can be shown to increase with the size. For example, in the test of a binomial hypothesis with two classes and one degree of freedom, the parameter of non-centrality is of the form, $N\phi^2$, where $N$ is the sample size. Without such mathematical considerations, Bartlett (1844) could say (see §18.6) that the certainty of the results depended on the uniformity of the comparisons and on the "greatness of their numbers," that is, the sample size. We may add that the difference between the binomial parameters, $p^* - p$, may be small but if the number of observations, $N$, is sufficiently large, $N(p^* - p)$ may be large, so that almost always the treated (under $H_1$) come out with more successes than the controls (under $H_0$).

A modern observer is surprised by the few observations made by the pioneers of anesthesia such as C.W. Long, W.T. Morton, and J.Y. Simpson before launching their respective forms of anesthesia. However, J. Snow made a more careful investigation of ether and chloroform [see the entry by K. Bryn Thomas (*DSB*, XII, pp. 502–503)]. The small sampling size considered by experimenters seems to be due to an idea, akin to Quetelet's *homme moyen* (see §14.4), that a representative could be chosen from a class and that almost all others in the class would have reacted to the medicine or treatment in the same way as the member chosen. The most notorious example of this idea is provided by J. Johnson (see §17.4), who rejected quinine as a therapy for malaria after a trial of one patient. Lind and Jenner escape this criticism because they obtained positive results on their respective trials of citrus fruits and vaccination and further observations confirmed their original trials. Liebermeister (1877) tested the Listerian treatment of wounds in a fourfold table with a total of 24 cases, a sample too small to have enough power to be of value, although this kind of error has been repeated in modern times by exponents of the small sample exact theory.

## 18.6 Choice of Treated and Controls

The word *control* or *contre-rolle* originally meant duplicate register. The use of controls is an essential feature of clinical trials. Bernard (1865/1957) writes (p. 194 in the translation) that the need for comparative experiments, especially in therapeutics, has always seemed important; in particular the question of whether the special therapy has any advantage over expectant treatment needs to be answered. He cites Gall (1800) as concluding that it is difficult to apportion credit to nature and physician in therapeutic trials (see also §17.8). Bernard believes that it is necessary to know the natural course and outcome of the disease, so that there must be controls in observations on the effect of therapy. He cites an enquiry by Béclard (1863), with controls, showing that the value of bleeding in the treatment of pneumonia is a mere therapeutic delusion. A similar conclusion was reached by Bennett (1865).

Bartlett (1844), the American student of Louis, remarks that cases and controls that are to be compared in therapeutic investigations must have equal disturbing factors of location, social class, and the like; they should be susceptible of a clear and positive diagnosis; there must be no selection of cases; the method of treatment must be clearly defined. The certainty of results will be "in proportion to the fixed and uniform character of the compared facts and to the greatness of their numbers."

A modern example of Bartlett's requirement is provided by a fourfold table discussed by Greenwood and Yule (1915). The experience of the troops, vaccinated against enteric fever, was compared with that of "controls," who had not been so vaccinated; but the vaccinated contained a smaller proportion of troops who had seen service in the field and so had been under lighter risk of infection than the controls. Vaccination was confounded with risk of infection. The choice of controls was thus not valid. Mathematically, the population specified by the null hypothesis was a mixture and so the usual analysis of the fourfold table was inapplicable (see Lancaster 1969, p. 248).

An unsatisfactory method was given by van Helmont (1662) and by Amberson, McMahon, and Pinner (1931), who suggested that the set of subjects entering the trial could be divided into two subsets and then treatment could be assigned to one of these subsets by lot. We can see that this is an unbiased method but, with a pair of unbalanced subsets, the lack of balance persists and its effect may dominate the experiment or trial. These difficulties are now solved by the use of randomization after Fisher (1935b).

According to Armitage (1983), Francis Bisset Hawkins (1796–1894) compared a case fatality rate of less than one in seven for fever patients treated actively in London during 1825 with the deaths of 21 out of 37 cases treated by Hippocrates (Hawkins, 1829). This comparison is absurd, but difficulties have arisen from using historical cases more nearly contemporary than the controls of Hawkins, for the cause of a disease such as diphtheria may not be homogeneous over time (see §17.9).

Claude Bernard* (1813–1878) mentions that Pinel recommended that the disease be observed without treatment one year and treated the following year (Bernard, 1865/1957). This procedure assumes that the disease is constant in its effects from year to year.

A.M. Lilienfeld (1982, p. 7) writes that "probably the most sophisticated clinical trial of a preventive type was conducted by Ignaz Semmelweis." Let us see how he dealt with the control problem. That Semmelweis's observations were made just after the peak of a severe epidemic of puerperal (childbed) fever was recognized by C.E.M. Levy in Semmelweis (1861, p. 300). Levy believed that it was inadmissible to choose as control a peak period of maternal mortality in the hospital, when Semmelweis's paper showed that much lower figures had occurred in the Vienna Hospital in the past.

Fibiger (1898) assigned every second case to the trial controls as did Pearson (1904b).

To avoid criticism of the methods and to follow procedures worked out in the agricultural or biometric fields, modern medical practice has almost invariably followed Fisher's ideals, namely the controls are selected from the subjects (individuals) available for trial by a random procedure (randomization), so that every subject has the same probabilities of becoming either a case or a control. Sometimes the practical setup forces some variation from these ideals, for example, some forms of restricted randomization are sometimes used to avoid concentrations of one form of treatment in some hospitals during a cooperative type of trial. Indeed, in the trial of §18.1, the statistician was careful to randomize, not by sets, but by individuals within each hospital set.

## 18.7 Further Testing of the Hypothesis and Interpretation of the Results of Experiment and Observation

There are few crucial experiments, although popular opinion has often been to the contrary. Usually it has been felt that confirmation of the results of a clinical trial or survey by another trial or survey is necessary, especially since the conclusion is based on a stochastic or chance procedure. Sometimes it is found desirable to have variations on the original trial, perhaps varying the form of disease tested, the special therapy used, or the age-sex groups treated.

Compared with the agricultural trials, there is much more difficulty with the interpretation of the results of a clinical trial or survey. This is notably so with the interpretations of the surveys of Mueller (1939), Wynder and Graham (1950), and Doll and Hill (1950) on the relation between cancer of the lung and cigarette smoking. In such cases, it is often said that a clinical trial or statistical survey determines only an association but not a causation. [See the discussion in Fisher (1953) on the value of observation.] The criteria for belief, that a causal relation exists, can be considered in detail taking note of the remarks of Hill (1965).

### Strength of Association

Percivall Pott (1775) (see §13.3) reported an unusually high incidence of scrotal cancer among London chimney sweeps, the death rates being more than 200 times higher than in workers in other trades. Similarly, Härting and

Hesse (1879) found that the death rates from cancer of the lung were very high among the miners at Schneeberg; the cause was held to be inhaled arsenic together with poor nutrition and respiratory disease until it was proved that the cancers were due to radioactivity (see §13.5). Another example is provided by Snow (1855) (see §8.9); cholera deaths occurred in 7 per thousand of houses supplied by one water company in London as against 0.5 per thousand of houses supplied by another company.

## Consistency

It will strengthen the experimenter's opinion that the association is due to causation if others carry or have carried out similar experiments and obtained similar results; but there are traps to be avoided. The hypothesis to be tested may be "in the news"; many laboratories may test it, some with positive, others with negative results; those with positive results may publish them. It would often be against journal policy to publish the negative. This form of selection will lead to false views on confirmation. Some experiments or surveys may be thought quite conclusive on their own.

## Temporality

The presumed cause must precede the presumed effect. Thus a high incidence of tuberculosis in a sedentary trade might occur because physical strength was not required in it.

## Biological Gradient

The existence of a biological gradient between presumed cause and effect strengthens the belief in the hypothesis. Thus a monotonic relation between the risk of lung cancer and the number of cigarettes smoked per day would be suggestive of a causal relation. [See §13.5 for difficulties in securing correct data.]

## Plausibility and Coherence

As Hill (1965) points out, what is biologically plausible depends on the biological knowledge of the day. It may be comforting to the critic to find that the result of the experiment is in keeping with accepted notions; but this argument cannot be carried too far as it would act against the acceptance of new conclusions. A conclusion is said to be coherent if it does not seriously conflict with the generally known facts; a conclusion may help to explain other as yet unexplained observations and this would strengthen belief in it (see §13.5).

## Further Experiment

Belief in causation would be strengthened by some experimental evidence, for example, the existence of several carcinogens in cigarette smoke. After the experimental work on niacin as the anti-pellagra vitamin, belief in causation would be strengthened when a change in cooking methods reduced the incidence of pellagra in the community. Pellagra was found to be cured by appropriate doses of niacin (see §13.1). We recall the work of Gregg (1941) on rubella deafness and, although it was an observation and not a clinical trial, the line of reasoning here was relevant to its interpretation. It was established by Lancaster (1951, 1954) and Lancaster and Pickering (1952) that other rubella epidemics had appeared in Australia and New Zealand and deafness and congenital cataract had followed.

## Analogy

It is easier to accept a hypothesis if it is similar to an already well-established one, for example, since rubella is well accepted as a cause of congenital deafness, observers might well be disposed to accept measles also as a cause; but care must be taken in using this argument.

## 18.8  Failure to Develop the Theory and Practice of Clinical Trials

In the early nineteenth century, the great French mathematicians had developed mathematical tests adequate for most clinical trials and had done excellent work in demography. Pinel and Louis (§16.2) had considered the gathering of hospital statistics and Gavarret (§18.4) had specialized the mathematics suitable for clinical trials. Some clinicians had carried out trials on new remedies but then no further progress in human clinical trials had been made from about 1850 to 1930. Yet in agricultural work similar problems had been solved, commencing around the period 1900 to 1930 (see §18.1).

We may discuss difficulties experienced in human clinical trials as follows.

## Choice of Hypothesis

### Disease

Diseases were not well defined and classified; they were gathered together in groups such as fevers; within such nosological groups there would be many diverse causes. For example, the pneumonias and the common septic conditions are now known to be due to a number of different infective agents. The drugs or therapy applied would have been set to act against different causes.

Drugs

There were few drugs acting specifically against given diseases; moreover, there were fewer in a pure state so that comparisons between different dosages would be hard to generalize. Perhaps, some credit should be given here to William Withering* (1741–1799) whose *Account of the Foxglove, and Some of its Medical Uses* in 1785 showed that the drug digitalis must be given with care and that its effective use was limited to only a few diseases. Withering worked with no formal controls but he drew his conclusions from his clinical experience.

The need for proper quantification of physiological drugs became more evident to the physicians when new drugs were introduced, commencing with the alkaloids in the early nineteenth century. This need was greatly increased when Paul Ehrlich* (1854–1915) began to test many chemical substances for their effectiveness against various diseases (see for example, C.E. Dolman, *DSB*, IV, p. 295*b*), sacrificing "hecatombs" of guinea pigs. He devised many methods of quantitative analysis of drugs and gave a dictum that the minimum curative dose must not be too close to the maximum tolerated dose—a very useful concept.

Homogeneity of Cases Treated

Too little of the "natural history" of the diseases was known, although Louis had begun such work; diseases varied with age, sex, and perhaps other features of the patient. Lack of homogeneity will usually weaken the power of the test.

## Infrastructure

Clinical trials would perhaps have been carried out in the great hospitals of the time; there would have been a shortage of skilled persons; nosocomial diseases would help to confuse the results of experiments.

## Randomization

With the heterogeneity of the patients themselves, proper randomization procedures would have to be used. Naive methods were often used or suggested, for example, divide the subjects into two equal classes and declare the members of one class to be the treated and the other the controls, but this was an unsatisfactory method. Here the randomization methods of Fisher (1935*b*) solved many of the difficulties.

The "numerical method" of Louis consisted of determining the mean values of statistics from two populations and comparing them by some method, as a rule not determined beforehand. The investigator might determine how they could be compared, for example judge that the numbers in the comparison were adequate in size. Gavarret (1840) gave rules as to how the comparison was to be carried out; but few investigators were prepared to use his methods; as Coleman (1987) has pointed out, some investigators deemed statistics to be irrelevant, deceptive, or false; and perhaps all three of these judgments. Medical investigators thought such analysis would divert attention from the individual case in all its particularities. Yet it was important to test why therapy had failed to make progress, as had diagnosis in the early part of the century. Gustav Radicke (1810–1883), a physicist and mathematician writing for practical and academic physicians, outlined a typical experiment comparing two series of treatments, but his mathematical techniques are greatly inferior to those of Gavarret (1840); for example, he does not compute a proper estimate of the variance to be used in testing significance (Radicke, 1858/1861).

## Size of Experiment

Until the development of the Neyman-Pearson theory, the appropriate minimum size of experiment could not be estimated. Often the small size of the experiment was due to an idea, akin to Quetelet's *homme moyen* (see §14.4).

## Views of the Physicians

There were many entrenched ideas to be combatted, for example, the great value of venesection, the individuality of the patient, the irrelevance of number in biology. Otherwise, the physicians had their own views on the minutiae of therapy and would be reluctant to pass control of their patient to another physician or even to a committee.

## Absence of Effective Drugs and Therapy

If too many negative results occurred in the trials, the charge of therapeutic nihilism would be raised.

The time was evidently not ripe for clinical trials. For the trials to be an effective aid to medicine, many things had to happen, which we can group together as "mathematization."

# 19
# Applications of Mathematics to Biology and Medicine

## 19.1 The Development of Mathematics

Mathematics in the ancient Greek world was highly developed in some fields as we have shown in Chapter 1 but the knowledge passed out of use in the Dark Ages of the Western Christian world. A principal mathematician in the revival of the West was Fibonacci (floruit 1225), who was skilled in algebra and arithmetic and whose problems of the population of rabbits we have discussed in §6.6. Our next authors are Kepler (floruit 1600), an astronomer, with his analysis of the physics of sight (see §15.1), and Harvey (floruit 1625), who, although not a mathematician, attached neat experimental proofs to his statements of new observations. By the time of Descartes (floruit 1625) and Leibniz (floruit 1675), mathematics could be said to have become highly developed.

Perhaps, we have agreed too closely with Fourier's view (see §3.4) that new mathematics was needed. What was needed was the broad interest of some biologists who would begin to mathematize biology. It is difficult to avoid giving principal credit to Galton (floruit 1875), who remained throughout his life a biologist but gave biological meanings to means, variance, correlation, and joint distributions, which the mathematicians had overlooked.

For biologists, the use by Gregor Mendel of combinatorial analysis has been important for the study of genetics, a rather neglected branch of mathematics in his time. Since the rediscovery of Mendel's work in 1900, this branch of mathematics has become highly developed, especially after it was realized that discrete mathematical functions abound in biology. R. A. Fisher (floruit 1940) was a leader in this field.

## 19.2 The Necessity of Mathematics in Biology

In §3.5 we have given the views of some biologists who, without being professional mathematicians, have called upon mathematics in their biological work; a notable example is D'Arcy Thompson.

Weldon (1901–1902) expressed his views on the necessity of the introduction of mathematics into biology in the editorial of the first issue of *Biometrika*:

The starting point of Darwin's theory of evolution is precisely the existence of those differences between individual members of a race or species which morphologists for the most part rightly neglect. The first condition necessary, in order that any process of Natual Selection may begin among a race, or species, is the existence of differences among its members; and the first step in an enquiry into the possible effect of a selective process upon any character of a race must be an estimate of the frequency with which individuals, exhibiting any given degree of abnormality with respect to that character, occur. The unit, with which such an enquiry must deal, is not an individual but a race, or a statistically representative sample of a race; and the result must take the form of a numerical statement, showing the relative frequency with which the various kinds of individuals composing the race occur.

The first tables to be made with such an explicit aim as Weldon wrote in his editorial were given in Weldon (1890, 1892, 1893, 1895) of a shrimp *Crangon vulgaris* and of *Carcina moenas*. Other biologists, encouraged by his example, made similar observations on external characters such as height, weight, lengths of limbs, length of bones, and so on, and other authors have made use of such tables in studies on other fields (see §7.3).

Because of these tables and his formulations on the future of mathematical theory in biology, Weldon can be considered an important pioneer in applications of mathematics to biology. He appears to have accepted the existence of normality in biology as a fact of observation.

Weldon's views given above were part of an *apologium* by him to justify the publication of the new journal *Biometrika*, undertaken as a result of a report of an editorial committee of the Royal Society. One can be sympathetic with the committee's views (see §4.12), even though Karl Pearson was introducing a new branch of applied mathematics; for he was very wasteful of space, for example, writing out the complete contents of a large matrix rather than stating the name of the matrix, **A**, and giving the value of its general term, $a_{ij}$, where available. Further, algebraists would recognize that he was not up to date in his treatment of matrices. Although in the famous paper (Pearson, 1900) he has obtained the inverse of a matrix, **B**, and gives several pages as to how to obtain it, all he needs to do is to verify that $\mathbf{AB} = \mathbf{1}$ or equivalently $\mathbf{BA} = \mathbf{1}$ as a check on his analysis.

Provine (1978) pointed out that the main roles of mathematical models were the following: (a) they showed that Mendelism and natural selection plus known processes in nature were sufficient to account for microevolution at the population level (see Chapter 7); (b) some mathematical or other models would lead to no fruitful results (see methods of criticizing the genetic theories of Galton in §4.4); (c) they complemented field research and gave useful criticism; (d) they stimulated further researches. Mayr and Provine (1980) give many historical and biographic details. Ewens (1979) gives a useful summary of the mathematical details, together with historical and some biological explanations. We may add that mathematics has been valuable in making statements more precise. (See §7.7 for an example on Neanderthal man.)

## 19.3 Normal Distributions

The most general form of the univariate normal distribution is given by the density function,

$$g(x) = \frac{1}{\sigma\sqrt{(2\pi)}} e^{-(x-\alpha)^2/2\sigma^2}, \qquad -\infty < x < \infty. \tag{19.3.1}$$

The standard normal density is given by

$$g(x) = \frac{1}{\sqrt{(2\pi)}} e^{-1/2x^2}, \qquad -\infty < x < \infty. \tag{19.3.2}$$

This second form is due to K. Pearson, who introduced the term *normal* and popularized the use of the "$\frac{1}{2}$" in the exponent. The mean and standard deviation of the variable in equation (19.3.2) are zero and unity, respectively; the corresponding values in (19.3.1) are $\alpha$ and $\sigma$.

We give an elegant proof given by Arthur Cayley* (1821–1895) in a postscript to a paper on the generalized central limit theorem by Isaac Todhunter* (1820–1884) in 1869. The postscript was not known to K. Pearson and Francis Ysidro Edgeworth (1845–1926) and we now give a brief sketch of it, longer than Cayley's because we have to prove that the square root of a positive definite matrix can be defined, which would not have been necessary if Cayley had been writing for algebraists.

Suppose that

$$f(\mathbf{x}) \equiv f(x_1, x_2, \dots, x_n) = C \exp(-\tfrac{1}{2}\mathbf{x}^T\mathbf{V}^{-1}\mathbf{x}) \tag{19.3.3}$$

is a probability density function; it is positive everywhere in the $n$-dimensional space supporting it. To be integrable over the whole space $\mathbf{V}^{-1}$ and equivalently $\mathbf{V}$ are both positive definite. There is an orthogonal $\mathbf{H}$ such that

$$\mathbf{H}^T\mathbf{V}\mathbf{H} = \mathbf{A} \tag{19.3.4}$$

is a diagonal matrix with each $a_{ii} > 0$ and we can find, in particular, a matrix $\mathbf{A}^{1/2}$ with positive diagonal terms. We now define

$$\mathbf{V}^{1/2} = \mathbf{H}\mathbf{A}^{1/2}\mathbf{H}^T \tag{19.3.5}$$

and find that $\mathbf{V}^{1/2}\mathbf{V}^{1/2} = \mathbf{V}$ and further $\mathbf{V}^{1/2}\mathbf{V}^{-1}\mathbf{V}^{1/2} = \mathbf{I}$. Now the transformation

$$\mathbf{x} = \mathbf{V}^{1/2}\mathbf{y} \tag{19.3.6}$$

has Jacobian $|\mathbf{V}^{1/2}| = |\mathbf{V}|^{1/2}$.

It follows that

$$\int \cdots \int f(\mathbf{x})\,dx_1\,dx_2\ldots dx_n = C|\mathbf{V}|^{1/2}\prod_1^n \int_{-\infty}^{\infty} \exp(-\tfrac{1}{2}y_k^2)\,dy_k = 1$$

so $\quad C = \dfrac{1}{(2\pi)^{1/2n}|\mathbf{V}|^{1/2}}$ $\qquad\qquad\qquad\qquad$ (19.3.7)

and f($\mathbf{x}$) is a probability density.

We conclude that

$$f(\mathbf{x}) = f(x_1, x_2, \ldots, x_n) = (2\pi)^{-1/2n}|\mathbf{V}|^{-1/2}\exp(-\tfrac{1}{2}\mathbf{x}^T\mathbf{V}^{-1}\mathbf{x}) \quad (19.3.8)$$

is a density function of $n$ variables, usually referred to as the joint normal distribution. Cayley has made his point and has no particular reason for proceeding further for he is only assisting Todhunter.

We note the identity

$$\mathbf{x}^T\mathbf{V}^{-1}\mathbf{x} - 2\mathbf{x}^T\mathbf{t} = (\mathbf{V}^{-1/2}\mathbf{x} - \mathbf{V}^{1/2}\mathbf{t})^T(\mathbf{V}^{-1/2}\mathbf{x} - \mathbf{V}^{1/2}\mathbf{t}) - \mathbf{t}^T\mathbf{V}\mathbf{t} \quad (19.3.9)$$

and so we can integrate $\mathbf{x}^T\mathbf{t}$ over the infinite space,

$$\int_{-\infty}^{\infty} \cdots \int f(\mathbf{x})e^{\mathbf{x}^T\mathbf{t}}\,dx_1, dx_2, \ldots, dx_n = \exp(\tfrac{1}{2}\mathbf{t}^T\mathbf{V}\mathbf{t}) \quad (19.3.10)$$

so $\exp(\tfrac{1}{2}\mathbf{t}^T\mathbf{V}\mathbf{t})$ is the moment generating function of the $n$ variables. Further, setting all but one of the elements of $t$ equal to zero, we have the moment generating function of a normal distribution. Every $X_j$ is normal. Similarly, setting all but two elements of $t$ equal to zero, we have the joint moment generating function of two normal variables and so on.

In particular, the joint bivariate normal distribution is given by

$$f(x_1, x_2) = \frac{1}{2\pi|\mathbf{V}|}e^{-1/2}\frac{(x_1^2 - 2\rho x_1 x_2 + x_2^2)}{(1 - \rho^2)} \quad (19.3.11)$$

in which the random variables are centered.

Now it is convenient to give a general form: $\mathbf{V}$ is called the variance-covariance matrix,

$$v_{ij} = \mathrm{E}(X_i X_j). \quad (19.3.12)$$

If the variables are standardized to have unit variance by the linear transformation

$$Y_j = X_j v_{jj}^{-1/2} \quad (19.3.13)$$

we obtain the correlation matrix

$$\mathbf{R} = (\rho_{ij}) \quad (19.3.14)$$

in which

$$\rho_{jj} = 1, \quad \rho_{ij} = \rho_{ji} = \mathrm{E}(Y_i Y_j) \quad (19.3.15)$$

and $\rho_{ij}$ is termed the coefficient of correlation between the variables, $Y_i$ and $Y_j$, and conventionally following Galton between $X_i$ and $X_j$ (see §§4.7, 4.11, 14.5, and 19.5).

From (19.3.10) all cumulants above the second order vanish.

$$\kappa_{kl} = 0 \qquad \text{for } k \text{ and/or } l > 2. \tag{19.3.16}$$

Much work has been done by the mathematical theorists on the central limit theorem to obtain the necessary and sufficient conditions that the standardized sum of a series of mutually independent random variables should converge to a normal random variable with zero mean and unit variance. In particular, it is now permissible to drop the condition that the summands should each be symmetrical. A graphical example is given by Whitaker (1914–1915) where the distribution of the sum of $n$ mutually independent Poisson variables is given for increasing (integer) values of $n$; with $n = 11$, the distribution of the sum displayed as a histogram is well fitted by a normal distribution with the same mean and variance.

Notwithstanding its analytical form given in equations (19.3.1) and (19.3.2), formidable to the beginner, the normal distribution is, in many ways, the simplest of all distributions. It has the following properties:

1. *Stability.* The sum of independently distributed normal random variables is also normal.
2. *Central limit theorem.* By this theorem, sums of independent random variables, having a finite variance and obeying a few other conditions, are normal. Thus if $X_i$, $i = 1, 2, \ldots$, has the uniform distribution on the interval $(-1, 1)$, the sum of even as few as six variables is so close to the theoretical normal that it is used as a method of generating normal variables. From the days of the theory of errors (or of least squares) many unnecessary rules were adopted, such as requiring symmetry of the distributions of the summands and identity of distributions of the summands. It is difficult to lay down such requirements but there should be no summand dominating the sum.
3. *Normal approximations.* Many statistical tests are based on normal theory, which is only an approximation in many cases; yet the normal theory gives rejection rates and other properties agreeing remarkably well with the discrete distribution. This is possibly often because the linear forms or standardized sums of independent random variables tend to have a form approximating to the normal.

It is of interest that Galton (1877) obtained some of these properties experimentally from his quincunx and pointed to the sum of independent binomial variables $X_1 + X_2$ with parameters, $p$ and $n_1$ and $n_2$, being also normal approximately, an early instance of simulation methods in statistical theory.

## 19.4  Normal Distribution and Theory of Errors

Abraham de Moivre* (1667–1754) in 1733 showed that for the binomial distribution with parameters, $N$ and $p$,

$$P(x|N,p) \simeq \frac{1}{\sqrt{(2\pi Npq)}} e^{-[(x-Np)^2/2Npq]} \tag{19.4.1}$$

was a good approximation if $Npq$ was not too small and if $x$ was not too far from the mean, that is, if $(x - Np)/(Npq)^{1/2}$ was in the range $\pm 4$ say.

The normal distribution was first considered as a possible density function for errors by Thomas Simpson* (1710–1761). It was to play an important role in the development of the theory of errors (or theory of least squares) as developed by Carl Friedrich Gauss* (1777–1855), Adrien-Marie Legendre* (1752–1833), and Pierre Simon Laplace* (1749–1827). Laplace (1812) in his *Théorie Analytique des Probabilités* made much use of the normal distribution for approximations, and Gavarret (1840) showed how it could be applied in medicine. Laplace developed this idea and showed that the approximation could be used not only for the calculation of the probability of the observation in the sense of its relative frequency, but also for summing the frequencies of the observation itself together with those of all the more extreme values. Laplace could then use the variable, $y = (x - Np)(Npq)^{-1/2}$, as a standardized normal variable and so consult tabulated values to give a significance test or approximation after a small correction.

Many empirical distributions had been collected before 1840, for example, of heights, weights, and other measurements of persons; heights of restricted populations; and other anthropometric measurements. Quetelet (1835/1842) noted that many such distributions could be approximated very closely by a normal distribution; but the normal curve has an infinite range on both sides and so he was inclined always to use the binomial curve $(\frac{1}{2} + \frac{1}{2})^N$ where the index $N$ was large, on the grounds that this discontinuous curve gave no part of its range to negative values of the variable being approximated (see §14.4). It may be that aesthetics took some part in this choice; Quetelet thought that his empirical frequency curve was approaching an actual theoretical curve, whereas modern workers would be content to determine a curve that would approximate closely to the empirical curve. Indeed, Galton (1877) saw that proper generality would only be attained by treating the normal distribution as an existent distribution; he states, "Quetelet, apparently from habit rather than theory, always adopted the binomial law of error, basing his tables on a binomial of high power. It is absolutely necessary to the theory of the present paper to get rid of binomial limitations and to consider the law of deviation or error in its exponential form" (p. 289). Thus, in one-dimensional distributions, Galton wishes to introduce the normal density determined by its mean and standard deviation, free of any reference to a limiting process by which it could be obtained. Galton could see that the normal distribution in any case would give an extremely small weight to negative values, for example, the mean of heights of human adult males is usually about 35 standard deviations. The density assigned to the region of zero or negative height is of the order of $\exp(-\frac{1}{2} \times 35^2)$, which has a common logarithm of $-264$, approximately, so Quetelet's fears were not justified.

This normal approximation brought some generality into describing the empirical distributions, for the number of observations and estimates of the mean and variance completely determines the distribution approximating to the empirical distribution. For the individual distribution, the estimates above enabled an estimate of the numbers of the empirical distribution in the tails, for example, about 5% of the distribution lies in the tails more than 1.96 standard deviations from the mean. Further, for comparisons between distributions, there are two parameters to test for differences, namely the mean and the variance.

Quetelet's choice noticed above had an unfortunate result. It prevented him from seeing that the variance was an important parameter in the analysis of observed data and in the description of theoretical distributions, so that it was Galton who first began to use the analysis of variance in a broad sense and not Quetelet.

## 19.5 Some Normal Distributions in Biology

There is a famous aphorism of Gabriel Jonas Lippmann* (1845–1921) that the mathematicians think that the normal distribution is a property of nature and the biologists think it is a mathematical theorem; but in any case, its usefulness, as shown by Weldon and others, is undoubted. We have introduced the multivariate normal by the shortest method in §19.3.

The reader may regret with K. Pearson that it was a mistake for Galton to invite James Douglas Hamilton Dickson (1849–1931) to give a mathematical form to Galton's hypotheses, namely that

1. the contour lines of equal density were concentric and similar ellipses; and
2. the regression lines were diameters conjugate to the two axes of stature;

derived from his experimental work and careful smoothing.

Dickson obtained the joint normal distribution from these conditions given by Galton so that

$$f(x, y) = g(x)h(y|x) \qquad \text{for every } x, y \qquad (19.5.1)$$

and further $h(y - \rho x|x)$ is a constant for every $x$, where $\rho$ is the correlation coefficient.

Indeed, Dickson writes that the logarithm of the density is, except for multiplication by a constant,

$$\frac{y^2}{(1.22)^2} + \frac{(3x - 2y)^2}{9(1.50)^2} \qquad (19.5.2)$$

without any general term. Dickson makes no attempt at multivariate form and seems not to have known Cayley's proof given in §19.3.

Pearson (1914–1930, IIIA, p. 3) notes that Galton did not at first see that the problems of inheritance and of correlations in the narrow biological sense had identical solutions. In about 1875 (Pearson, II, p. 392), Galton plotted the means of the grades (or ranks) of seed of the daughter, $Y$, against the grades (or ranks) of seed of the mother, $X$. The rank was defined to be the cumulative probability of the normal distribution, and the regression was found to be nonlinear. Later in 1875 (Pearson, II, p. 187), Galton found that the mean weight of the $Y$'s conditional on $X = x$ was a linear function of $x$; the regression was linear. Pearson (IIIA, p. 4) gives this regression line, and he makes comments that Galton first came to the bivariate normal distribution via regression (pp. 3–7).

It is explained (Pearson, IIIA, p. 8) that Galton found certain relations between two normal variables in joint distributions. The analyses convinced him that the variables should be expressed in standard measure. We assume this has been done. If the regression coefficient is $\rho$, the regression is linear and the conditional variance of $Y$ given $X$ is $v$, then

$$v = 1 - \rho^2. \tag{19.5.3}$$

Galton generalized his results by making linear transformations on the marginal variables. The standard measure is a necessity when dealing with joint distributions of several variables.

The Mendelian theory introduces difficulties for the effect of each gene has discrete values whereas most observed effects possessed continuous distributions. Yule (1902, p. 235) showed that if the character was determined by a sum

$$L = a_1 X_1 + a_2 X_2 + \cdots + a_n X_n + b Y \tag{19.5.4}$$

where the $X$'s are the effects of genes $1, 2, \ldots, n$, $Y$ is a nonheritable variable and the $a$'s and $b$ are certain constants, then there would be at least $2^{n+1}$ points of increase in $(n + 1)$-dimensional space and so $L$ would have many points of increase and would approximate closely to some continuous variable. Fisher (1918) made a stronger statement that the continuous variable would be approximately normal. [See Moran and Smith (1966) and Yates and Mather (1963).]

Before 1900, there were in existence many tables showing the frequencies of heights of army recruits, the number of sibships with a given size containing $0, 1, 2, \ldots$ boys, the days of fine weather per month, and so on. Sometimes there were models to test which distribution specified the frequencies in the classes. At other times, the observed distribution appeared to be tending toward some distribution such as the normal. The chi-squared statistics of Pearson (1900) were designed to test the closeness of the observed distribution to some other, theoretical or postulated, distribution. In many uses, the

Pearson chi-squared test takes the form,

$$\chi^2 = \sum_1^n \frac{((\text{Observed})_j - (\text{Expected})_j)^2}{(\text{Observed})_j} \tag{19.5.5}$$

where the summation is over all classes (cells) of the distribution being tested. Equation (19.5.5) can often be represented as the sum of squares of $n$ standardized normal variables, subject to a single restriction that

$$\sum_1^n (\text{Observed} - \text{Expected})_j = 0.$$

It has been used in such varied backgrounds as counts of yeast cells (Student, 1907), bacterial colony counting (Fisher, Thornton, and MacKenzie, 1922), and sex ratios in sibships (Fisher, 1925). [See also Lancaster (1974) for many examples.]

## 19.6  History and Philosophy of Science

A reading of "The Changing Intellectual Milieu of Biology" (Mayr, 1982, pp. 83–132) should convince the reader that biology should be discussed in "any history and philosophy of science" on equal terms with the traditional mathematics, physics, and chemistry (see §15.1).

Mayr concludes his chapter with a statement that biology has become so diversified that its history can no longer be dominated by any one of its specialties; nevertheless, he says, there is a spirit of unity not excelled for several centuries.

Let us see some problems of those historians educated in mathematics, physics, and astronomy. For example, they like to see results expressed in mathematical form, preferably closed or at least in a set of differential equations. They have been fortunate in their studies of the planetary motions within the solar system, where hypotheses can be simplified; for example, a planet can be shown to be equivalent to a point containing a mass equal to that of the planet. Similarly, it is often possible in classic problems to treat discrete masses (molecules) as being part of a continuum. Such methods brought about remarkable solutions in the field of astronomy.

In biological problems there are dangers for the classical theorists in making continuous approximations to all-or-nothing events, whereby in the theory of epidemics the individual passes from the susceptible class to death or to the immune class. A mathematical flaw now appears, as we have pointed out in §12.8. In the statements of the McKendrick equations (12.4.4 et seq.) there is no suggestion that the total size of the populations given is relevant to permissible values of the parameters; yet this is easily determined by a discretization of the theory. The whole of the literature of the theory of epidemics has contributed very little to understanding epidemics.

We may pass to another example from epidemiology. The introduction of continuity into the discussion has led the mathematicians to the idea of differentiation and the question of what are the "forces" and ideas of the ebb and flow of an epidemic, for example, the notion that bubonic plague disappeared from Europe as a single process, with a boundary line separating a zone of active plague from one of freedom from plague. An interest in the occurrence of the great pestilences of history was spurred by the pandemic of bubonic plague commencing in 1894 (see Lancaster, 1990, especially p. 98). The solution appears to be that there are four species of the genus *Meriones* with variable susceptibility to the plague bacillus living in contact with one another, which might suggest the possibility of some difficult mathematics.

The classically educated applied mathematician will expect every problem to have a simple solution. [For example, see §4.12 for Karl Pearson and §7.3 for Raphael Weldon, who, in the years after the rediscovery of Mendelism, both expected that any solution of the genetics problem must be simple.]

In physics and astronomy, there are given laws to summarize knowledge obtained by observation or experiment but in biology there are contingencies also. [See Gould (1989, p. 280 passim), who cites Charles Darwin (p. 282) among biologists and Leo Tolstoy the great novelist (p. 285) as being aware of contingencies.] Mathematically, we might write that the classical history of science deals with determined or stochastic processes, such as the revolution of the planets around the sun, usually determined, whereas a contingency is described as an event in a discrete stochastic process such as perhaps a mixture of an infinite set of Poisson variables, usually with a small parameter and acting over a short period of time and perhaps mutually independent. Now a contingency may be unimportant but such may affect the evolution of man, first by a mutation, an accident including attacks by wild beasts, an infection, floods, volcanic events, extraterrestrial events, and so on. But mankind was not numerous in prehistoric times.

We have shown above in this section that there are limits to what mathematics can do in biology, but with modern methods there are new problems such as topology in the movements of the chromosomes in mitosis and meiosis, and in applying physics and chemistry to biological problems.

Broadbent (1952) wrote a review of the theory of transforms by some four authors. The transform in which statisticians are interested is named after Laplace,

$$\phi(p) = \int_{-\infty}^{\infty} K(p, x) f(x) \, dx \qquad (19.6.1)$$

of which a multivariate example is given in equation (19.3.10), where $p$ and $x$ are to be regarded as vectors. Broadbent spoke of the domains of the functions f and $\phi$. Our problem in the domain of f has been transformed into the domain of $\phi$ and a result obtained; the result is then transformed into the domain of f. Broadbent points out that there are pairs of domains in which some results are more readily obtainable than in the other. It is as if there were two languages. Some thoughts are more readily expressed in one language than in another. Let us generalize his idea to scientific thought. We do not accept mathematics as a science but as a language. We assume that there is but one science and that the various so-called sciences—physics, chemistry, geology, biology, biochemistry— are but arbitrary or convenient methods of classifying the methodology of science. Man has at various times made discoveries about the natural world and these have been refined by later observations and/or experiment. Let us examine briefly some case histories.

1. *Genome* can be variously defined. In this context, it means the totality of the genetic material that controls the heredity of an organism. The objects can be observed by the naked eye, light microscope, or electronic microscope, or by chemical or other means. Maupertuis in §4.1 showed that hereditary material was passed down by either sex. Mendel showed that heredity was conveyed by discrete material. Mendel's work was ignored by Naegeli (see §4.5), who no doubt felt that mathematics and botany did not mix, and it was only rediscovered by de Vries, Correns, and Tschermak in 1902. A great feature of Mendel's contribution was his introduction of combinatorial methods, actively studied by academics in Vienna at that time; Mendel and R.A. Fisher are sometimes cited as having raised interest in combinatorial analysis, which has since solved many genetic problems. In 1875, in a letter to Charles Darwin, Francis Galton suggested that heredity might be based on a discrete or a continuous theory. Unfortunately, Darwin chose the continuous theory. Later, Galton (1908) (see §3.6) praised Mendel for the clarity of his papers and for his steady labor to develop his views on heredity. Fisher showed that Darwin's and Galton's theory of continuous heredity could not be supported, arguing that their theory suggested a loss of variance, which would have required mutations to stabilize. Authoritative voices now tell us that there can only be *Mendelian* heredity. Note that Galton's statement of a law of heredity, unhappily accepted by the *Biometrika* school, includes a phrase that implies infinite divisibility of matter (see §4.4). We acknowledge the importance of the Human Genome Project but it is marginal to our main interests in this book.

2. *Double helix and molecular biology.* Work in this field has revolutionized the scientific approach to biology. There are many applications in it that depend on combinatorial mathematics, but the field is too large for satisfactory treatment in this book. There are several popularizing books for both these subjects.

# Epilogue

From the examples given in this book and from miscellaneous readings, it is clear that there are many important, even unavoidable, applications of mathematics to biology. My original plan was to give examples of the use of mathematics in the biological, including medical, sciences that would not amount to a history of the topic.

We have found it desirable to give some brief details of the developments of mathematics and science from ancient Greek times, because these traditions were revived with the coming of the Renaissance after the Dark Ages of the Western world. For example, we show how the ancient Greek problems of atomism have been resolved in modern times by John Dalton, Marian Smoluchowski, and Albert Einstein. Aristotle's views on classification of animals are still much admired. On the other hand, Plato wished to separate universals from sensible particulars, forms from ideas, and believed that causation in nature was regulated by laws that could be stated in mathematical terms. These notions have persisted through the ages into modern times, whereby man was supposed by Plato to be able to describe the natural world without observation or experiment. This was a disaster and so Plato has become the antihero of modern science. Happily there was another tradition, whereby the scientist checked whether his conclusions were consistent with nature, and we have suggested that this tradition should be termed Stratonic. With these definitions the application of mathematics to genetic and evolutionary theory has been Stratonic, whereas the application of higher mathematics to epidemiology has been Platonic.

Chapters 8 through 12, dealing with epidemiology, have incidentally shown how little interest or ability in mathematics the educated classes had; this was a leftover from the Dark Ages, when only in the Church and the legal world was there facility even in reading and writing; however, the coming of commerce required skill in writing and arithmetic. We may consider Fibonacci as an example of the broader use of these skills. Nevertheless, the old views on exact arithmetic held sway to a late period. Possibly the education system and its specialization in subject matter led to the notion of distinct sciences and the success of mathematical methods led to such phrases as "mathematics, the queen of the sciences," but there was no appreciation of its value in the biological sciences, especially in medicine. In particular, there was little progress in the utilization of mathematics for the proof of contagion, a key step in the control of the infective diseases. Statistical methods have been rather more effective in the study of the noninfectious diseases: (a) avitaminoses, especially scurvy caused by lack of vitamin C, and pellagra caused by lack of niacin; (b) genetic diseases, for example, sickle-cell anemia and the hemoglobinopathies, where genetic theory has given very satisfactory answers to their existence; (c) cancers caused by ionizing radiation and ultraviolet light; and (d) cancer of the lung—unfortunately we have here an example of an "adversary" discussion, which so often distorts the truth.

We give in Chapters 14 and 15 examples of the development of physiology and other related fields of research. The demographic side of biostatistics has been discussed more briefly in this monograph than in my *Expectations*. The classical fields of death rates, life tables, fertility, and marriage are still of value and have been applied to many other fields. In particular, the methods have played an important part in working out the consequences of theories of evolution. In Chapter 16, we have moved on to the classification of disease, a difficult field for the mathematician. In Chapter 17 we deal with the numerical analysis of clinical experience, and in Chapter 18 with a quick overview of clinical trials. Comparisons have often given rise to the need for mathematics, a point stressed by Gavarret and Galton.

Chapter 19 is more mathematical. Since the normal distribution has played such a dominant role in biology, we give a rather full account of it and consider mathematical background and biological applications. An important point is raised; in many theories biologists and mathematicians give an implicit reference to uniformitarianism. In many applications this is useful but we must admit to contingencies that may have important effects in nature, especially when numbers are small, for example in the evolution of man, in the passing of the dinosaurs, and so on.

What is the role of the biostatistician (the medical statistician in prewar terms)? It is my opinion that his field has passed into a professional stage. It is possible that the former study of medical statistics will become a branch of epidemiology, with emphasis on the herd rather than the individual.

Perhaps this book will aid in the process of recognizing that biology must be represented more effectively in the history and philosophy of science.

We consider that most of the great discoveries in biology have been made by biologists, although they may often have been assisted by mathematicians or a knowledge of mathematics. Now that there have been so many developments in biology, a proper knowledge of the language will be essential. We look forward to many trained mathematicians becoming interested in the field of biology.

# References

Abbe, E. 1879. Ueber Blutkörper-Zählung. *Jenäische Z. Med. Naturwiss.* 13(NS 6): 98–105.

Abbey, H. 1952. An examination of the Reed-Frost theory of epidemics. *Hum. Biol.* 3:201–233.

Ackerknecht, E.H. 1948. Anticontagionism between 1821 and 1867. *Bull. Hist. Med.* 22:562–593.

Ackerknecht, E.H. 1953. *Rudolf Virchow, doctor, statesman, anthropologist.* University of Wisconsin Press, Madison.

Ackerknecht, E.H. 1967. *Medicine at the Paris Hospital, 1794–1848.* Johns Hopkins, Baltimore.

Ackerknecht, E.H., and Vallois, H.V. 1956. *Franz Joseph Gall, inventor of phrenology and his collection.* Transl. from the French by C. St. Léon. Wisconsin Studies in Medical History, No. 1. University of Wisconsin Medical School, Madison.

Adams, F. 1844–1847. *The seven books of Paulus Aegineta. Translation and commentary.* Sydenham Society, London. 3 vols.

Adams, J. 1814. *A treatise on the supposed hereditary properties of disease.* J. Callow, London.

Ainsworth, G.C. 1976. *Introduction to the history of mycology.* Cambridge University Press, London.

Allee, W.C., and Schmidt, K.P. [Hesse, R.] 1951. *Ecological animal geography. An authorized edition rewritten and revised, based on "Tiergeographie auf ökologischer Grundlage" by the late Richard Hesse.* 2nd edition. Wiley, New York; Chapman and Hall, London.

Allen, G.E. 1978. *Thomas Hunt Morgan: The man and his science.* Princeton University Press, Princeton, NJ.

Allen G.E. 1979. Naturalists and experimentalists: The genotype and the phenotype. In Coleman, W. and Limoges, C. (Eds.) 1979. *Studies in history of biology,* Johns Hopkins, Baltimore, 3:179–209.

Allison, A.C. 1954a. The distribution of the sickle cell trait in East Africa and elsewhere and its apparent relationship to the incidence of subtertian malaria. *Trans. Roy. Soc. Trop. Med. Hyg.* 48:312–318.

Allison, A.C. 1954b. Notes on sickle-cell polymorphism. With statistical appendix by S. Maynard Smith. *Ann. Hum. Genet.* 19:39–57.

Allison, A.C. 1954c. Protection afforded by sickle-cell trait against subtertian malarial infection. *Br. Med. J.* 1:290–294.

Allison, A.C. 1955. Aspects of polymorphism in man. *Cold Spring Harbor Symp. Quant. Biol.* 20:239–255.

Allison, A.C. 1956. The sickle and hemoglobin C-genes in some African populations. *Ann. Hum. Genet.* 21:67–89.

Allison, A.C. 1964. Polymorphism and natural selection in human populations. *Cold Spring Harbor Symp. Quant. Biol.* 24:137–149.

Amberson, J.B., Jr., McMahon, B.T., and Pinner, M. 1931. A clinical trial of sano-crysin in pulmonary tuberculosis. *Am. Rev. Tuberc.* 24:401–435.

Anderson, J.S., Cooper, K.E., McLeod, J.W., and Thomson, J.G. 1931. On the existence of two forms of diphtheria bacillus—*B. diphtheriae gravis* and *B. diphtheriae mitis*—and a new medium for their differentiation and for the bacteriological diagnosis of diphtheria. *J. Pathol. Bacteriol.* 34:667–681.

Andvord, K.F. 1921. Is tuberculosis to be regarded from the etiological standpoint as an acute disease of childhood? *Tubercle* 3:97–116.

Andvord, K.F. 1930. What can we learn by studying tuberculosis by generation? *Norsk. Mag. for Laegevid.* 91:642–660.

Angerstein, W. 1865. *Die Massverhältnisse des männlichen Körpers und das Wachstum der Knaben dargestellt und erläutert.* Selbsverlag, Cologne.

Anonymous, 1880. Ovariotomy. (Editorial) *Br. Med. J.* 1:931–932.

Antonelli, P.L. (Ed.) 1985. *Mathematical essays on growth and the emergence of form.* University of Alberta Press, Edmonton.

Armitage, P. 1983. Trials and errors: The emergence of clinical statistics. Presidential address delivered to the Royal Statistical Society, June 22nd, 1983. *J. Roy. Stat. Soc.* A146:321–334.

Aurelian, L., Manak, M.M., McKinlay, M., Smith, C.C., Klacsmann, K.T., and Gupta, P.K. 1981. "The Herpesvirus hypothesis"—are Koch's postulates satisfied? *Gynecol. Oncology* 12:S56–S87.

Ayala, F.J., and Dobzhansky, T. (Eds.) 1972. *Studies in the philosophy of biology: Reduction and related problems.* Macmillan, London.

Aycock, W.L., and Ingalls, T.H. 1946. Maternal disease as a principle in the epidemiology of congenital anomalies: With a review of rubella. *Am. J. Med. Sci.* 212:366–379.

Bailey, N.T.J. 1975. *The mathematical theory of infectious diseases and its applications.* 2nd edition. Griffin, London.

Balfour, G.W. 1847. Notes on the practice of Skoda. *Edinburgh Med. Surg. J.* 68:397–405.

Baltazard, M., Bahmanyar, M., Mostachfi, P., Eftekhari, M., and Mofidi, Ch. 1960. Recherches sur la peste en Iran. *Bull. WHO* 23:141–155.

Bariéty, M. 1972. Louis et la méthode numérique. *Clio Medica* 7:177–183.

Barinaga, M. 1991. How long is the human life-span? *Science* 254:936–938.

Bartlett, E. 1844. *An essay on the philosophy of medical science.* Lea and Blanchard, Philadelphia. Cited in Bull (1959).

Bartlett, M.S. 1960. The critical community size for measles in the United States. *J. Roy. Stat. Soc.* A123:37–44.

Bassi, A. 1835–1836/1958. *Del mal del segno.* Pts. I & II. Tipografia Orcesi, Lodi. Transl. by P.J. Yarrow, edited by G.C. Ainsworth, Phytopathological Classics No. 10, American Phytopathological Society, Baltimore, 1958.

Bateson, W. 1894. *Materials for the study of variation.* Macmillan, London.

Bateson, W. 1900. Problems of heredity as a subject for horticultural investigation. *J. Roy. Hort. Soc.* 25(1 & 2):54–63. Reprinted in B. Bateson, *William Bateson, F.R.S., naturalist: His essays and addresses together with a short account of his life*, Cambridge University Press, Cambridge, 1928.

Bateson, W. 1902/1909/1913. *Mendel's principles of heredity.* Cambridge University Press, Cambridge.

Beaver, P.C., Jung, R.C., and Cupp, E.W. 1984. *Clinical parasitology.* Lea and Febiger, Philadelphia.

Béclard, P.A. 1863. Rapport général sur les prix décernés en 1862. *Mém. Acad. Méd. Paris* XXVI:xxiii.

Beddall, B.G. 1957. Historical notes on avian classification. *Systematic Zool.* 6:129–136.

Beet, E.A. 1946. Sickle-cell disease in the Balovale district of North Rhodesia. *East Afr. Med. J.* 23:75.

Beet, E.A. 1947. Sickle-cell disease in Northern Rhodesia. *East Afr. Med. J.* 24:212–222.

Bennett, J.H. 1865. *The restorative treatment of pneumonia.* Adam, Edinburgh. 3rd edition, 1866, Adam & Charles Black, Edinburgh.

Bennett, J.H. (Ed.) 1983. *Natural selection, heredity and eugenics. Including selected correspondence of R.A. Fisher with Leonard Darwin and others.* Clarendon (Oxford University Press), New York.

Bennett, J.H. 1991. R.A. Fisher and the role of a statistical consultant. *J. Roy. Stat. Soc.* A154:443–445.

Berkson, J., Magath, T.B., and Hurn, M. 1935. Laboratory standards in relation to chance fluctuations of the erythrocyte count as estimated with the haemocytometer. *J. Am. Stat. Assoc.* 30:414–426.

Berkson, J., Magath, T.B., and Hurn, M. 1940. The error of estimate of the red blood cell as made with the haemocytometer. *Am. J. Physiol.* 128:309–323.

Bernard, C. 1865/1957. *An introduction to the study of experimental medicine.* Transl. by H.C. Greene. 1st edition, 1865. Dover Publications, New York, 1957.

Bernoulli, D. 1721. *Dissertatio inauguralis physico-medica de respiratione.* J.L. Brandmüller, Basel.

Bernoulli, D. 1735. *Recherches physiques et astronomiques sur le problème proposé pour la seconde fois par l'Académie royale des sciences de Paris. Quelle est la cause physique de l'inclinaison des plans des orbites des planètes par rapport au plan de l'équateur de la révolution du soleil autour de son axe? Prix (Paris Academy) 1734.* 2nd edition. Bachelier, Paris, 1808.

Bernoulli, D. 1760. Réflexions sur les avantages de l'inoculation. *Mercure de France,* June 1760.

Bernoulli, D. 1766. Essai d'une nouvelle analyse de la mortalité causée par la petite vérole, & des avantages de l'inoculation pour la prévenir. *Mém. math. phys. Acad. roy. sci., Paris,* 1760.

Bernstein, F. 1925. Zusammenfassende Betrachtungen über die erblichen Blutstrukturen des Menschen. *Z. Induktive Abstammungs- und Vererbungslehre* 37:237–270.

Bernstein, F. 1930a. Ueber die Erblichkeit der Blutgruppen. *Z. Induktive Abstammungs- und Vererbungslehre* 54:400–426.

Bernstein, F. 1930b. Fortgesetzte Untersuchungen aus der Theorie der Blutgruppen. *Z. Induktive Abstammungs- und Vererbungslehre* 56:233–273.

Bernstein, S.N. 1932. Sur les liaisons entre les grandeurs aléatoires. *Verh. Math. Kongr. Zürich* 1:288–309.

Bertillon, J. 1883. *La vie et les oeuvres du Dr. L.A. Bertillon.* Paris.

Bertillon, S. 1941. *Vie d'Alphonse Bertillon, inventeur de l'anthropométrie.* 2nd edition. Gallimard, Paris.

Bienaymé, I.J. 1845. De la loi de multiplication et de la durée des familles. *Soc. Philomat. Paris Extraits,* Sér. 5, 37–39; *L'Institut* 589, 13:131–132. Reprinted in D.G. Kendall, "The genealogy of genealogy: Branching processes before (and after) 1873," *Bull. London Math. Soc.* 7:225–253 (1975).

Biewener, A.A. 1990. Biomechanics of mammalian terrestrial locomotion. *Science* 250:1097–1103.

Biochemical Society Symposium No. 4 1950. *Biochemical aspects of genetics.* R.T. Williams (Ed.). Cambridge University Press, Cambridge.

Birdsell, J.B. 1981. *Human evolution: An introduction to the new physical anthropology.* 3rd edition. Houghton Mifflin, Boston.

Black, F.L. 1966. Measles endemicity in insular populations: Critical community size and its evolutionary implication. *J. Theor. Biol.* 11:207–211.

Boeckh, R. 1886. *Statistisches Jahrbuch der Stadt Berlin. Vol. 12. Statistik des Jahres 1884.* Berlin.

Boeckh, R. 1893. Halley als Statistiker. *Bull. Inst. Int. Stat.* 7:1–24.

Bogin, B. 1988. *Patterns of human growth.* Cambridge Studies in Biological Anthropology, vol. 3. Cambridge University Press, New York.

Boice, J.D., Jr., and Land, C.E. 1982. Ionizing radiation. In Schottenfeld, D. and Fraumeni, J.F. Jr. (Eds) 1982. *Cancer epidemiology and prevention.* W.B. Saunders Co., Philadelphia, pp. 231–253.

Bókai, J. 1892. Varicella and herpes. [Anonymous report.] *Lancet* 2:679.

Bonner, T.N. 1963. *American doctors and German universities: A chapter in international intellectual relations, 1870–1914.* University of Nebraska, Lincoln.

Bonney, V. 1918–1919. The continued high maternal mortality of child-bearing: The reason and the remedy. *Proc. Roy. Soc. Med.* 12:75–107.

Bora, K.C., Douglas, G.R., and Nestleman, E.R. (Eds.) 1982. *Chemical mutagenesis, human population monitoring and genetic risk assessment.* Prog. Med. Research, vol. 3. Elsevier, Amsterdam.

Bortkiewicz, L. 1898. *Das Gesetz der kleinen Zahlen.* Teubner, Leipzig.

Boveri, T. 1902/1964. Ueber mehrpolige Mitosen als Mittel zur Analyse des Zellkerns. *Verh. physikal. -medizin. Gesellschaft zu Würzburg* NS 35:67–90. Transl. in *Foundations of experimental embryology,* B.H. Willier and J.M. Oppenheimer (Eds.), Prentice-Hall, Englewood Cliffs, NJ, 1964, pp. 76–97.

Boveri, T. 1904. *Ergebnisse über die Konstitution der chromatischen Substanz des Zellkerns.* Fischer, Jena.

Box, J.F. 1978. *Fisher, the life of a scientist.* Wiley, New York.

Boxall, R. 1893. The mortality of childbirth. *Lancet* 2:9–15.

Bracegirdle, B. 1978. *A history of microtechnique: The evolution of the microtome and the development of tissue preparation.* Heinemann, London.

Bradbury, S. 1967. *The evolution of the microscope.* Pergamon Press, Oxford.

Bradbury, S., and Turner, G.L'E. (Eds.) 1967. *Historical aspects of microscopy.* W. Heffer & Sons, Cambridge, for the Royal Microscopical Society.

Bretonneau, P.F. 1826. *Des inflammations spéciales du tissu muqueux et en particulier de la diphthérie ou inflammation pelliculaire ...* Crevot, Paris.

Bretonneau, P.F. 1922. *Traités de la dothinentérie et de la spécificité, publiés pour la première fois d'après les manuscrits originaux par L. Dubreuil-Chambardel.* Vigot, Paris.

Broadbent, T.A.A. 1952. Integral transforms. *Nature* 169:1027–1028.

Brown, D.R., and Smith, J.E.K. (Eds.) 1991. *Frontiers of mathematical psychology.* Springer-Verlag, New York.

Brown, F. 1984. The nature of viruses. In Topley and Wilson, vol. 4, pp. 1–4.

Brownlee, J. 1907. Statistical studies in immunity: The theory of an epidemic. *Proc. Roy. Soc. Edinburgh* 26(for 1905–1906):484–521.

Brownlee, J. 1916a. Certain considerations regarding the epidemiology of phthisis pulmonalis. *Public Health* 29:130–145.

Brownlee, J. 1916b. The history of the birth—and death—rates in England and Wales taken as a whole, from 1570 to the present time. *Public Health* 29:211–222, 228–238.

Brownlee, J. 1918. An investigation into the periodicity of measles epidemics in London from 1703 to the present day by the method of the periodogram. *Philos. Trans. Roy. Soc. London* B208:225–250.

Brownlee, J. 1919. Notes on the biology of a life table. *J. Roy. Stat. Soc.* 82:34–77.

Budd, W. 1873/1931. *Typhoid fever: Its nature, mode of spreading, and prevention.* Longmans, London. Reprinted 1931 by The American Public Health Assoc. for the Delta Omega Society, New York.

Buffon, G.-L.L. 1774–1789. *Histoire naturelle—Supplément.* 7 vols. Imprimerie Royale, Paris. [Vol. 4, 1777, contains Montbeillard's measurements of his son's growth.]

Bull, J.P. 1959. The historical development of clinical therapeutic trials. *J. Chron. Dis.* 10:218–248.

Bulloch, W. 1938/1960. *The history of bacteriology. University of London, Heath Clark Lectures, 1936.* Oxford University Press, London.

Bumpus, H.C. 1899. The elimination of the unfit as illustrated by the introduced sparrow, *Passer domesticus. Biological Lectures, Marine Biological Laboratory, Wood's Hole,* 6:209–226 (1898).

Bundesen, H.N., Connelly, J.I., Gorman, A.E., Hardy, A.V., McCoy, G.W., and Rawlings, I.D. 1936. *Epidemic amebic dysentery. The Chicago outbreak of 1933.* Natl. Inst. of Health Bull. No. 166. Govt. Printing Office, Washington, DC.

Burdett, H. 1882. On the relative mortality after amputations of large and small hospitals and the influence of the antiseptic (Listerian) system upon such mortality. *J. Stat. Soc. (London)* 45:444–483.

Burnet, F. 1941. The biological approach to infectious disease. *Med. J. Austral.* 2:607–612.

Busvine, J.R. 1976. *Insects, hygiene and history.* Athlone Press, London.

Busvine, J.R. 1980. *Insects and hygiene: The biology and control of insect pests of medical and domestic importance.* 3rd edition. Chapman and Hall, London.

Butlin, H.T. 1892. Three lectures on cancer of the scrotum in chimney sweeps and others. *Br. Med. J.* 1:1341–1346; 2:1–6, 66–71.

Bylebyl, J.J. 1977. Nutrition, quantification and circulation. *Bull. Hist. Med.* 51:369–385.

Campbell, J.M. 1924. *Maternal mortality.* U.K. Ministry of Health, Reports on Public Health and Medical Subjects No. 25. HMSO, London.

Campbell, J.M. 1927. *The protection of motherhood.* U.K. Ministry of Health, Reports on Public Health and Medical Subjects No. 48. HMSO, London.

Cannon, W.B. 1929. Organization for physiological homeostasis. *Physiol. Rev.* 9:399–431.

Cantlie, N. 1974a. *A history of the Army Medical Department.* Vol. I. Churchill-Livingstone, Edinburgh-London.

Cantlie, N. 1974b. *A history of the Army Medical Department.* Vol. II. Churchill-Livingstone, Edinburgh-London.

Carpenter, K.J. (Ed.) 1981. *Pellagra.* Benchmark Papers in Biochemistry, Vol. 2. Hutchinson Ross, Stroudsburg, PA.

Cartwright, F.F. 1967. *The development of modern surgery.* Arthur Barker, London.

Castle, W.E. 1903–1904. The laws of heredity of Galton and Mendel, and some laws governing race improvement by selection. *Proc. Am. Acad. Arts Sci. (Daedalus)* 39:223–242.

Chauveau, J-B.A. 1868. Nature du virus vaccin. Détermination expérimentale des éléments qui constituent le principe actif de la sérosité vaccinale virulente. *C.R. Acad. Sci.* 66:289–293.

Chetverikov (Četverikov), S.S. 1915/1920. The fundamental factor of insect evolution. *Smithsonian Report for 1918*, No. 2566, 441–449, Washington, 1920. Russian original in *Izv. Moskovsk. Entomol. Obščestva* 1:14–24 (1915).

Chetverikov (Četverikov), S.S. 1926/1961. On certain aspects of the evolutionary process from the standpoint of modern genetics. Transl. by M. Barker, edited by I.M. Lerner. *Proc. Am. Philos. Soc.* 105:167–195 (1961). Russian original in *Žurnal Eksperimental. Biologiĭ* A2:3–54 (1926).

Chittenden, R.H., and Underhill, F.P. 1917. The production in dogs of a pathological condition which closely resembles human pellagra. *Am. J. Physiol.* 44:13–66.

Churchill, F. 1850. *Essays on the puerperal fever and other diseases peculiar to women: Selected from the writings of British authors previous to the close of the eighteenth century.* By request of the Sydenham Society. Lea and Blanchard, Philadelphia.

Clark, P.F. 1961. *Pioneer microbiologists of America.* University of Wisconsin Press, Madison.

Clark, R.W. 1984. *The survival of Charles Darwin: A biography of a man and an idea.* Weidenfeld & Nicolson, London.

Clark, W.E. le G., and Medawar, P.B. (Eds.) 1945. *Essays on growth and form. Presented to D'Arcy Wentworth Thompson.* Clarendon Press, Oxford.

Clarke, C. 1985. Robert Russell Race. 28th November 1907–15th April 1984. Elected F.R.S. 1952. *Biog. Mem. Fell. Roy. Soc.* 31:453–492.

Cockayne, E.A. 1933. *Inherited abnormalities of the skin and its appendages.* Oxford University Press, London.

Coleman, W. 1964. *Georges Cuvier, zoologist. A study in the history of evolution theory.* Harvard University Press, Cambridge, MA.

Coleman, W. 1965. Cell, nucleus, and inheritance: An historical study. *Proc. Am. Philos. Soc.* 109:124–158.

Coleman, W. (Ed.) 1967. *The interpretation of animal form. Essays by J. Wyman, C. Gegenbauer, E.R. Lankester, H.L. Duthiers, W. His, and H.N. Martin.* Sources of Science, No. 15. Johnson Reprint Corp., New York-London.

Coleman, W. 1979. Bergmann's rule: Animal heat as a biological phenomenon. In Coleman, W., and Limoges, C. (Eds.) 1979. *Studies in history of biology.* Johns Hopkins, Baltimore, 3:67–88.

Coleman, W. 1987. Experimental physiology and statistical inference: The therapeutic trial in nineteenth century Germany. In Krüger, L., Gigerenzer, G., and Morgan, M.S. (Eds.) 1987. *The probabilistic revolution, 2: Ideas in the sciences*. Bradford (MIT Press), Cambridge, MA, pp. 201–226.

Constable, G. 1973. *The Neanderthals: The emergence of man*. With an introduction by R.S. Solecki. Time-Life Books, New York.

Corney, B.G. 1883–1884. The behaviour of certain epidemic diseases in natives of Polynesia, with especial reference to the Fiji Islands. *Trans. Epidemiol. Soc. London* 3:76–95.

Creighton, C. 1965. *A history of epidemics in Britain; with additional material by D.E.C. Eversley (and others)*. 2nd edition. Cass, London. Barnes & Noble, New York. 2 vols. (First published 1891/1894 as *A history of epidemics in Britain from A.D. 664 to the extinction of plague*, Cambridge University Press.)

Crombie, A.C. 1967. The mechanistic hypothesis and the scientific study of vision: Some optical ideas as a background to the invention of the microscope. In Bradbury and Turner, pp. 3–112.

Crow, J.F., and Denniston, C. 1985. Mutation in human populations. *Adv. Hum. Genet.* 14:59–123.

Cuénot, L. 1905. Les races pures et leurs combinaisons chez les souris. *Arch. Zool. Exp. Génét.* 3:123–132.

Cumpston, J.H.L. 1914. *The history of smallpox in Australia, 1788–1908*. Serv. Publ. No. 3, Quarantine Service, Commonwealth of Australia. Govt. Printer, Melbourne.

Cumpston, J.H.L. 1927. *The history of diphtheria, scarlet fever, measles and whooping cough in Australia*. Serv. Publ. No. 37, Dept. Health, Commonwealth of Australia. Govt. Printer, Canberra.

Cumpston, J.H.L., and McCallum, F. 1926. *The history of plague in Australia, 1900–1925*. Serv. Publ. No. 32, Dept. Health, Commonwealth of Australia. Govt. Printer, Melbourne.

Cushing, H. 1940. *The life of Sir William Osler*. Oxford University Press, London.

Dacie, J. 1988. Hermann Lehmann, 1910–1985. *Biog. Mem. Fell. Roy. Soc.* 34:407–449.

Dahlberg, G. 1929. Inbreeding in man. *Genetics* 14:421–454.

Dareste, C. 1874. De la duplicité monstrueuse. *Bull. Soc. d'Anthrop. Paris* 9(Sér. 2): 321–338.

Davenport, C.B. 1926. Human metamorphosis. *Am. J. Phys. Anthrop.* 9:205–226.

Daw, S.F. 1970. Age of boys' puberty in Leipzig 1727–49 as indicated by voice breaking in J.S. Bach's choir members. *Hum. Biol.* 42:87–89.

Dean, G. 1971. *The porphyrias: A story of inheritance and environment*. Pitman, London.

von Decastello, A., and Sturli, A. 1902. Ueber die Isoagglutinine im Serum gesunder und kranker Menschen. *München. Med. Wochenschr.* 49:1090–1095.

Decouflé, P. 1982. Occupation. In Schottenfeld, D., and Fraumeni, J.F. Jr. (Eds.) 1982. *Cancer epidemiology and prevention*. W.B. Saunders Co., Philadelphia, pp. 318–335.

Deniker, J. 1900/1924. *The races of man: An outline of anthropology and ethnography*. Scribner's, New York; W. Scott, London. Extracts reprinted in Kroeber, A.L., and Waterman, T.T. 1924; *Source book in anthropology*. University of California Press, Berkeley, CA, pp. 166–180.

Denniston, C. 1982. Low level radiation and genetic risk estimation in man. *Annu. Rev. Genet.* 16:329–355.

Deparcieux, A. 1746. *Essai sur les probabilités de la durée de la vie humaine* ... Frères Guerin, Paris.

Derrick, V.P.A. 1927. Observations on (1) errors in age in the population statistics of England and Wales, and (2) the changes in mortality indicated by the national records. With discussion. *J. Inst. Actuar.* 58:117–159.

Dittrich, F.L. 1963. *Biophysics of the ear.* In collaboration with R.C. Extermann. C.C. Thomas, Springfield, IL.

Dobell, C. 1914. Mendelism in the seventeenth century. *Nature* 94:588–589.

Dobell, C. 1932. *Antony van Leeuwenhoek and his "little animals": Being some account of the father of protozoology and bacteriology and his multifarious discoveries in these disciplines* ... John Bale, Sons & Danielsson, London.

Dobzhansky, T. 1965. Mendelism, Darwinism, and evolutionism. *Proc. Am. Philos. Soc.* 109:205–215.

Doll, R., and Hill, A.B. 1950. Smoking and carcinoma of the lung. *Br. Med. J.* 2:739–748.

Donné, A. 1837. *Recherches microscopiques sur la nature des mucus et la matière des divers écoulemens des organes génito-urinaires chez l'homme et chez la femme: Description des nouveaux animalcules découverts dans quelques-uns de ces fluides: Observations sur un nouveau mode de traitement de la blennorrhagie.* Paris.

Donovan, J.W. 1970. Measles in Australia and New Zealand, 1834–1835. *Med. J. Austral.* 1:5–10.

Double, F.J. 1837. The inapplicability of statistics to the practice of medicine. *London Med. Gaz.* 20:361–364.

Drayna, D., and White, R. 1985. The genetic linkage map of the human X chromosome. *Science* 230:753–758.

Dubois, D., and Dubois, F.E. 1916. A formula to estimate the approximate surface area if height and weight are known. *Arch. Intern. Med.* 17:863–871.

Dudley, S.F. 1923. *The Schick test, diphtheria and scarlet fever.* Med. Res. Council Spec. Rep. Series, No. 75. HMSO, London.

Dudley, S.F. 1926. *The spread of droplet infection in semi-isolated communities.* Med. Res. Council Spec. Rep. Series, No. 111. HMSO, London.

Dudley, S.F. 1931. Lessons on the infectious diseases in the Royal Navy. The Milroy Lecture. *Lancet* 1:509–517, 570–578.

Dudley, S.F. 1936. The ecological outlook on epidemiology. *Proc. Roy. Soc. Med.* 30:57–70.

von Dungern, E., and Hirszfeld, L. 1910. Ueber Vererbung gruppenspezifischer Strukturen des Blutes. II. *Z. Immunitätsforsch.* 6:284–292. Transl. by G.P. Pohlmann in *Transfusion (Philadelphia)* 2:70–74 (1962).

von Dungern, E., and Hirszfeld, L. 1911. Ueber gruppenspezifische Strukturen des Blutes. III. *Z. Immunitätsforsch.* 8:526–562.

Dunn, L.C. 1965. Mendel, his work and his place in history. *Proc. Am. Philos. Soc.* 109:189–198.

Dunn, L.C., Sturtevant, A.H., Dobzhansky, T., Stern, C., Glass, B., Sonnenborn, T.M., and Mangelsdorf, P.C. 1965. Commemoration of the publication of Gregor Mendel's pioneer experiments in genetics. *Proc. Am. Philos. Soc.* 109:189–248.

Edmonds, T.R. 1832. *Life tables, founded upon the discovery of a numerical law regulating the existence of every human being* ... J. Duncan, London.

Ehrlich, P.R., and Wilson, E.O. 1991. Biodiversity studies: Science and policy. *Science* 253:758–762.

Einstein, A. 1905. Die von der molekularkinetischen Theorie der Wärme geforderte Bewegung von in ruhenden Flüssigkeiten suspendierten Teilchen. *Ann. der Physik* (4th ser.) 17:549–560.

Einstein, A. 1922. *Untersuchungen über die Theorie der Brownschen Bewegung.* [Contains 5 papers by Einstein on Brownian movement.] Oswalds Klassiker der Exakten Wissenschaften, No. 199. Akademische Verlagsgesellschaft, (R. Fürth), Leipzig.

Eisenhart, C., and Wilson, P.W. 1943. Statistical methods and control in bacteriology. *Bacteriol. Rev.* 7:57–137.

Elderton, W.P. 1922–1923. Review: Is tuberculosis to be regarded from the aetiological standpoint as an acute disease of childhood. *Tubercle* 3(3), 1921. *Biometrika* 14:191–192.

Elvehjem, C.A., Madden, R.J., Strong, F.M., and Woolley, D.W. 1937. Relation of nicotinic acid and nicotinic acid amide to canine black tongue. *J. Am. Chem. Soc.* 59:1767–1768.

Epstein, J.H. 1978. Photocarcinogenesis: A review. *Natl. Cancer Inst. Monog.* 50:13–25.

von Ettingshausen, A. 1826. *Die combinatorische Analysis als Vorbereitungslehre zum Studium der theoretischen höhern Mathematik.* Wallishausser, Vienna.

Euler, L. 1761. An illustration of population growth. In Suessmilch, vol. 1 1761, pp. 291–299. Transl. in Smith and Keyfitz, 1977, pp. 79–82.

Euler, L. 1767. Recherches générales sur la mortalité et la multiplication du genre humain. *Hist. Acad. Roy. Sci. Belles Lettres*, Berlin, Année 1760;144–164. Reprinted in *Opera omnia* (1911– ) 1st ser., 7:79–100. Transl. in Smith and Keyfitz, 1977, pp. 83–91.

Evans, A.S. 1976. Causation and disease: The Henle-Koch postulates revisited. *Yale J. Biol. Med.* 49:175–195.

Eveleth, P.B., and Tanner, J.M. 1991. *Worldwide variation in human growth.* 2nd edition. Cambridge University Press, New York.

Ewens, W.J. 1979. *Mathematical population genetics.* Biomathematics, vol. 9. Springer-Verlag, New York.

Eyler, J.M. 1980. The conceptual origins of William Farr's epidemiology: Numerical methods and social thought in the 1830s. With discussion. In A.M. Lilienfeld (Ed.), *Times, places and persons: Aspects of the history of epidemiology.* The Henry F. Sigerist Supplements to the *Bulletin of the History of Medicine*, New Ser. No. 4, Johns Hopkins, Baltimore, pp. 1–27.

Farabee, W.C. 1905. *Inheritance of digital malformations in man.* [Extract from a Ph.D. thesis, *Hereditary and sexual influences in meristic variation: A study of digital malformations in man*, Harvard, 1903.] Papers of the Peabody Museum of American Archaeology and Ethnology, Cambridge, MA., Vol. 3, plates 23–27, text 65–77.

Farley, J. 1977. *The spontaneous generation controversy from Descartes to Oparin.* Johns Hopkins University Press, Baltimore-London. First published 1974.

Farr, W. 1839. First Annual Report of the Registrar General of England and Wales, p. 99. Cited in Introduction to *International Classification of Diseases*, 1955 revision, WHO, Geneva, 1957.

Farr, W. 1843. *Fifth Annual Report of the Registrar General of Births, Deaths and Marriages in England.* Parliamentary Papers, XXI, 1843.

Farr, W. 1885/1975. *Vital statistics: A memorial volume of selections from the reports and writings of William Farr.* N.A. Humphreys (Ed.). Offices of the Sanitary Institute, London. Reprinted 1975 under the auspices of the Library of the N.Y. Academy of Medicine, with introd. by M. Susser and A. Adelstein. History of Medicine Series No. 46, Scarecrow Press, Metuchen, NJ.

Faust, E.C. 1955. History of human parasitic infections. *Public Health Rep.* 70:958–965.

Feinstein, A.R. 1967. *Clinical judgment.* Williams & Wilkins, Baltimore.

Feller, W. 1941. On the integral equation of renewal theory. *Ann. Math. Stat.* 12:243–268. Excerpts reprinted in Smith and Keyfitz, 1977.

Fenaroli, G., Garibaldi, U., and Penco, M.A. 1981. Games, life wagers, mortality tables: Birth of the calculus of probability, statistics and population theory. (Italian) *Arch. Hist. Exact Sci.* 25(4):329–341.

Fenner, F., Henderson, D.A., Arita, I., Jezek, Z., and Ladnyi, I.D. 1988. *Smallpox and its eradication.* WHO, Geneva.

Fenner, F., and Ratcliffe, F.N. 1965. *Myxomatosis.* Cambridge University Press, London.

Fibiger, J. 1898. Om Serumbehandlung af Difteri. *Hospitalstidende* 4(Ser. 6):309–325, 337–350.

Filipčenko, J.A. 1923. *Variation and methods for its study.* (Russian) [*Izmenchivost' i metody ee izucheniia.*] Petrograd. 2nd edition, Leningrad, 1926; 3rd edition, 1927; 4th edition, Moscow and Leningrad, 1929.

Findlay, G.M. 1928. Ultraviolet light and skin cancer. *Lancet* 2:1070–1073.

Fisher, R.A. 1918. The correlation between relatives on the supposition of Mendelian inheritance. *Trans. Roy. Soc. Edinburgh* 52:399–433.

Fisher, R.A. 1925. *Statistical methods for research workers.* Oliver and Boyd, Edinburgh. 13th edition, 1958.

Fisher, R.A. 1930. *The genetical theory of natural selection.* Oxford University Press, Oxford.

Fisher, R.A. 1935a. Eugenics, academic and practical. *Eugen. Rev.* 27:95–100.

Fisher, R.A. 1935b. *The design of experiments.* Oliver & Boyd, Edinburgh-London.

Fisher, R.A. 1943. Personal communication cited in R.R. Race, "An 'incomplete' antibody in human serum." *Nature* 153:771–772 (1944).

Fisher, R.A. 1945. G.L. Taylor, M.D., Ph.D., F.R.C.P. *Br. Med. J.* 1:463.

Fisher, R.A. 1947. The Rhesus factor—a study in scientific method. *Am. Scientist* 35:95–103.

Fisher, R.A. 1953. The expansion of statistics: The inaugural address of the President, Sir Ronald Fisher, F.R.S., delivered to the Royal Statistical Society on November 19th, 1952. *J. Roy. Stat. Soc.* A116:1–9.

Fisher, R.A. 1959. *Smoking; the cancer controversy: Some attempts to assess the evidence.* Oliver and Boyd, Edinburgh.

Fisher, R.A. 1971–1974. *Collected papers of R.A. Fisher.* J.H. Bennett, Ed. University of Adelaide, Adelaide, South Australia. 5 vols.

Fisher, R.A., Thornton, H.G., and MacKenzie, W.A. 1922. The accuracy of the plating method of estimating the density of bacterial populations. *Ann. Appl. Biol.* 9:325–359.

Florkin, M. 1972. *A history of biochemistry.* Pt I—*Proto-biochemistry*; Pt II—*From proto-biochemistry to biochemistry.* Vol. 30 of *Comprehensive biochemistry* (M. Florkin, Ed.). Elsevier, Amsterdam.

Forbes, T.R. 1971. *Chronicle from Aldgate. Life and death in Shakespeare's London.* Yale University Press, New Haven-London.

Ford, W.W. 1939. *Bacteriology.* Hoeber, New York-London. (Reprinted 1964 as Clio Medica Series, vol. 22. Hafner, New York.)

Foster, W.D. 1961. *A short history of clinical pathology.* E. & S. Livingstone, Edinburgh-London.

Foster, W.D. 1965. *A history of parasitology.* E. & S. Livingstone, Edinburgh-London.

Foster, W.D. 1970. *A history of medical bacteriology and immunology.* Heinemann, London.

Freind, J. 1717. *De Febribus Commentarii novem ad Librum Epidemiorum primum et tertium accommodati.* Transl. by T. Dale as *Nine commentaries upon fevers ...* T. Cox, London, 1730.

Frost, W.H. 1939. The age selection of the mortality from tuberculosis in successive decades. *Am. J. Hyg.* 30 (Nov. issue):91–96. (Reprinted in *Papers of Wade Hampton Frost, M.D.,* K.F. Maxcy, Ed., The Commonwealth Fund, New York, 1941.)

Fruton, J.S. 1972. *Molecules and life: Historical essays on the interplay of chemistry and biology.* Wiley-Interscience, New York.

Fuller, T. 1730. *Exanthematologia; or an attempt to give a rational account of eruptive fevers, especially of the measles and small-pox.* C. Rivington & S. Austen, London. Our citation from Rolleston (1937).

Galilei, G. 1638. *Discorsi e dimostrazioni matematiche, intorno à due nuove scienze attenenti alla mecanica ed i muovimenti locali ...* Appresso gli Elzevirii, Leiden. *Opere,* Ed. Favaro, VIII, p. 169 et seq. Transl. by H. Crew and A. de Salvio, Macmillan, New York, 1914.

Gall, F.J. 1800. *Philosophisch-medicinische Untersuchungen über Natur und Kunst im kranken und gesunden Zustande des Menschen.* Baumgärtner, Leipzig.

Galton, F. 1865. Hereditary talent and character, I. & II. *Macmillan's Mag.* 12:157–166, 318–327.

Galton, F. 1874. *Notes and queries on anthropology for the use of travellers and residents in uncivilised lands, published by the Committee of the British Association for the Advancement of Science ...* E. Stanford, London. [Section on statistics by Galton.]

Galton, F. 1875. Statistics by intercomparison with remarks on the law of frequency of error. *Philos. Mag.* (Ser. 4) 49:33–46.

Galton, F. 1876. The history of twins as a criterium of the relative powers of nature and nurture. *J. (Roy.) Anthropol. Inst.* 5:391–406.

Galton, F. 1877. Typical laws of heredity. *Proc. Roy. Instn. Gt. Britain* 8:282–301.

Galton, F. 1886. Family likeness in eye-colour. *Proc. Roy. Soc. London.* 40:402–416.

Galton, F. 1889a. On the advisability of assigning marks for bodily efficiency in the examination of candidates for those public services in which bodily efficiency is of importance. (Reply to H.M. Civil Service Commissioners, xxxi, p. 15.) Paper read before the Anthropological Section of the British Association, see Pearson (1914–1930), vol. II, p. 386.

Galton, F. 1889b. Human variety. *J. (Roy.) Anthropol. Inst.* 18:401–419. [This address became Chapter II of F.Galton's *Anthropometric Laboratory, Notes & Memoirs, No. 1,* 1890, published at his own expense. Hence the pagination of quotes in Pearson (1914–1930).]

Galton, F. 1889c. *Natural inheritance.* Macmillan, London.

Galton, F. 1897. The average contribution of each several ancestor to the total heritage of the offspring. *Proc. Roy. Soc. London* 61:401–413.

Galton, F. 1908. *Memories of my life.* E.P. Dutton, New York; Methuen, London.

Galton, F., and Watson, H.W. 1874. On the probability of extinction of families. *J. (Roy.) Anthropol. Inst.* 4:138–144.

Garland, J. 1943. Varicella following exposure to herpes zoster. *N. Engl. J. Med.* 228:336–337.

Garnham, P.C.C. 1971. *Progress in parasitology.* University of London Heath Clark Lectures 1968. Athlone Press, University of London.

Garrod, A.E. 1902. The incidence of alkaptonuria: A study in chemical individuality. *Lancet* 2:1616–1620.

Gavarret, J. 1840. *Principes généraux de statistique médicale, ou, Développement des règles qui doivent présider à son emploi.* Bechet jeune et Labé, Paris.

Geissler, A. 1889. Beiträge zur Frage des Geschlechtsverhältnisses der Geborenen. *Z. K. Sächs. Statist. Bur.* 35:1–24, 56.

George, W. 1964. *Biologist philosopher: A study of the life and writings of Alfred Russel Wallace.* Life of Science Library. Abelard-Schuman, London.

Gerhard, W.W. 1837. On the typhus fever which occurred at Philadelphia in the spring and summer of 1836. *Am. J. Med. Sci.* 19–20:289–322.

Giddings, L.V., Kaneshiro, K.Y., and Anderson, W.W. (Eds.) 1989. *Genetics, speciation, and the founder principle.* Oxford University Press, New York-Oxford.

Gilbert, S.F. (Ed.) 1991. *Developmental biology: A comprehensive synthesis. Vol. 7, A conceptual history of modern embryology.* Plenum, New York.

Gini, C. 1951. Combinations and sequences of sexes in human families and mammal litters. *Acta Genet. Stat. Med.* 2:220–244.

Glass, B. 1959. Maupertuis, pioneer of genetics and evolution. In Glass, B., Temkin, O., and Strauss, W.L. Jr. (Eds.) 1959. *Forerunners of Darwin: 1745–1859.* Johns Hopkins, Baltimore, pp. 51–83.

Glass, B. 1963. The establishment of modern genetical theory as an example of the interaction of different models, techniques and inferences. In Crombie, A.C. (Ed.). *Scientific change.* Heinemann, London, pp. 521–541.

Glass, B. 1965. A century of biochemical genetics. *Proc. Am. Philos. Soc.* 109:227–236.

Goldberger, J. 1914. The etiology of pellagra: The significance of certain epidemiological observations with respect thereto. *Public Health Rep.* 29:1683–1686.

Goldberger, J. 1916. The transmissibility of pellagra: Experimental attempts at transmission to the human subject. *Public Health Rep.* 31:3159–3173.

Goldberger, J., and Sebrell, W.H. 1930. The Blacktongue preventive value of Minot's Liver Extract. *Public Health Rep.* 45:3064–3070.

Goldberger, J., Waring, C.H., and Willets, D.G. 1915. The prevention of pellagra: A test of diet among institutional inmates. *Public Health Rep.* 30:3117–3131.

Goldberger, J., and Wheeler, G.A. 1920. The experimental production of pellagra in human subjects by means of diet. *Bull. Hygienic Laboratory,* No. 120, 7–116.

Goldberger, J., Wheeler, G.A., Lillie, R.D., and Rogers, L.M. 1928. A study of the blacktongue-preventive action of 16 foodstuffs, with special reference to the identity of blacktongue of dogs and pellagra of man. *Public Health Rep.* 43:1385–1454.

Goldberger, J., Wheeler, G.A., and Sydenstricker, E. 1920. A study of the relation of diet to pellagra incidence in seven textile-mill communities of South Carolina in 1916. *Public Health Rep.* 35:648–713. [Reprinted in *The challenge of facts: Selected public health papers of Edgar Sydenstricker.* Kasius, R.V. (Ed.). Prodist (Neale Watson Academic Publications), New York, 1974;345–369.

Goldschmid, E. 1953. Nosologia naturalis. In Underwood, E.A. (Ed.). *Science, medicine and history: Essays on the evolution of scientific thought and medical practice, written in honour of Charles Singer*, vol. II. Oxford University Press, London, pp. 103–122.

Goodall, A. 1908. Review of Student 1907. [Biometrika 5(3):351.] *Folia Haemat., Leipzig* 5:430.

Gordon, A. 1795. *A treatise on the epidemic puerperal fever of Aberdeen*. London. (Reprinted by the Sydenham Society in Churchill, 1850, pp. 377–422.)

Gordon, J.E. 1962. Chickenpox: An epidemiological review. *Am. J. Med. Sci.* 244: 362–389.

Goubert, J.-P. 1989. *The conquest of water. The advent of health in the industrial age.* Transl. by A. Wilson from the French edition, Paris, 1986. Princeton University Press, Princeton.

Gould, S.J. 1970. Evolutionary paleontology and the science of form. *Earth-Science Reviews* 6:77–119.

Gould, S.J. 1989. *Wonderful life: The Burgess shale and the nature of history.* W.W. Norton, New York-London.

Graetzer, J. 1883. *Edmund Halley und Caspar Neumann. Ein Beitrag zur Geschichte der Bevölkerungs-Statistik.* S. Schottlaender, Breslau.

Graham, H. 1950. *Eternal Eve.* Heinemann, Altrincham. Rev. 2nd edition, 1960.

Graunt, J. 1662. *Natural and political observations mentioned in a following index, and made upon the Bills of Mortality ... by John Graunt, fellow of the Royal Society. With reference to the government, religion, trade, growth, ayre, diseases, and the several changes of the said city ...* J. Martin, J. Allestry, and T. Dicas, London. [Reprinted in Hull, C.H. (Ed.). *The economic writings of Sir William Petty.* Cambridge University Press, 1899; vol. 2, pp. 315–435 (reprinted 1964 by Augustus M. Kelly, New York); *Natural and Political Observations made upon the Bills of Mortality*, W.F. Willcox (Ed.), Johns Hopkins, Baltimore, 1939; *J. Inst. Actuar.* 90(384): 4–61 (1964).]

Gray, J. 1929. The kinetics of growth. (*Brit.*) *J. Exper. Biol.* 6:248–274.

Greenbaum, L.S. 1972. The humanitarianism of Antoine Laurent Lavoisier. *Studies on Voltaire and the Eighteenth Century* 88:651–675.

Greenhow, E.H. 1857–1858. The results of an inquiry into the different proportions of death produced by certain diseases in different districts of England. In *Papers Relating to the Sanitary State of the People of England. Parliamentary Papers*, 1857–58; vol. XXIII, pp. 1–64. (With a popular introduction and prefatory comment to this Report, entitled "Illustrations", ibid. 365–387.)

Greenwood, M. 1921. Galen as an epidemiologist. *Proc. Roy. Soc. Med.* 14 (Sect. Hist. Med.):3–16.

Greenwood, M. 1927a. In discussion of Derrick (1927). *J. Inst. Actuar.* 58:117–159.

Greenwood, M. 1927b. Abstract of Dudley (1926), "The spread of droplet infection in semi-isolated communities." *Bull. Hyg.* 2:417–418.

Greenwood, M. 1928. 'Laws' of mortality from the biological point of view. *J. Hyg. Camb.* 28:267–294.

Greenwood, M. 1931. On the statistical measure of infectiousness. *J. Hyg. Camb.* 31:336–351.

Greenwood, M. 1932. *Epidemiology, historical and experimental. The Herter Lectures for 1931.* Johns Hopkins, Baltimore.

Greenwood, M. 1936a. English death rates, past, present and future. A valedictory address. *J. Roy. Stat. Soc.* 99:674–715.

Greenwood, M. 1936b. *The medical dictator and other biographical studies.* Williams and Norgate, London.

Greenwood, M. 1948. *Medical statistics from Graunt to Farr. The Fitzpatrick Lectures for the years, 1941 and 1943.* Reprinted from *Biometrika.* Cambridge University Press, Cambridge.

Greenwood, M., Hill, A.B., Topley, W.W.C., and Wilson, G.S. 1936. *Experimental epidemiology.* Med. Res. Council Spec. Rep. Series No. 209. HMSO, London.

Greenwood, M., and Topley, W.W.C. 1925. A further contribution to the experimental study of epidemiology. *J. Hyg. Camb.* 24:45–110.

Greenwood, M., and Yule, G.U. 1915. The statistics of anti-typhoid and anti-cholera inoculations and the interpretation of such statistics in general. *Proc. Roy. Soc. Med.* 8:113–194.

Greenwood, M., and Yule, G.U. 1917. On the statistical interpretation of some bacteriological methods employed in water analysis. *J. Hyg. Camb.* 16:36–54.

Gregg, N.M. 1941. Congenital cataract following German measles in the mother. *Trans. Ophthalmol. Soc. Aust.* 3:35–46.

Guy, W.A. 1847. On the duration of life of sovereigns. *J. Stat. Soc. (London)* 11:62–69.

H., J.H., and McC., R.M.S. 1969. William Norman Pickles. *Lancet* 1:581–582.

H.-S., R.E., and H., J.H. 1969. W.N. Pickles, C.B.E., M.D., D.Sc., F.R.C.P., F.R.C.P.E.D., F.R.C.G.P. *Br. Med. J.* 1:719–720.

Haeckel, E.H.P.A. 1866. *Generelle Morphologie der Organismen.* G. Reimer, Berlin. 2 vols.

Härting, F.H., and Hesse, W. 1879. Der Lungenkrebs, die Bergkrankheit in den Schneeberger Gruben. *Vierteljahrs schrift gerichtl. Med. u. öffentl. Gesundheitswesen* (N.F.) 30:296–309; 31:102–129, 313–337.

Haight, F.A. 1967. *Handbook of the Poisson distribution.* Wiley, New York.

Hald, A. 1990. *A history of probability and statistics and their applications before 1750.* Wiley, New York.

Haldane, J.B.S. 1924. A mathematical theory of natural and artificial selection. Part 1. *Trans. Camb. Philos. Soc.* 23:19–41.

Haldane, J.B.S. 1957. Karl Pearson, 1857–1957. Centenary. *Biometrika* 44:303–313.

Hales, S. 1733. *Haemastaticks; or an account of some hydraulic and hydrostatical experiments made on the blood and blood-vessels of animals.* Comprises vol. 2 of 2nd edition of *Statical essays.* W. Innys, R. Manby and T. Woodward, London. 2nd edition, 1738–1740.

Hall, T.S. 1969. *Ideas of life and matter.* University of Chicago Press, Chicago. 2 vols.

Halley, E. 1693/1942. An estimate of the degrees of the mortality of mankind, drawn from curious tables of the births and funerals at the city of Breslau; with an attempt to ascertain the price of annuities upon lives. *Philos. Trans. Roy. Soc. London* 17: 596–610, 654–656. Reprinted 1942, L.J. Reed (Ed.), Johns Hopkins, Baltimore.

Halley, E., Haygarth, J., Wigglesworth, E., Barton, W., and Milne, J. 1973. *Mortality in pre-industrial times: The contemporary verdict.* Containing reprints of the above authors, 1694–1837. Gregg International, London.

Halliday, J.L. 1928. *An inquiry into the relationship between housing conditions and the incidence and fatality of measles.* Med. Res. Council Spec. Rep. Series No. 120. HMSO, London.

Ham, B.B. 1907. *Report on plague in Queensland, 1900–1907.* Govt. Printer, Brisbane.

Hamer, W.H. 1906. The Milroy Lectures on epidemic disease in England—the evidence of variability and persistency of type. *Lancet* 1:569–574, 655–662, 733–739.

Hammond, E.C. 1966. Smoking in relation to the death rates of 1 million men and women. *Natl. Cancer Inst. Monog.* 19:127–204.

Hammond, E.C. 1972. Smoking habits and air pollution in relation to lung cancer. In Lee, D.H.K. (Ed.). *Environmental factors in respiratory disease*, Academic Press, New York, pp. 177–198.

Hardy, G.H. 1908. Mendelian proportions in a mixed population. *Science* 28:49–50.

Harris, H. 1969. Enzyme and protein polymorphism in human populations. *Br. Med. Bull.* 25:5–13.

Harris, H. 1973. Lionel Sharples Penrose, 11th June 1898–12th May 1972. Elected F.R.S. 1953. *Biog. Mem. Fell. Roy. Soc.* 19:521–561.

Harvey, C. 1936. Cancer of the lung: A clinical survey of one hundred cases. *Med. J. Austral.* 1:565–572.

Harvey, P.H., and Pagel, M.D. 1991. *The comparative method in evolutionary biology*. Oxford University Press, New York.

Hawkins, F.B. 1829. *Elements of medical statistics*. Longman, Rees, Orme, Brown and Green, London.

Heath, T.L. 1921. *A history of Greek mathematics*. Reprinted 1960, 1965. Clarendon Press, Oxford. 2 vols.

Hebra, F., and Kaposi, M. 1866–1880. *On diseases of the skin, including the exanthemata*. Transl. by Fagge, Pye-Smith and Tay. Publ. Nos. 30, 37, 61, 64, 86. New Sydenham Society, London. 5 vols.

Heckscher, E.F. 1950. Swedish population trends before the Industrial Revolution. *Econ. Hist. Rev.* 2(2):266–277.

von Helmholtz, H. 1924–1925. *Helmholtz's treatise on physiological optics*. Transl. from the 3rd German edition. J.P.C. Southall (Ed.), Optical Society of America. 3 vols. Reprinted by Dover Publications, New York, 1962. 3 vols. in 2.

van Helmont, J.B. 1662. *Oriatrike or Physik Refined*. Transl. by J. Chandler. Lodowick Loyd, London. [Cited in Debus, A.G. *The chemical dream of the Renaissance*, Heffer, Cambridge, 1968, p. 27.]

Henle, F.G.J. 1840. *Pathologische Untersuchungen, Pt. 1*. Berlin. Reprinted as *Von den Miasmen und Kontagien*, in Sudhoffs Klassiker der Medizin, No. 3, F. Marchand, Leipzig, 1910; and as *On miasmata and contagia*, transl. by G. Rosen, *Bull. Hist. Med.* 6:907–983 (1938).

Heyde, C.C., and Seneta, E. 1977. *I.J. Bienaymé: Statistical theory anticipated*. Stud. Hist. Math. Phys. Sci., vol. 3. Springer-Verlag, Berlin-Heidelberg-New York.

Hill, A.B. 1960. Conclusion: The statistician. In *Controlled clinical trials. Papers from a conference, Vienna, 1959*. Council for International Organizations of Medical Sciences (CIOMS). Blackwell, Oxford, pp. 168–171.

Hill, A.B. 1962. *Statistical methods in clinical and preventive medicine*. E. & S. Livingstone, Edinburgh-London.

Hill, A.B. 1965. The environment and disease: Association or causation? President's Address, Section of Occupational Medicine. *Proc. Roy. Soc. Med.* 58:295–300.

Hillis, D.M., and Moritz, C. (Eds.) 1990. *Molecular systematics*. Sinauer, Sunderland, MA.

Hilts, V.L. 1980. Epidemiology and the statistical movement. With discussion. In Lilienfeld, A.M. (Ed.). *Times, places and persons: Aspects of the history of epidemiology*. The Henry F. Sigerist Supplements to the *Bulletin of the History of Medicine*, New Ser. No.4, Johns Hopkins, Baltimore, pp. 43–61.

Hirst, L.F. 1953. *The conquest of plague*: *A study of the evolution of epidemiology*. Clarendon Press, Oxford.

Holmes, F.L. 1984. Carl Voit and the quantitative tradition in biology. In Mendelsohn, E. (Ed.) 1984. *Transformation and tradition in the sciences*: *Essays in honor of I.B. Cohen*. Cambridge University Press, Cambridge, pp. 455–470.

Holmes, O.W. 1843. The contagiousness of puerperal fever. *N. Engl. Quart. J. Med. Surg.* Reprinted in *Medical Classics* 1:211–243 (1936).

Holmes, O.W. 1855. *Puerperal fever as a private pestilence*. [Holmes, 1843, reprinted with additions.] Ticknor and Fields, Boston. Reprinted in *Medical Classics* 1:247–268 (1936).

Holmes, O.W. 1891. *Medical essays, 1842–1882*. [Contains reprints of 1843 and 1855 papers, above.] Vol. 9 of *The Works of Oliver Wendell Holmes*, Houghton, Mifflin, Boston–New York. 13 vols.

Holmes, T. 1863. *Report by Mr. Holmes and Dr. Bristowe forming part of the Sixth Report of the Medical Officer of the Privy Council*. Govt. Blue Book for 1863, published 1864, London. [Cited in J.Y. Simpson (1871), p. 340.]

Holmes, T. 1869. On "Hospitalism." *Lancet* 2:194–196, 229–230.

Hopkins, D.R. 1983. *Princes and peasants*: *Smallpox in history*. University of Chicago Press, Chicago-London.

Hopwood, D.A., and Chater, K.F. (Eds.) 1989. *Genetics of bacterial diversity*. Academic Press, London-San Diego.

Howells, W.W. 1973. *Cranial variation in man*: *A study by multivariate analysis of patterns of difference among recent human populations*. Papers of the Peabody Museum, vol. 67. Peabody Museum of Archaeol. and Ethnol., Harvard University, Cambridge, MA.

Huber, F. 1959. *Daniel Bernoulli als Physiologe und Statistiker*. Basler Veröffent. Geschichte Med. Biol. No. 8. B. Schwabe, Basel.

Hueper, W.C. 1966. *Occupational and environmental cancers of the respiratory system*. Springer-Verlag, Berlin.

Hughes, A. 1959. *A history of cytology*. Abelard-Schuman, London.

Humphreys, N.A. 1885. Biographical sketch of William Farr, M.D., D.C.L., C.B., F.R.S., etc. In Farr (1885/1975), pp. vii–xxiv.

Hyde, J.N. 1906. On the influence of light in the production of cancer of the skin. *Am. J. Med. Sci.* 131:1–22.

Imhof, A.E. (in collaboration with R. Gehrmann, I.E. Kloke, M. Roycroft, and H. Wintrich). 1990. *Lebenserwartungen in Deutschland vom 17. bis 19. Jahrhundert*. German/English text. VCH—Acta humaniora, Weinheim, D-6940.

Ingalls, T.H., Babbott, F.L., Hampson, K.W., and Gordon, J.E. 1960. Rubella: Its epidemiology and teratology. *Am. J. Med. Sci.* 239:363–383.

Ingalls, T.H., Plotkin, S.A., Meyer, H.M., and Parkman, P.D. 1967. Rubella: Epidemiology, virology and immunology. *Am. J. Med. Sci.* 253:349–373.

Irgens, L.M. 1981. Epidemiological aspects and implications of the disappearance of leprosy from Norway: Some factors contributing to the decline. *Leprosy Rev.* 52(Suppl.):147–165.

Jackson, J. 1836. Preface and Appendix. In Louis (1836), pp. v–xxviii, 99–171.

Jampert, C.F. 1754. *De causis incrementum corporis animalis limitantes*. Fürstenia, Halle.

Jardine, N., and Sibson, R. 1971. *Mathematical taxonomy*. Wiley, London–New York.

Jenkin, H.C.F. 1867. The origin of species. *North British Rev.* 46(92):277–318.

Jenner, E. 1798. *An Inquiry into the Causes and Effects of the Variolae Vaccinae* ... Published by the author, London.

Johannsen, W. 1903/1959. Ueber Erblichkeit in Populationen und in reinen Linien. K. *Danske Videnskabernes Selskabs Forhandlinger* No. 3 (1903). Transl. and reprinted in Peters, 1959, pp. 20–26.

Johannsen, W.(L.) 1905. *Arvelighedslaerens elementer.* (*The elements of heredity.*) Gyldendalske Boghandel, Copenhagen. An enlargement of *On heredity and variation* (1896); rewritten and enlarged again (1909).

Johannsen, W. 1929. Biology and statistics. *Nordic Stat. J.* 1:351–361. [An English version of "Biologi og Statistik", *Norsk. Stat. Tidskr.* 1:71–80 (1922).]

Johanson, D. and Edey, M.A. 1981. *Lucy, the beginning of humankind.* Granada, London.

Johanson, D., and Shreeve, J. 1989. *Lucy's child: The discovery of a human ancestor.* William Morrow, New York.

Johnson, J. 1813. *The influence of tropical climate, more especially the climate of India, on European constitutions ... an essay.* J.J. Stockdale, London.

Kandel, E.R., Schwartz, J.H., and Jessell, T.M. 1991. *Principles of neural science.* 3rd edition. Elsevier, New York.

Keen, H. 1982. Problems in the definition of diabetes mellitus and its subtypes. In Koebberling and Tattersall, pp. 1–11.

Kennaway, E.L., and Waller, R.E. 1953/1954. Studies on cancer of the lung. In Clemmesen, J. (Ed.). *Cancer of the lung (endemiology): A symposium.* Council Internat. Org. Med. Sci. (CIOMS), Louvain, Belgium, 1954. Reprinted from *Acta Un. Int. c. Cancr.* 9:485–494 (1953).

Kepler, J. 1604. *Ad Vitellionem paralipomena or Astronomiae pars optica.* Frankfurt. [See *Johannes Kepler Gesammelte Werke, Vol. II.* W. von Dyck, M. Caspar and F. Hammer (Eds.), Munich, 1937.]

Kermack, W.O., and McKendrick, A.G. 1927. Contributions to the mathematical theory of epidemics. Pt 1. *Proc. Roy. Soc. London* A138:55–83.

Kermack, W.O., McKendrick, A.G., and McKinlay, P.L. 1934a. Death rates in Great Britain and Sweden: Some general regularities and their significance. *Lancet* 1: 698–703.

Kermack, W.O., McKendrick, A.G., and McKinlay, P.L. 1934b. Death rates in Great Britain and Sweden: Expression of specific mortality rates as products of two factors and some consequences thereof. *J. Hyg. Camb.* 34:433–457.

Kingman, J.F.C. 1980. *Mathematics of genetic diversity.* CBMS-NSF Regional Conf. Ser. in Appl. Math. 34. Society for Industrial and Applied Mathematics, Philadelphia.

Kisch, B. 1954. David Gruby, 1810–1898. Forgotten Leaders in Modern Medicine. *Trans. Am. Philos. Soc.* 44:193–226.

Knapp, G.F. 1868. *Ueber die Ermittlung der Sterblichkeit aus den Aufzeichnungen der Bevölkerungs-Statistik.* J.C. Hinrichs, Leipzig.

Knapp, G.F. 1869. *Die Sterblichkeit in Sachsen nach amtlichen Quellen dargestellt.* 2 Parts. Duncker & Humblot, Leipzig.

Koch, R. 1876. Die Aetiologie der Milzbrand-Krankheit, begründet auf die Entwicklungsgeschichte des Bacillus Anthracis. *Beiträge zur Biologie der Pflanzen* 2:277–311. Reprinted in K. Sudhoff's *Klassiker der Medizin* No. 9 (1910) and transl. in *Medical Classics* 2:787–820 (1937–1938) as "The etiology of anthrax, based on the ontogeny of the anthrax bacillus."

Koch, R. 1877. Verfahren zur Untersuchung, zum Conservieren und Photographieren der Bakterien. *Beiträge zur Biologie der Pflanzen* 2:399–434. Transl. and abstr. as "Methods for studying, preserving and photographing bacteria." In Doetsch, R.N. Jr. (Ed.). *Microbiology: Historical contributions from 1776 to 1908.* Rutgers University Press, New Brunswick, NJ, 1960, pp. 67–73.

Koch, R. 1878. *Untersuchungen über die Aetiologie der Wundinfectionskrankheiten.* Leipzig. Transl. by W.W. Cheyne as *Investigations into the etiology of traumatic infective diseases.* Publ. No. 88, New Sydenham Society, London, 1880.

Koch, R. 1882. Die Aetiologie der Tuberculose. *Berliner klin. Wochenschr.* 19:221–230. Transl. by B. Pinner and M. Pinner as "The aetiology of tuberculosis", *Am. Rev. Tuberc.* 25:285–323 (1932). Issued as a pamphlet by National Tuberculosis Assoc., New York, 1932. Transl. by W. de Rouville in *Medical Classics* 2:853–880 (1937–1938).

Koch, R. 1890. Ueber bakteriologische Forschung. *Verh. X Internat. Med. Kongr., Berlin* I:35–47. Transl. as "An address on bacteriological research", *Br. Med. J.* 2:380–383 (1890).

Koebberling, J., and Tattersall, R. (Eds.) 1982. *The genetics of diabetes mellitus.* Proc. of the Serono Symposia, vol. 47. Academic Press, London.

Koenigsberger, L. 1906. *Hermann von Helmholtz.* Transl. by F.A. Welby with a preface by Lord Kelvin. Clarendon Press, Oxford.

Kopf, E.W. 1918. Florence Nightingale as statistician. *Publ. Am. Stat. Assoc.* 15:388–404. Reprinted in Kendall, M.G., and Plackett, R.L. *Historical studies in probability and statistics.* Griffin, London, 1977, pp. 310–326.

Kreyberg, L. 1962. Histological lung cancer types: A morphological and biological correlation. *Acta Pathol. Microbiol. Scand. Suppl.* 152:1–92.

Kuczynski, R.R. 1928. *The balance of births and deaths. Vol. 1, Western and Northern Europe.* Macmillan, New York.

Kuczynski, R.R. 1930. A reply to Dr. Lotka's review of "The Balance of Births and Deaths." *J. Am. Stat. Assoc.* 25:84–85.

Lancaster, H.O. 1950a. Statistical control in haematology. *J. Hyg. Camb.* 48:402–417.

Lancaster, H.O. 1950b. The sex ratios in sibships with special reference to Geissler's data. *Ann. Eugen. London* 15:153–158.

Lancaster, H.O. 1951. Deafness as an epidemic disease in Australia: A note on census and institutional data. *Br. Med. J.* 2:1429–1432.

Lancaster, H.O. 1952. The mortality in Australia from measles, scarlatina and diphtheria. *Med. J. Austral.* 2:272–276.

Lancaster, H.O. 1954. Deafness due to rubella. *Med. J. Austral.* 2:323–324.

Lancaster, H.O. 1956. Some geographical aspects of the mortality from melanoma in Europeans. *Med. J. Austral.* 1:1082–1087.

Lancaster, H.O. 1960. Australian mortality in the late nineteenth century. *Med. J. Austral.* 2:84–87.

Lancaster, H.O. 1964. Bibliography of vital statistics in Australia and New Zealand. *Austral. J. Stat.* 6:33–99.

Lancaster, H.O. 1967. The infections and population size in Australia. *Bull. Int. Stat. Inst.* 42:459–471.

Lancaster, H.O. 1969. *The chi-squared distribution.* Wiley, New York.

Lancaster, H.O. 1973. Bibliography of vital statistics in Australia: A second list. *Austral. J. Stat.* 15:1–26.

Lancaster, H.O. 1974. *An introduction to medical statistics.* Wiley, New York.

Lancaster, H.O. 1982a. Bibliography of vital statistics in Australia: A third list. *Austral. J. Stat.* 24:361–380.

Lancaster, H.O. 1982b. A bibliography of statistical bibliographies: A fourteenth list. *Int. Stat. Rev.* 50:195–217.

Lancaster, H.O. 1990. *Expectations of life: A study in the demography, statistics, and history of world mortality.* Springer-Verlag. New York.

Lancaster, H.O. 1994a. Genetics before Mendel. Maupertuis and Réaumur: Pedigrees. *J. Med. Biog.*: (in press).

Lancaster, H.O. 1994b. Semmelweis: A rereading of *Die Aetiologie. J. Med. Biog.* 2(1): 12–21;2(2):84–88.

Lancaster, H.O., and Nelson, J. 1957. Sunlight as a cause of melanoma: A clinical survey. *Med. J. Austral.* 1:452–456.

Lancaster, H.O., and Pickering, H. 1952. The incidence of births of the deaf in New Zealand. *N.Z. Med. J.* 51:184–189.

Lancisi, G.M. 1717. *De noxiis paludum effluviis eorumque remediis. Libri duo.* Jo. Mariae Salvioni, Rome.

Landsteiner, K. 1901. Ueber Agglutinationserscheinungen normalen menschlichen Blutes. *Wiener klin. Wochenschr.* 14:1132–1134.

Laplace, P.S. 1812. *Théorie analytique des probabilités.* Courcier, Paris. Reprinted in *Oeuvres complètes* (1878–1912), vol. VII, with 4 supplements published in 1816, 1818, 1820 and 1825. Facsimile of first edition published by Éditions Culture et Civilisation, Brussels, 1967.

Laplace, P.S. 1814/1951. *A philosophical essay on probabilities.* Transl. from the sixth French edition by F.W. Truscott and F.L. Emory with introd. by E.T. Bell, Dover, New York, 1951.

Laughlin, W.S., and Osborne, R.H. (Compilers) 1967. *Human variation and origins: An introduction to human biology and evolution. Readings from the Scientific American.* W.H. Freeman & Co., San Francisco-London.

Lavoisier, A.-L. 1784. Cited in Hill, A.B. 1960, pp. 170–171. [No details of Lavoisier reference given.]

Lavoisier, A.-L. 1787. *Méthode de nomenclature chimique, proposée par MM. de Morveau, Lavoisier, Bertholet et de Fourcroy.* Cuchet, Paris. Augmented edition, 1789. Also in *Oeuvres*, 1892, Imprimerie Nationale, Paris, 6 vols. Facsimile reprod. by Johnson Reprint Co., New York, 1965. Citation in *Oeuvres* 5:358.

Lechevalier, H.A., and Solotorovsky, M. 1965. *Three centuries of microbiology.* McGraw-Hill, New York.

Lee, J.A.H. 1982. Melanoma and exposure to sunlight. *Epidemiol. Rev.* 4:110–136.

van Leeuwenhoek, A. 1679. Letter to Viscount Brouncker. November 1677. *Philos. Trans. Roy. Soc. London* 12:1040–1043.

Lesky, E. 1976. *The Vienna Medical School of the 19th century.* Transl. from the German by L. Williams and I.S. Levij. Johns Hopkins, Baltimore-London.

Leslie, P.H. 1945. On the uses of matrices in certain population mathematics. *Biometrika* 33:183–212. Excerpts reprinted in Smith and Keyfitz, 1977.

Levin, M.L., Goldstein, H., and Gerhardt, P.R. 1950. Cancer and tobacco smoking: A preliminary report. *J. Am. Med. Assoc.* 143:336–338.

Levine, P., and Stetson, R.E. 1939. An unusual case of intragroup agglutination. *J. Am. Med. Assoc.* 113:126–127.

Levy, C.E.M. 1847. Report on lying-in establishments and on practical instruction in delivery in London and Dublin. (Danish). J.C. Scharling, Copenhagen. Reprinted from *Bibliothek for Laeger* (1847). See Semmelweis, 1861, pp. 153–169.

Liebermeister, C. 1877. Ueber Wahrscheinlichkeitsrechnung in Anwendung auf thera-peutische Statistik. In Volkmann, R. (Ed.). *Sammlung Klinischer Vorträge*, No. 110 (Innere Medizin No. 39), pp. 935–962.

von Liebig, J. 1846. *Animal chemistry or chemistry in its applications to physiology and pathology*. Part I. 3rd edition, W. Gregory (Ed.). Taylor and Walton, London.

Lilienfeld, A.M. 1982. Ceteris paribus: The evolution of the clinical trial. *Bull. Hist. Med.* 56:1–18.

Lilienfeld, D.E. 1977. Contagium vivum and the development of water filtration: The beginning of the sanitary movement. *Prev. Med.* 6:361–375.

Lister, J. 1878. On lactic fermentation. *Trans. Pathol. Soc. London.* 29:425–467.

Lloyd, C. 1961. The introduction of lemon juice as a cure for scurvy. *Bull. Hist. Med.* 35:123–132.

Lloyd, C., and Coulter, J.L.S. 1961. *Medicine and the Navy 1200–1900. Vol. 3, 1714–1815.* E. & S. Livingstone, Edinburgh-London.

Lloyd, G.E.R. 1991. *Methods and problems in Greek science. Selected papers.* Cambridge University Press, Cambridge.

London, W.P., and Yorke, J.A. 1973. Recurrent outbreaks of measles, chickenpox and mumps. I, Seasonal variation in contact rates. II, Systematic differences in contact rates and stochastic effects. *Am J. Epidemiol.* 98:453–468, 469–482.

Lorenz, E. 1944. Radioactivity and lung cancer: A critical review of lung cancers in the miners of Schneeberg and Joachimsthal. *J. Natl. Cancer Inst.* 5:1–15.

Lorimer, F. 1959. The development of demography. In Hauser, P.M., Duncan, O.D. (Eds.). *The study of population: An inventory and appraisal*, University of Chicago Press, Chicago, IL, pp. 124–179.

Lotka, A.J. 1907. Studies on the mode of growth of material aggregates. *Am. J. Sci.* 24(141):199–216.

Lotka, A.J. 1934–1939. *Théorie analytique des associations biologiques. Part I: Principes. Part II: Analyse démographique avec application particulière à l'espèce humaine.* Actualités scientifiques et industrielles, Nos. 187 and 780. Hermann et Cie, Paris. 2 vols.

Louis, P.C.A. 1825/1844. *Recherches anatomico-pathologiques sur la phthisie, etc.* Gabon, Paris, 1825. 2nd edition, 1844. Transl. by W.H. Walshe as *Researches on phthisis, anatomical, pathological and therapeutical.* Sydenham Society, London, 1844.

Louis, P.C.A. 1829/1836. *Recherches anatomiques, pathologiques et thérapeutiques sur la maladie connue sous les noms de gastro-entérite, fièvre putride, adynamique, ataxique, typhoïde, etc. etc., comparée avec les maladies aigues les plus ordinaires.* J.-B. Baillière, Paris, 1829. 2nd (augmented) edition, 1841. Transl. by Henry I. Bowditch as *Anatomical, pathological and therapeutic researches upon the disease known under the name of gastro-enterite, putrid, adynamique, ataxic, or typhoid fever, etc. compared with the most common acute diseases.* 2 vols. 1836. Vol. 1, I.R. Butts, Boston; vol. 2, Hilliard, Gray & Co., Boston.

Louis, P.C.A. 1831/1834. *Généralités sur l'enseignement de la médecine clinique ...* Paris, 1831. Transl. by P. Martin as *Essay on clinical instruction.* S. Highley, London, 1834.

Louis, P.C.A. 1835/1836. *Recherches sur les effets de la saignée dans quelques maladies inflammatoires et sur l'action de l'émétique et les vésicatoires dans la pneumonie.* J.-B. Baillière, Paris, 1835. Transl. by C.G. Putnam as *Researches on the effects of blood-letting in some inflammatory diseases, and on the influence of tartarised antimony and vesication in pneumonitis.* With preface and appendix by J. Jackson. Hilliard, Gray & Co., Boston, 1836.

Louis, P.C.A. 1837. The applicability of statistics to the practice of medicine. *London Med. Gaz.* 20:488–491.

Lussana, F., and Frua, C. 1856. Su la pellagra. *Su la pellagra*, Giuseppi Bernardoni, Milan, pp. 1–25. Translated and reprinted in Carpenter, 1981, pp. 13–18.

Lyell, C. 1830–1833. *Principles of geology. Being an attempt to explain the former changes of the earth's surface by reference to causes now in operation.* John Murray, London. 3 vols.

Lyon, J.F., and Thoma, R. 1881. Ueber die Methode der Blutkörperzählung. *Virchows Arch.* 84:131–154.

Lyon, J.L., Gardner, J.W., and West, D.W. 1980. Cancer risk and life style among Mormons from 1967–1975. In Cairns, J., Lyon, J.L., and Skolnick, M. (Eds.). *Cancer incidence in defined populations.* Banbury Rep. No. 4, Cold Spring Harbor Laboratory, Cold Spring Harbor, NY, pp. 3–30.

McGovern, V.J. 1952. Melanoblastoma. *Med. J. Austral.* 1:139–142.

MacIntyre, D. 1926. The serum treatment of diphtheria. *Lancet* 1:855–858.

McKendrick, A.G. 1926. Applications of mathematics to medical problems. *Proc. Edinburgh Math. Soc.* 44:98–130.

McKendrick, A.G. 1940. The dynamics of crowd infections. *Edinburgh Med. J.* (N.S.) (IV) 47:117–136.

McKinney, H.L. 1972. *Wallace and natural selection.* Yale University Press, New Haven-London.

McLeod, J.W. 1943. The types *mitis, intermedius* and *gravis* of *Corynebacterium diphtheriae*: A review of observations during the past ten years. *Bacteriol. Rev.* 7:1–56.

Macmichael, W. (? anonymously) 1825. Review of "The Progress of Opinion on the Subject of Contagion." *Quart. Rev.* 33:218–257.

Madsen, Th., and Madsen, S. 1956. Diphtheria in Denmark. From 23,695 to 1 case— Post or propter. I. Serum therapy. II. Diphtheria immunization. *Dan. Med. Bull.* 3:112–121.

Magath, T.B., Berkson, J., and Hurn, M. 1936. The error determination of the erythrocyte count. *Am. J. Clin. Pathol.* 6:568–579.

Magner, L.N. 1979. *A history of the life sciences.* Marcel Dekker, New York-Basel.

Maiocchi, R. 1990. The case of Brownian motion. *Br. J. Hist. Sci.* 23:257–283.

Malgaigne, J.-F. 1842. Études statistiques sur les résultats des grandes opérations dans les hôpitaux de Paris. *Arch. Gén. de Méd.* 13:389–418;14:50–81.

Mangelsdorf, P.C. 1965. Genetics, agriculture and the world food problem. *Proc. Am. Philos. Soc.* 109:242–248.

Marshall, H., and Tulloch, A.M. 1838. *Statistical report on the sickness, mortality and invaliding in the West Indies. Prepared from the records of the Army Medical Department and War Office returns. Presented to both houses of Parliament by command of Her Majesty.* W. Clowes & Sons, London for HMSO.

Martin, C. ca. 1740. *De animalium calori.*

Martland, H.S., and Humphries, R.E. 1929. Osteogenic sarcoma in dial painters using luminous paint. *Arch. Pathol.* 7:406–417.

Matumoto, M. 1969. Mechanism of perpetuation of animal viruses in nature. *Bacteriol. Rev.* 33:404–418.

de Maupertuis, P.L.M. 1744–1745/1980. *Vénus physique, contenant deux dissertations, l'une sur l'origine des hommes et des animaux; et l'autre sur l'origine des noirs.* La Haye. Also in *Oeuvres* (Lyons, 1756) 2:1–134. Republished as *Vénus physique, suivi de la Lettre sur le progrès des sciences, précédé d'un essai de Patrick Tort.* Collection Palimpseste, Aubier-Montaigne, Paris, 1980.

de Maupertuis, P.L.M. 1752. *Lettres*. [No. XIV cited by Glass, 1959.] Also in *Oeuvres* (Lyons, 1756) 2:185–340.

Mayr, E. 1982. *The growth of biological thought: Diversity, evolution, and inheritance*. Belknap Press (Harvard), Cambridge, MA.

Mayr, E. 1991. *One long argument: Charles Darwin and the genesis of modern evolutionary thought*. Harvard University Press, Cambridge, MA.

Mayr, E., and Provine, W.B. 1980. *The evolutionary synthesis: Perspectives on the unification of biology*. Harvard University Press, Cambridge, MA.

Medawar, P.B. 1962. D'Arcy Thompson and "Growth and Form." *Perspect. Biol. Med.* 5:220–232.

Medical Research Council 1948. Streptomycin treatment of pulmonary tuberculosis. *Br. Med. J.* 2:769–782.

Meigs, C.D. 1856. *Obstetrics: The science and the art*. 3rd edition. Blanchard and Lea, Philadelphia.

Meiklejohn, A.P. 1954. The curious obscurity of Dr. James Lind. *J. Hist. Med. Allied Sci.* 9:304–310.

Mendel, G.J. 1865. Versuche über Pflanzenhybriden. *Verh. des Naturforschenden Vereins (Brünn)*, 4:1–47. Transl. by the Royal Horticultural Society as "Experiments in plant hybridisation" and published in Bateson, 1913, pp. 335–379. Also published as *Experiments in plant hybridisation*, with commentary by Sir Ronald Fisher and reprint of "Notice of Mendel" by W. Bateson, Oliver & Boyd, Edinburgh, 1965. Also published in Peters (1959).

Merrill, M. 1931. The relationship of individual to average growth. *Hum. Biol.* 3:37–70.

Merson, M.H. 1993. Slowing the spread of HIV: Agenda for the 1990s. *Science* 260: 1266–1268.

Mettler, C.C. 1947. *History of medicine. A correlative text, arranged according to subjects*. F.A. Mettler (Ed.). Blakiston, Philadelphia-Toronto.

Meyer, H.M., Brooks, B.E., Douglas, R.D., and Rogers, N.G. 1962. Ecology of measles in monkeys. *Am. J. Dis. Child.* 103:307–313.

Michell, J. 1767. An Inquiry into the probable Parallax, and Magnitude of the fixed Stars, from the Quantity of Light which they afford us, and the particular Circumstances of their Situation. *Philos. Trans. Roy. Soc. London* 57(Pt. 1): 234–264.

Middleton, W.E. Knowles. 1966. *A history of the thermometer and its use in meteorology*. Johns Hopkins, Baltimore.

Milne, J. 1815. *A treatise on the valuation of annuities and assurances on lives and survivorships; on the contruction of tables of mortality; and other probabilities and expectations of life ... With a variety of new tables*. Longman, Hurst, Rees, Orme & Brown, London. 2 vols.

Milne-Edwards, H., and Villermé, L.R. 1829. Mémoire sur l'influence de température sur la mortalité des enfants nouveaux-nés. *Ann. d'Hyg.* 2:291–307.

Mitra, S.K. 1947. *The upper atmosphere*. Royal Asiatic Society of Bengal, Calcutta.

de Moivre, A. 1733. *Approximatio ad Summam Terminorum Binomii $(a + b)^n$ in Seriem expansi*. Printed for private circulation. Photographically reprinted in R.C. Archibald, "A rare pamphlet of Moivre and some of his discoveries." *Isis* 8:671–683 (1926).

Moore, J.A. (Ed.) 1972. *Readings in heredity and development*. Oxford University Press, New York-London.

Moran, P.A.P., and Smith, C.A.B. 1966. *Commentary on R.A. Fisher's paper on the correlation between relatives on the supposition of Mendelian inheritance.* Cambridge University Press, London.

Morgan, T.H., Sturtevant, A.H., Muller, H.J., and Bridges, C.B. 1915. *The mechanism of Mendelian heredity.* Constable & Co., London.

Moser, L.F. 1839. *Die Gesetze der Lebensdauer. Nebst Untersuchungen über Dauer, Fruchtbarkeit der Ehen, über Tödlichkeit der Krankheiten ... etc.* Veit & Co., Berlin.

Motulsky, A.G. 1959. Joseph Adams (1756–1818). *AMA Arch. Intern. Med.* 104:490–496.

Mrosovsky, N. 1990. *Rheostasis. The physiology of change.* Oxford University Press, New York.

Mueller, F.H. 1939. Tabakmissbrauch und Lungencarcinom. *Z. Krebsforsch.* 49:57–85.

Mullett, C.F. 1956. *The bubonic plague and England. An essay in the history of preventive medicine.* University of Kentucky Press, Lexington.

Munro, A. Campbell. 1890–1891. Measles, an epidemiological study. *Trans. Epidemiol. Soc. London* 10(N.S.):94–109.

Mustacchi, P. 1961. Ramazzini and Rigoni-Stern on parity and breast cancer. *AMA Arch. Intern. Med.* 108:639–642.

Nasse, C.F. 1820. Von einer erblichen Neigung zu tödlichen Blutungen. *Arch. Med. Erfahr.* 1:385–434.

Neel, J.V. 1977. Health and disease in unacculturated Amerindian populations. With discussion. In Elliott, K., and Whelan, J. (Eds.). *Health and disease in tribal societies.* Ciba Found. Symp. 49. Elsevier/Excerpta Medica/North-Holland, Amsterdam, pp. 155–177.

Neyman, J., and Pearson, E.S. 1967. *Joint statistical papers.* Cambridge University Press, Cambridge.

Nora, J.J., and Fraser, F.C. 1974. *Medical genetics: Principles and practice.* Lea & Febiger, Philadelphia.

Norton, H.T.J. 1928. Natural selection and Mendelian variation. *Proc. London Math. Soc.* Ser. 2, 28:1–45.

Nunneley, T. 1841. *A treatise on the nature, causes, and treatment of erysipelas.* John Churchill, London.

Nuttall, G.H.F. 1904. *Blood immunity and blood relationship: A demonstration of certain blood relationships amongst animals by means of the precipitin test for blood, including original researches by G.S. Graham-Smith and T.S.P. Strangeways.* Cambridge University Press, Cambridge.

Ogle, W. 1892. An inquiry into the trustworthiness of the old Bills of Mortality. *J. Roy. Stat. Soc.* 55:437–460.

Ollivier, A.-F. 1822. *Traité expérimental du typhus traumatique, gangrène ou pourriture des hôpitaux; contenant des observations nouvelles sur diverses gangrènes, épidémies, contagions; sur les antiseptiques, les désinfectans, etc. ... Ouvrage ampliatif de deux mémoires adressés in 1810 et 1811 au Conseil de santé des armées, suivi de pièces justificatives.* Seignot, Paris.

Osiander, F.C. 1795. Resultate von Beobachtungen und Nachrichten über die erste Erscheinung des Monatliches. *Denkwürdigkeiten für das Heilkunde und Geburtshilfe* 2:380–388.

Ottenberg, R. 1908. Transfusion and arterial anastomosis. *Ann. Surg.* 47:486–502.

Ottenberg, R. 1937. Reminiscences of the history of blood transfusion. *Mt. Sinai J. Med. N.Y.* 4:264–271.

Oxnard, C.E. 1984. *The order of man: A biomathematical anatomy of the primates.* Yale University Press, New Haven-London.

Palladino, P. 1990. Stereochemistry and the nature of life: Mechanist, vitalist, and evolutionary perspectives. *Isis* 81:44–67.

Panum, P.L. 1847/1940. *Iagttagelser, anstillende under Maeslinge-Epidemien paa Faeröerne i Aaret 1846.* Transl. by A.S. Hatcher as *Observations made upon the epidemic of measles in the Faroe Islands in the year 1846.* With a biographical memoir by J.J. Petersen, transl. by J. Dimont. Delta Omega Soc. and Am. Public Health Assoc., New York, 1940.

Passey, R.D. 1922. Experimental soot cancer. *Br. Med. J.* 2:1112–1113.

Pasteur, L. (with collab. of J.F. Joubert and C.E. Chamberland). 1878. La théorie des germes et ses applications à la médecine et à la chirurgie. *C.R. Acad. Sci.* 86:1037–1043 (*Oeuvres* 6:112–130).

Pearl, R., and Reed, L.J. 1920. On the rate of growth of the population of the United States since 1790 and its mathematical representation. *Proc. Natl. Acad. Sci.* 6:275–288.

Pearson, E.S. 1990. *"Student": A statistical biography of William Sealy Gosset.* Edited and augmented by R.L. Plackett with the assistance of G.A. Barnard. Clarendon Press, Oxford.

Pearson, K. 1892/1900. *The grammar of science.* Contemp. Science Series. W. Scott, London. 2nd edition, Adam & Charles Black, London, 1900.

Pearson, K. 1900. On a criterion that a given system of deviations from the probable in the case of a correlated system of variables is such that it can be reasonably supposed to have arisen from random sampling. *Philos. Mag.* (Ser. 5)50:157–175.

Pearson, K. 1903. Mathematical contributions to the theory of evolution. XI. On the influence of natural selection on the variability of organs. *Philos. Trans. Roy. Soc. London* A200:1–66.

Pearson, K. 1904a. On the generalized theory of alternative inheritance with special references to Mendel's law. *Philos. Trans. Roy. Soc. London* A203:53–86.

Pearson, K. 1904b. Report on certain enteric fever inoculation statistics. *Br. Med. J.* 2: 1243–1246.

Pearson, K. 1906. Walter Frank Raphael Weldon, 1860–1906. *Biometrika* 5:1–52.

Pearson, K. 1908–1909. Note on inheritance in man. *Biometrika* 6:327–328.

Pearson, K. 1914–1930. *The life, letters and labours of Francis Galton.* Cambridge University Press, Cambridge, U.K. 4 vols.

Pearson, K., Nettleship, E., and Usher, C.H. 1911–1913. *A monograph on albinism in man. Parts I, II and IV.* Drapers' Co. Research Memoirs, Biometric Series 6, 8, and 9. Dulau and Co., London (Cambridge University Press). 6 vols. [Part III not published.]

Pell, M.B. 1878. Rates of mortality and increase of the population of New South Wales. *The Australian Practitioner* 3:145–175.

Peller, S. 1965. Births and deaths among Europe's ruling families since 1500. [Revised from three papers in *Bull. Hist. Med.* 13 (1943), 16 (1944), and 21 (1947).] In Glass, D.V., and Eversley, D.E.C. (Eds.). *Population in history.* Edward Arnold, London, pp. 87–100. Reprinted 1969.

Peller, S. 1967. *Quantitative research in human biology and medicine.* John Wright, Bristol.

Pemberton, J. 1970. *Will Pickles of Wensleydale: The life of a country doctor*. Bles, London.

Penrose, L.S. 1953. The genetical background of common diseases. *Acta Genet. Stat. Med.* 4:257–265.

Penrose, L.S., and Haldane, J.B.S. 1935. Mutation rates in man. *Nature* 135:907–908.

Penrose, L.S., Mackenzie, H.J., and Karn, M.N. 1948. A genetic study of human mammary cancer. *Ann. Eugen. London* 14:234–266.

Penrose, L.S., and Stern, C. 1958. A reconsideration of the Lambert pedigree. *Ann. Hum. Genet.* 22:258–283.

van Pesch, A.J. 1866. No details available. [Presumably cited in his "Sterftetafels voor Nederland afgeleid uit de waarnemingen over het tijdvak 1870–1880." *Bijdragen van het Statistisch Instituut* No. 3, Haarlem, 1885.]

Peters, J.A. (Ed.) 1959. *Classic papers in genetics*. Prentice Hall. Englewood Cliffs, NJ.

Philip, R. 1924. The effects of the anti-tuberculosis campaign on the diminution of the mortality from tuberculosis. *Edinburgh Med. J.* 31:482–526.

Phillips, R.L., Kuzma, J.W., and Lotz, T.M. 1980. Cancer mortality among comparable members versus nonmembers of the Seventh-day Adventist Church. In Cairns, J., Lyon, J.L., and Skolnick, M. (Eds.) 1980. *Cancer incidence in defined populations*. Banbury Rep. No. 4, Cold Spring Harbor Laboratory, Cold Spring Harbor, NY, pp. 93–108.

Pickles, W.N. 1939. *Epidemiology in country practice*. John Wright, Bristol.

Pinel, P. 1798. *Nosographie philosophique ou la méthode de l'analyse appliquée à la médecine*. Richard, Caille et Ravier, Paris. 2 vols.

Pinel, P. 1807. Résultats d'observations et construction des tables pour servir à déterminer le degré de probabilité de la guérison des aliénés. *Mém. Cl. Sci. Math. Phys. Institut Nat. de France*, 1er Sem. 1807;169–205. Baudouin, Paris.

Plackett, R.L. 1988. Data analysis before 1750. *Int. Stat. Rev.* 56:181–195.

Plum, P. 1936. Accuracy of haematological counting methods. *Acta Med. Scand.* 90:342–364.

Possion, S.-D. 1837. *Recherches sur la probabilité des jugements en matière criminelle et en matière civile, précédées des règles générales du calcul des probabilités*. Bachelier, Paris.

Pott, P. 1775. *Chirurgical observations relative to the cataract, the polypus of the nose, the cancer of the scrotum, the different kinds of ruptures, and the mortification of the toes and feet*. Hawes, Clarke and Collins, London.

Potter, M. 1963. Percivall Pott's contribution to cancer research. *Natl. Cancer Inst. Monog.* 10:1–13.

Pressat, R. 1972. *Demographic analysis: Methods, results, applications*. Transl. by J. Matras from the French editions of 1961 and 1969. Edward Arnold, London.

Provine, W.B. 1971. *The origins of theoretical population genetics*. Chicago University Press, Chicago-London.

Provine, W.B. 1978. The role of mathematical population geneticists in the evolutionary synthesis of the 1930s and 1940s. In Coleman, W., and Limoges, C. (Eds.) 1978. *Studies in history of biology*. Johns Hopkins, Baltimore, vol. 2, pp. 167–192.

Punnett, R.C. 1915. *Mimicry in butterflies*. Cambridge University Press, Cambridge.

Quetelet, L.A.J. 1826. *Correspondance mathématique et physique*. Vol. 2. Garnier, Gand-Bruxelles.

Quetelet, L.A.J. 1827. *Recherches sur la population, les naissances, les décès, les prisons, les dépôts de mendicité, etc. dans le royaume des Pays-Bas*. H. Tarlier, Bruxelles.

Quetelet, L.A.J. 1835/1842. *Sur l'homme et le développement de ses facultés. Essai sur physique sociale.* Bachelier, Paris, 1835. Transl. as *A treatise on man and the development of his faculties*, W. & R. Chambers, Edinburgh, 1842. Facsimile reprod. of English transl. with introd. by S. Diamond, Scholars' Facsimiles and Reprints, Gainesville, FL, 1969, and in *Comparative statistics in the 19th century*, introd. by R. Wall, Gregg International, Farnborough, UK, 1973.

Quetelet, L.A.J. 1845. Sur l'appréciation des documents statistiques et en particulier sur l'appréciation des moyennes. *Bull. Commission Stat. de Belgique* 205–286.

Quine, M.P., and Seneta, E. 1987. Bortkiewicz's data and the law of small numbers. *Int. Stat. Rev.* 55:173–181.

Race, R.R., and Sanger, R. 1975. *Blood groups in man.* 6th edition. Blackwell Scientific, Oxford.

Radicke, G. 1858/1861. Die Bedeutung und Werth arithmetischer Mittel mit besonderer Beziehung auf die neueren physiologischen Versuche zur Bestimmung des Einflusses gegebener Momente auf den Stoffwechsel, und Regeln zur exacten Beurtheilung dieses Einflusses. *Arch. für physiol. Heilkunde* (N.F.) 2:145–219 (1858). Transl. by F.T. Bond as "On the importance and value of arithmetic means ..." New Sydenham Society, London, vol. XI, 1861.

Ramazzini, B. 1713. *De morbis artificum diatriba.* Jo. Baptistam Conzattum, Padua. First published, Modena, 1700. Transl. by Wilmer Cave Wright, as *Diseases of workers*, University of Chicago Press, Chicago, 1940. Reprinted in History of Medicine Series, Library of the New York Academy of Medicine, Hafner, New York-London, 1964.

Ransome, A. 1868. On epidemics, studied by means of statistics of disease. *Br. Med. J.* 2:386–388.

Ransome, A. 1880. On epidemic cycles. *Manchester Lit. Philos. Soc. Proc.* 19:75–95.

Ransome, A. 1881–1882. On the form of the epidemic wave, and some of its probable causes. *Trans. Epidemiol. Soc. London* 1:96–111.

Ratliff, F. 1976. Georg von Békésy. *Biog. Mem. Nat. Acad. Sci.* 48:25–49.

Raven, C.E. 1950. *John Ray, naturalist: His life and works.* Cambridge University Press, Cambridge.

Ray, J. 1686–1704. *Historia plantarum.* London. 3 vols.

de Réaumur, R.A.F. 1751. *L'art de faire eclorre et d'enlever en toute saison des oiseaux domestiques des toutes esspèces.* 2nd edition. First publ. 1749. Impr. Royale, Paris. 2 vols.

Redi, F. 1684. *Osservazioni intorno agli animali viventi che si trovano negli animali viventi.* P. Matini, Florence.

Redmond, D.E. 1970. Tobacco and cancer: The first clinical report, 1761. *N. Engl. J. Med.* 282:18–23.

Reed, T.E., and Neel, J.V. 1959. Huntington's chorea in Michigan. 2, Selection and mutation. *Am. J. Hum. Genet.* 11:107–136.

Reeve, E.C.R., and Huxley, J.S. 1945. Some problems in the study of allometric growth. In Clark, W.E. le G., and Medawar, P.B. (Eds.). *Essays on growth and form. Presented to D'Arcy Wentworth Thompson.* Clarendon Press, Oxford, pp. 121–156.

Remmert, R. 1991. *Theory of complex functions.* Transl. from the German by R.B. Burckel. Springer-Verlag, New York.

Renbourn, E.T. 1960. The natural history of insensible perspiration: A forgotten doctrine of health and disease. *Med. Hist.* 4:135–152.

Rényi, A. 1965. On the theory of random search. *Bull. Am. Math. Soc.* 71:809–828.

Rényi, A. 1970a. *Probability theory.* North-Holland, Amsterdam and Akadémiai Kiadó, Budapest.

Rényi, A. 1970b. *Foundations of probability.* Holden-Day, San Francisco.

Retzius, A. 1845. Ueber die Schädelformen der Nordbewohner. *Arch. für Anatomie, Physiologie und wissenschaftliche Medizin* 84–129.

Retzius, G. 1909. A review of, and views on, the development of some anthropological questions. Huxley Lecture for 1909. *J. (Roy.) Anthropol. Inst.* 39:279–295. Reprinted in Kroeber, A.L., and Waterman, T.T. 1924. *Source book in anthropology.* University of California Press, Berkeley, CA, pp. 162–169.

Richardson, B.W. 1887/1936. John Snow, M.D. A representative of medical science and art of the Victorian era. *The Asclepiad, London,* 4:274–300 (1887). Reprinted in Snow, 1936, pp. xxiii–xlviii.

Rivers, T.M. 1937. Viruses and Koch's postulates. *J. Bacteriol.* 33:1–12.

Roberts, D.F. (Ed.) 1975. *Human variation and natural selection.* Symp. Society for Study of Human Biology, vol. 13. Taylor & Francis, London.

Rochoux, J.A. 1982. Sur la contagion et les maladies contagieuses. [Summary.] *Arch. Gén. de Méd.* 28:285–286.

Roddis, L.H. 1950. *James Lind: Founder of nautical medicine.* Henry Schuman, New York.

Roederer, J.G. 1754. De pondere et longitudine infantum recens natorum. *Comment. Soc. Reg. Scient. Göttingensis* 3:410–424.

von Rokitansky, K. 1842–1846. *Handbuch der (allgemeinen) pathologischen Anatomie.* Braumüller and Seidel, Vienna. 3 vols. [English edition, London, 1854.]

Rolleston, J.D. 1924. Bretonneau: His life and work. *Proc. Roy. Soc. Med. (Sect. Hist. Med.)* 18:1–12.

Rolleston, J.D. 1937. *History of the acute exanthemata.* Heinemann, London.

Rosen, G. 1938. Jacob Henle: On miasmata and contagia. Transl. of *Von den Miasmen und Kontagien* (Henle, 1840). *Bull. Hist. Med.* 6:907–983.

Rosen, G. 1955. Problems in the application of statistical analysis to questions of health: 1700–1880. *Bull. Hist. Med.* 29:27–45.

Rosen, G. 1968. Louis, P.C.A. *Int. Encyc. Soc. Sci.* 9:478–479.

Rosenbaum, S. 1988. 100 years of heights and weights. *J. Roy. Stat. Soc.* A151:276–309.

Rosenberg, C.E. 1979. Florence Nightingale on contagion: The hospital as moral universe. In Rosenberg, C.E. (Ed.). *Healing and history: Essays for George Rosen.* Science History Publ., New York; W. Dawson, Folkestone, UK, pp. 116–136.

Rosenbloom, J. 1922. The history of pulse timing with some remarks on Sir John Floyer and his physician's pulse watch. *Ann. Med. Hist.* 4:97–99.

Rothschuh, K.E. 1973. *History of physiology.* Transl. and edited by G.B. Risse. Krieger, Huntington, NY.

Roux, E. 1897. Sur la peste bubonique. Essais de traitement par le sérum antipesteux, à propos d'une note du Dr. Yersin, médecin de 2e classe des Colonies, directeur de l'Institut Pasteur de Nha-Trang. *Bull. Acad. Méd. Paris* 37:91–99.

Royal College of Physicians 1971. *Smoking and health now.* Pitman Medical & Scientific, London.

Roychoudhury, A.K., and Nei, M. 1988. *Human polymorphic genes: World distribution.* Oxford University Press, New York.

Royston, E. 1956. Studies in the history of probability and statistics III. A note on the history of the graphical presentation of data. *Biometrika* 43:241–247.

Russell, J.C. 1948. Demographic pattern in history. *Popul. Studies* 1:388–404.

Russell, P.F. 1955. *Man's mastery of malaria. University of London Heath Clark Lectures 1953.* Oxford University Press, London.

Salas, I. 1863. Etiology and prophylaxis of pellagra. From *Etiologie et prophylaxie de la pellagra.* M.D. Thesis, Fac. de Médecine de Paris, pp. 26–35. Transl. in Carpenter 1981, pp. 19–24.

Sambon, L.W. 1910. Progress report on the investigation of pellagra. *J. Trop. Med. Hyg.* 13:271–282, 287–300, 305–315, 319–321. Excerpts reprinted in Carpenter, 1981, pp. 28–30.

Samuelson, P.A. 1976. Resolving a historical confusion in population analysis. *Hum. Biol.* 48:559–580. Reprinted in Smith and Keyfitz 1977; pp. 109–129.

Sarton, G. 1935. Preface to volume XXIII of *Isis.* (Quetelet) *Isis* 23:6–24.

Schadow, J.-G. 1834–1835. *Polyclète, ou théorie des mesures de l'homme, selon le sexe et l'âge avec indication des grandeurs réelles d'après le pied du Rhin . . .* Sachse, Berlin. 2 vols.

von Schelling, H. 1949. Erste Ergebnisse einer Aufgliederung der Sterblichkeit nach Geburtsjahrgängen und Kalender jahren. *Metron* 15:359–374.

Schleiden, J.M. 1842–1843. *Grundzüge der wissenschaftlichen Botanik.* W. Engelmann, Leipzig. 2 vols.

Schleisner, P.A. 1851. Vital statistics of Iceland. *J. Stat. Soc. (London)* 14:1–10.

Schmidt, E.J. 1922. The breeding places of the eel. *Philos. Trans. Roy. Soc. London,* Ser. B., 211:179–208.

Schoenlein, J.L. 1839a. *Allgemeine und specielle Pathologie und Therapie.* Literatur-Comptoir, St. Gallen. 4 vols.

Schoenlein, J.L. 1839b. Zur Pathogenie der Impetigines. *Arch. für Anatomie, Physiologie und wissenschaftliche Medicin,* p. 82.

Schröer, H. 1967. *Carl Ludwig, Begründer der messenden Experimentalphysiologie, 1816–1895.* Grosse Naturforscher, Bd. 33. Wissenschaftliche Verlagsgesellschaft, Stuttgart.

Schwann, T.A.H. 1839. *Mikroskopische Untersuchungen über die Uebereinstimmung in der Struktur und dem Wachstum der Thiere und Pflanzen.* G.E. Reimer, Berlin.

Scott, H.H. 1939. *A history of tropical medicine. Based on the Fitzpatrick Lectures delivered before the Royal College of Physicians of London 1937–38.* Edward Arnold, London. 2 vols.

Scotto, J., and Bailar III, J.C. 1969. Rigoni-Stern and medical statistics. *J. Hist. Med.* 24:65–75.

Scotto, J., Fears, T.R., and Fraumeni, J.F., Jr. 1982. Solar radiation. In Schottenfeld, D., and Fraumeni, J.F., Jr (Eds.) 1982. *Cancer epidemiology and prevention.* W.B. Saunders Co., Philadelphia, pp. 254–276.

Semm, K., and Weichert-von Hassel, M. (Eds.) 1985. *Kiel University Hospital of Gynecology and Michaelis School of Midwifery–its contributions to gynecology from 1805 to 1985. A study in medical history on the occasion of its 180th anniversary.* University Hospital of Gynecology, Kiel.

Semmelweis, I.P. 1861/1941/1966. *Die Aetiologie, der Begriff und die Prophylaxis des Kindbettfiebers.* C.A. Hartleben, Pest-Vienna-Leipzig, 1861. Reprinted by Johnson Reprint Corp., Sources of Science No. 19, New York-London, 1966. Transl. by F.P. Murphy as *The etiology, the concept and prophylaxis of childbed fever. Medical Classics* 5:350–773 (1941).

Sharpe, F.R., and Lotka, A.J. 1911. A problem in age distribution. *Philos. Mag.* (Ser. 6) 21:435–438.

Sherwood, E.R. 1964. Bibliography of the publications of Wilhelm Weinberg. *Jahreshefte d. Vereins f. vaterländische Naturkunde in Württemberg* 118–119:61–67. [Original on deposit in Biology Library, University of California, Berkeley.]

Short, R. 1991. HIV infection and AIDS—the global scene. In *HIV infection and AIDS: Present status and future prospects for prevention, treatment and cure. Proc. 1991 Annual General Meeting of the Australian Academy of Science.* Australian Academy of Science, Canberra, pp. 15–29.

Simon, J. 1864. Sixth report of the Medical Officer of the Privy Council, 1863. [Incorporates a report by Mr. T. Holmes and Dr. Bristowe; see Simpson 1871, p. 340, footnote.] *Government Blue Book for 1863*, London.

Simpson, G.G. 1949/1966. *The meaning of evolution. A study of the history of life and of its significance for man.* Yale University Press, New Haven.

Simpson, G.G., Roe, A., and Lewontin, R.C. 1960. *Quantitative zoology.* Revised Edition. Harcourt, Brace and World, New York.

Simpson, J.Y. 1850. On the analogy between puerperal and surgical fever. *Edinburgh Mon. J. Med. Sci.* 11:414–429 (1850). Reprinted in *The obstetric memoirs and contributions of James Y. Simpson, M.D., F.R.S.E.* W.O.Priestley and H.R. Storer (Eds.) Adam and Charles Black, Edinburgh, 2:1–19 (1856).

Simpson, J.Y. 1851. On the communicability and propagation of puerperal fever. *Edinburgh Mon. J. Med. Sci.* 12:72. Reprinted in *The obstetric memoirs and contributions of James Y. Simpson, M.D., F.R.S.E.* W.O.Priestley and H.R. Storer (Eds.) Adam and Charles Black, Edinburgh, 2:20–33 (1856).

Simpson, J.Y. 1868–1869, 1869–1870. Our existing system of hospitalism and its effects. *Edinburgh Mon. J. Med. Sci.* 14:816–830, 1084–1115; 15:523–532. Reprinted in *Works* 2:289–405 (1871).

Simpson, J.Y. 1871. *The works of Sir J.Y. Simpson, Bart. Vol. 2: Anaesthesia, hospitalism, hermaphroditism and a proposal to stamp out smallpox and other contagious diseases.* W.G. Simpson (Ed.) Adam and Charles Black, Edinburgh.

Singer, C., and Singer, D. 1917–1918. The scientific position of Girolamo Fracastoro [1478?–1553] with especial reference to the sources, character and influence of his theory of infection. *Ann. Med. Hist.* 1:1–34.

Škoda, J. 1839/1853/1854. *Abhandlung über Perkussion und Auskultation.* 2nd edition. Seidel, Vienna. Transl. By W.O. Markham from the 4th ed. 1850 as *A treatise on auscultation and percussion.* Highley & Son, London, 1853, Lindsay & Blakiston, Philadelphia, 1854.

Smith, D.P., and Keyfitz, N. 1977. *Mathematical demography.* Springer-Verlag, Berlin-Heidelberg-New York.

Smith, E.L., Hill R.L., et al. 1983. Hemoglobin and the chemistry of respiration. In Smith, E.L., et al. *Principles of biochemistry: Mammalian biochemistry*, 7th edition. McGraw-Hill, New York, pp. 100–140.

Smith, T. (with Kilborne, F.L.) 1893. *Investigations into the nature, causation, and prevention of Texas or Southern cattle fever.* Government Printing Office, Washington. Reprinted in *Medical Classics* 1:372–597 (1936–1937).

Smoluchowski, M. 1906. Zur kinetischen Theorie der Brownschen Molekularbewegung und der Suspensionen. *Ann. der Physik* (4th ser.) 21:756–780.

Snow, J. 1853/1855/1936. *Snow on cholera: Being a reprint of two papers by John Snow. Together with a biographical memoir by B.W. Richardson and an introduction by W.H. Frost.* Commonwealth Fund, New York (1936).

Sonnenborn, T.M. 1965. Genetics and man's vision. *Proc. Am. Philos. Soc.* 109:237–241.

Soper, H.E. 1929. The interpretation of periodicity in disease prevalence. *J. Roy. Stat. Soc.* 92:34–61.

Spies, T.D., Cooper, C., and Blankenhorn, M.A. 1938. The use of nicotinic acid in the treatment of pellagra. *J. Am. Med. Assoc.* 110:622–627.

Spiess, O., and Verzár, F. 1941. Eine akademische Festrede von Daniel Bernoulli—Ueber das Leben (De Vita). *Verh. Naturforsch. Ges. Basel* 52:189–266.

Springett, V.H. 1950. A comparative study of tuberculosis mortality rates. *J. Hyg. Camb.* 48:361–395.

Springett, V.H. 1952. An interpretation of statistical trends in tuberculosis. *Lancet* 1:521–525, 575–579.

Squire, W. 1875–1877. On measles in Fiji. *Trans. Epidemiol. Soc. London* 4:72–74.

Stanley, N.F. 1980. Man's role in changing patterns of arbovirus infections. In Stanley, N.F., and Joske, R.A. (Eds.). *Changing disease patterns and human behaviour*. Academic Press, New York, pp. 151–173.

Stanley, S.M. 1987. *Extinction*. Scientific American Library No. 20, New York.

Steiner, W.R. 1939. Some distinguished American medical students of Pierre-Charles-Alexandre Louis of Paris. *Bull. Hist. Med.* 7:783–793.

Stern, C. 1962. Wilhelm Weinberg, 1862–1937. *Genetics* 47:1–5.

Stern, C. 1965. Mendel and human genetics. *Proc. Am. Philos. Soc.* 109:216–226.

Stigler, S.M. 1986. *The history of statistics. The measurement of uncertainty before 1900*. Belknap Press (Harvard), Cambridge, MA.

Stocks, P. 1942. Measles and whooping-cough incidence before and during the Dispersal of 1939–1941. *J. Roy. Stat. Soc.* 105:259–291.

Strauss, E. 1954. *Sir William Petty: Portrait of a genius*. Bodley Head, London.

Strong, T.B. (Ed.) 1906. *Lectures on the method of science*. Clarendon Press, Oxford.

Stubbe, H. 1972. *History of genetics from prehistoric times to the rediscovery of Mendel's laws*. Transl. from the German edition of 1965. MIT Press, Cambridge, MA.

Student (W.S. Gosset) 1907. On the error of counting with a haemocytometer. *Biometrika* 5:351–360.

Sturtevant, A.H. 1965a. The early Mendelians. *Proc. Am. Philos. Soc.* 109:199–204.

Sturtevant, A.H. 1965b. *A history of genetics*. Harper & Row, New York.

Suessmilch, J.P. 1761–1762. *Die göttliche Ordnung in den Veränderungen des menschlichen Geschlechts, aus der Geburt, dem Tode und der Fortpflanzung desselben erwiesen*. 2nd edition (original 1741–1742). Verlag des Buchladens der Realschule, Berlin. 2 vols.

Sutton, W.S. 1902. On the morphology of the chromosome group in *Brachystola magna*. *Biol. Bull.* 4:24–39.

Sutton, W.S. 1903. The chromosomes in heredity. *Biol. Bull.* 4:231–251. Reprinted in Peters (1959).

Swan, C., Tostevin, A.L., Moore, B., Mayo, H., and Black, G.H.B. 1943. Congenital defects in infants following infectious diseases during pregnancy. *Med. J. Austral.* 2:201–210.

Sydenstricker, E. 1927. The declining death rate from tuberculosis. *Trans. Natl. Tuberc. Assoc.* 23:102–124. Reprinted in Kasius, R.V. (Ed.). *The challenge of facts: Selected public health papers of Edgar Sydenstricker*. Prodist, New York, 1974.

Tait, R.L. 1886. One hundred and thirty-nine consecutive ovariotomies performed between January 1st, 1884 and December 31st, 1885 without a death. *Br. Med. J.* 1:921–924.

Tait, R.L. 1890. Address in surgery: Surgical training, surgical practice, surgical results. *Br. Med. J.* 2:267–273.

Tanner, J.M. 1981. *A history of the study of human growth.* Cambridge University Press, Cambridge.

Teissier, G., and Huxley, J.S. 1936. Terminologie et notation dans la description de la croissance relative. *C.R. Soc. de biologie* 121:934–936.

Teissier, G., and Lambert, R. 1927. Théorie de la similitude biologique. *Ann. physique et physico chimie biologique* 3:212–246.

Temkin, O. 1953. Remarks on the neurology of Gall and Spurzheim. In Underwood, E.A. (Ed.). *Science, medicine and history*, vol. II. Oxford University Press, London, pp. 282–289.

Tenon, J.R. 1788. *Mémoires sur les hôpitaux de Paris.* P-D. Pierres, Paris.

Thackray, A. 1970. *Atoms and powers. An essay on Newtonian matter-theory and the development of chemistry.* Harvard University Press, Cambridge, MA.

Thiersch, C. 1865. *Der Epithelialkrebs, namentlich der Haut. Eine anatomisch-klinische Untersuchung.* W. Engelmann, Leipzig.

Thompson, D'A.W. 1917/1942. *On growth and form.* Cambridge University Press, London; Macmillan, New York, (1942 edition).

Thompson, J.A. (with Tidswell, F.) 1903. *Report of the Board of Health on a second outbreak of plague at Sydney, 1902.* Govt. Printer, Sydney, Australia.

Thompson, J.N., and Thoday, J.M. (Eds.) 1979. *Quantitative genetic variation.* Academic Press, New York-London.

Thomsen, O., Friedenreich, V., and Worsaae, E. 1930a. Ueber die Möglichkeit der Existenz zweier neuer Blutgruppen; auch ein Beitrag zur Beleuchtung sogenannter Untergruppen. *Acta Pathol. Microbiol. Scand.* 7:157–190.

Thomsen, O., Friedenreich, V., and Worsaae, E. 1930b. Die wahrscheinliche Existenz eines neuen, mit den drei bekannten Blutgruppengenen (O, A, B) Allelomorphen, A' benannten Gens mit den daraus folgenden zwei neuen Blutgruppen: A' und A'B. *Klin. Wochenschr.* 9:67–69, 938.

Tilling, L. 1975. Early experimental graphs. *Br. J. Hist. Sci.* 8:193–213.

Todhunter, I. 1865. *A history of the mathematical theory of probability from the time of Pascal to that of Laplace.* Macmillan, London. Reprinted 1965, Chelsea, New York.

Todhunter, I. 1869. On the method of least squares. [Includes appendix by A. Cayley.] *Trans. Camb. Philos. Soc.* 11:219–238.

Topley, W.W.C. 1919. *The Goulstonian Lectures on the spread of bacterial infection. Lancet* 2:1–5, 45–49, 91–96.

Topley, W.W.C., and Wilson, G.S. 1975. *Topley and Wilson's "Principles of bacteriology, virology and immunity."* Wilson, G.S., Miles, A., et al. (Eds.). 6th edition. Edward Arnold, London. 2 vols.

Topley, W.W.C., and Wilson, G.S. 1983–1984. *Topley and Wilson's "Principles of bacteriology, virology and immunity."* Wilson, G.S., Miles, A., and Parker, M.T. (Gen. Eds.). 7th edition. Edward Arnold, London. Williams and Wilkins, Baltimore. 4 vols.

Topley, W.W.C., and Wilson, G.S. 1990. *Topley and Wilson's "Principles of bacteriology, virology and immunity."* Parker, M.T., and Collier, L.H. (Gen. Eds.). 8th edition. Edward Arnold, London. 5 vols.

Tousey, R. 1953. Solar work at high altitudes from rockets. In Kuiper, G.P. (Ed.). *The solar system, Vol. 1: The Sun.* University of Chicago Press, Chicago, pp. 658–676.

Tschermak von Seysenegg, E. 1900. Ueber künstliche Kreuzung von *Pisum sativum*. *Z. landwirtschaftliche Versuchswesen in Oesterreich* 3:465–555. Summarized in *Ber. Deutsch. Botan. Ges.* 18:232–239 (1900).

Uhland, R. 1953. *Geschichte der Hohen Karlsschule in Stuttgart.* Kohlhammer, Stuttgart.

US Office on Smoking and Health. 1983. *Bibliography on smoking and health.* (In collaboration with WHO.) Public Health Serv. Bibliog. Ser. No. 45, DHHS(PHS) 84–50196. US Govt. Printing Office, Rockville, MD.

US Surgeon-General. 1979. *Smoking and health: A report of the Surgeon-General.* DHEW(PHS) 79-50066. Supt. Documents, Washington, DC.

Unna, P.G. 1894. *Die Histopathologie der Hautkrankheiten.* August Hirschwald, Berlin.

Upton, A.C. 1975. Physical carcinogenesis. Radiation: history and sources. In Becker, F.F. (Ed.). *Cancer: A comprehensive treatise, Vol. 1,* Plenum, New York, pp. 387–404.

Vallery-Radot, R. 1919. *The life of Pasteur.* Transl. from the French by R.L. Devonshire. Constable, London.

van Vark, G.N., and Bilsborough, A. 1991. Shaking the family tree. Letter to the Editor. *Science* 253:834.

van Vark, G.N., and Howells, W.W. (Eds.) 1984. *Multivariate statistical methods in physical anthropology.* Reidel, Dordrecht.

Verhulst, P.F. 1838. Notice sur la loi que la population suit dans son accroissement. *Corresp. Math. Phys. publ. par A. Quételet* 10:113–121. Gand-Bruxelles.

Verhulst, P.F. 1845–1847. Recherches mathématiques sur la loi d'accroissement de la population. Deuxième mémoire sur la loi d'accroissement de la population. *Nouveaux Mém. Acad. Roy. Sci. Belles Lettres* (*Brussels*) 18:3–41 (1845); 20:4–32 (1847).

Viner, J. 1952. *International trade and economic development: Lectures delivered at the National University of Brazil.* The Free Press, Glencoe, IL.

Virchow, R. 1863. *Cellular pathology as based upon physiological and pathological histology. Twenty lectures delivered in the Pathological Institute of Berlin during the months of February, March and April, 1858.* Transl. from the 2nd edition by F. Chance. R.M. DeWitt, New York, 1863. Reprinted by Dover Publications, New York, 1971. Excerpts reprinted in Moore (1972).

Virchow, R. 1959. *Disease, life and man: Selected essays by Rudolf Virchow.* Transl. and with introd. by L.J. Rather. Stanford University Press, Stanford, CA.

Vittadini, C. 1852. Della natura del calcino o mal del segno. *Mem. dell. I.R. Istituto Lombardo* 3:447–512.

Vogel, F., and Motulsky, A.G. 1982. *Human genetics. Problems and approaches.* Springer-Verlag, Berlin-Heidelberg-New York. (1st edition, 1979).

Vorzimmer, P. 1963. Charles Darwin and blending inheritance. *Isis* 54:371–380.

de Vries, H. 1889. *Intracellulare Pangenesis.* G. Fischer, Jena.

de Vries, H. 1900. Sur la loi de disjonction des hybrides. *C.R. Acad. Sci.* 130:845–847.

Wallace, A.R. 1855. On the law which has regulated the introduction of new species. *Annals and Mag. of Nat. Hist.* 16:184–196.

Wangensteen, O.H., and Wangensteen, S.D. 1978. *The rise of surgery. From empiric craft to scientific discipline.* Wm. Dawson, Folkestone, U.K.; University of Minnesota Press, Minneapolis.

Wargentin, P.W. 1766/1930. Mortaliteten i Sverige, i anledning af Tabell-Verket. *Kongl. Vetenskaps Acad. Handl.* 27:1–25 (1766). Reprinted as *Tables of mortality based upon the Swedish population, prepared and presented in 1766 by Per Wilhelm Wargentin.* Thule Life Insurance Co., I. Haeggström, Stockholm, 1930. See also *Mémoires abrégés Acad. Roy. Sci. Stockholm,* I, 4, Paris, 1772.

Watt, J., Freeman, E.J., and Bynum, W.F. (Eds.) 1981. *Starving sailors: The influence of nutrition upon naval and maritime history.* National Maritime Museum, London.

Webster, L.T. 1946. Experimental epidemiology. *Medicine (Baltimore)* 25:77–109.

Weinberg, W. 1901. Beiträge zur Physiologie und Pathologie der Mehrlings-geburten beim Menschen. *Pflügers Arch. gesamte Physiol. d. Menschen u.d. Tiere* 88:346–430.

Weinberg, W. 1908. Ueber den Nachweis der Vererbung beim Menschen. *Jahreshefte d. Vereins f. vaterländische Naturkunde in Württemberg* 64:369–382.

Weinberg, W. 1909. Ueber Vererbungsgesetze beim Menschen, II. *Z. Induktive Abstammungs- und Vererbungslehre* 2:276–330.

Weinberg, W. 1910. Weitere Beiträge zur Theorie der Vererbung. *Arch. Rassen- und Gesellschaftsbiologie* 7:35–49, 169–173.

Weinberg, W. 1912. Weitere Beiträge zur Theorie der Vererbung. IV. Ueber Methode und Fehlerquellen der Untersuchung auf Mendelsche Zahlen beim Menschen. *Arch. Rassen- und Gesellschaftsbiologie* 9:165–174.

Weismann, A. 1885/1891. On the number of polar bodies and their significance in heredity. In *Essays upon heredity and kindred biological problems,* transl. by E.B. Poulton, S. Schönland, and A.E. Shipley. 2nd edition. Clarendon Press, Oxford, 1891–1892 (2 vols). vol. 1, p. 370.

Welch, W.H. 1925. *Public health in theory and practice: An historical view. The second William Thompson Sedgwick Memorial Lecture.* Yale University Press, New Haven, CT.

Weldon, W.F.R. 1890. The variations occurring in certain decapod crustacea. I. *Crangon vulgaris. Proc. Roy. Soc. London* 47:445–453.

Weldon, W.F.R. 1892. On certain correlated variations in *Crangon vulgaris. Proc. Roy. Soc. London* 51:2–21.

Weldon, W.F.R. 1893. On certain correlated variations in *Carcina moenas. Proc. Roy. Soc. London* 54:318–329.

Weldon, W.F.R. 1895. Report of the Committee for conducting Inquiries into the measurable Characteristics of Plants and Animals, Part I. Attempt to measure the deathrate due to the selective destruction of *Carcina moenas* with respect to a particular dimension. *Proc. Roy. Soc. London* 57:360–369.

Weldon, W.F.R. 1901–1902. Editorial. *Biometrika* 1:1–6.

Weldon, W.F.R. 1906. Inheritance in animals and plants. In Strong, pp. 81–109.

W(eldon), W.F.R, and P(earson), K. 1902–1903. Inheritance in *Phaseolus vulgaris. Biometrika* 2:499–503.

Wendt, G.G., and Drohm, D. 1972. *Fortschritte der allgemeinen und klinischen Humangenetik. Vol. IV: Die Huntingtonsche Chorea: eine populationsgenetische Studie.* Thieme, Stuttgart.

Westergaard, H. 1932. *Contributions to the history of statistics.* P.S. King & Son, London. Reprinted, Mouton, Paris, and S.R. Publishers, Wakefield, UK, 1969.

Whitaker, L. 1914–1915. On the Poisson law of small numbers. *Biometrika* 10:36–71.

Whitelegge, B.A. 1892–1893. Measles epidemics, major and minor. *Trans. Epidemiol. Soc. London* 12 (NS):37–54.

Wiener, A.S. 1939. *Blood groups and blood transfusion.* 2nd edition. C.C. Thomas, Springfield, IL-Baltimore, MD.

Willius, F.A., and Dry, T.J. 1948. *A history of the heart and the circulation.* W.B. Saunders, Philadelphia-London.

Wilson, E.B. 1914. The bearing of cytological research on heredity. *Proc. Roy. Soc. London* 88B:333–352.

Wilson, E.B. 1947. The spread of measles in the family. *Proc. Natl. Acad. Sci.* 33:68–72.

Wilson, E.B., Bennett, C., Allen, M., and Worcester, J. 1939. Measles and scarlet fever in Providence, R.I., 1929–1934, with respect to age and size of family. *Proc. Am. Philos. Soc.* 80:357–476.

Wilson, E.B., and Worcester, J. 1945a. The law of mass action in epidemiology. *Proc. Natl. Acad. Sci.* 31:24–34, 109–116.

Wilson, E.B., and Worcester, J. 1945b. Damping of epidemic waves. *Proc. Natl. Acad. Sci.* 31:294–298.

Wilson, G.S. 1962. Measles as a universal disease. *Am. J. Dis. Child.* 103:219–224.

Wilson, G.S. 1984. General epidemiology. In Topley and Wilson, vol. 3, pp. 1–9.

Wilson, G.S., and Smith, G. 1984. Diphtheria and other diseases due to *Corynebacteria.* In Topley and Wilson, vol. 3, pp. 73–101.

Wilson. L.G. 1978. Fevers and science in early nineteenth century medicine. *J. Hist. Med.* 33:386–407.

Winge, O. 1958. Wilhelm Johannsen: The creator of the terms gene, genotype, phenotype and pure line. *J. Heredity* 49:82–88.

Winslow, C.E.A. 1943. *The conquest of epidemic diseases.* Princeton University Press, Princeton. (1980 edition, University of Wisconsin Press.)

de Witt, J. 1671. *Waardye van Lyf-renten naer Proportie van Losrenten.* The Hague. Transl. in F. Hendriks, "Contributions to the history of insurance, with a restoration of the Grand Pensionary De Wit's treatise on life annuities," *Assurance Mag.* 2:121–150, 222–258 (1852); 3:93–120 (1853).

Wolff, C.F. 1759. *Theoria generationis.* [Dissertation.] Halle, 1759. Restated as *Theorie von der Generation in zwo Abhandlungen erklärt und bewiesen,* Berlin, 1764. Photographic reprint of both the above, with introd. by R. Herrlinger, Olms, Hildesheim, 1966.

Woolf, H. 1961a. The Conference on the History of Quantification in the Sciences. [Introducing papers presented at the conference held in November, 1959.] *Isis* 52:133–134.

Woolf, H. 1961b. *Quantification—a history of the meaning of measurement in the natural and social sciences.* Bobbs-Merrill, Indianapolis.

Wunderlich, C.A. 1871. *On the temperature in diseases: A manual of medical thermometry.* Transl. from the 2nd German edition by W.B. Woodman. New Sydenham Society, London.

Wynder, E.L. (Ed.) 1955. *The biologic effects of tobacco, with emphasis on the clinical and experimental aspects.* Little, Brown, Boston.

Wynder, E.L. 1972. Etiology of lung cancer. Reflections of two decades of research. *Cancer* 30:1332–1339.

Wynder, E.L., and Graham, E.A. 1950. Tobacco smoking as a possible etiologic factor in bronchiogenic carcinoma. *J. Am. Med. Assoc.* 143:329–336.

Wynder, E.L., Lemon, F.R., and Bross, I.J. 1959. Cancer and coronary artery disease among Seventh-Day Adventists. *Calif. Med.* 89:267–272. See also *Cancer* 12:1016–1028 (1959).

Wynn, H.P. 1982. Controlled versus random experimentation. *Statistician* 31:237–244.

Yates, F., and Mather, K. 1963. Ronald Aylmer Fisher: 1890–1962. *Biog. Mem. Fell. Roy. Soc.* 9:91–129.

Yorke, J.A., Nathanson, N., Pianigiani, G., and Martin, J. 1979. Seasonality and the requirements for perpetuation of viruses in populations. *Am. J. Epidemiol.* 109:103–123.

Young, J.Z. 1971. *An introduction to the study of man.* Clarendon Press, Oxford.

Yule, G.U. 1902. Mendel's laws and their probable relations to intra-racial heredity. *New Phytologist* 1:193–207, 222–238.

Yule, G.U. 1934. On some points relating to vital statistics, more especially statistics of occupational mortality. *J. Roy. Stat. Soc.* 97:1–84.

Zeuner, G. 1869. *Abhandlungen aus der mathematischen Statistik.* Arthur Felix, Leipzig.

# Subject Index

# Person Index

---

*to be found in Dictionary of Scientific Biography. For persons having multiple entries, page numbers in italics indicate the main entry.